DISTILLATION

AND

RECTIFICATION

by

DR. EMIL KIRSCHBAUM

Professor, Karlsruhe Institute of Technology

Translated by

M. WULFINGHOFF

Engineer, Sanderson & Porter

Published and Distributed in the Public Interest
by Authority of the Alien Property Custodian
under License No. A-1019

1948

CHEMICAL PUBLISHING CO., INC.

BROOKLYN NEW YORK

Distillation and Rectification

ISBN: 978-0-8206-0018-5

Chemical Publishing Company:
www.chemical-publishing.com
www.chemicalpublishing.net

First Edition:

Chemical Publishing Company, Inc. - New York 1948

Copyright Vested in the Alien Property Custodian, 1943, Pursuant to Law

Second Impression:

Chemical Publishing Company, Inc. - 2011

Printed in the United States of America

FOREWORD TO THE FIRST ENGLISH EDITION

Kirschbaum's book may be regarded as a classic of the distillation and rectification industry. This first edition in the English language should prove a great help to American engineers.

The translator, who has been engaged in distillation work, both in Europe and the United States, did justice to the original. He also included a résumé on the American practice and recent developments in distillation and rectification technique.

IRENE E. BERCK, PH.D.
Editor

ACKNOWLEDGMENTS

The translator is indebted to Dr. Donald F. Othmer, Head of the Department of Chemical Engineering, Brooklyn Polytechnic Institute, and to Professor H. J. Garber of the Department of Chemical Engineering, University of Knoxville, Ky., for many valuable suggestions which have improved the readability of the text.

Dedicated to My Parents

PREFACE

Although chemical engineering and the design of chemical equipment have been presented as a whole in the literature, in accordance with their increasing importance, the individual branches of equipment design have not as yet been treated with the thoroughness they deserve. One reason for this has been that because of certain necessities in the manufacturing of many commodities, industrial practice was well ahead of theory.

In Germany this condition was somewhat alleviated more than a decade ago, when chemical engineering was adopted as a branch of advanced technical study and much research work was devoted to distillation and rectification. The significance of this can be readily appreciated since the liquid fuels which are so essential to the life and economy of a nation are always obtained by distilling and rectifying. The same situation exists in the case of other products, and there can be no doubt that recent developments will make distilling and rectifying apparatus even more essential.

The purpose of this book is to close one of the gaps in the literature and to present results of research together with engineering data. This book is meant to aid the student as well as the practicing engineer. The fundamentals are presented in a strictly scientific way, and design and other factors, important to an engineer, have also been considered.

To prevent excessive bulk, it was not thought expedient to describe the specific industries such as alcohol or petroleum distillation, gas liquefaction, solvent recovery, Buna processing, etc. Instead, principles and generally suitable designs have been stressed and numerical and descriptive examples referred to specialized industries. Certain processes have been demonstrated for the case of binary mixtures only, although they may involve liquid mixtures of more than two individual substances. The fundamentals thus gained will also enable an engineer to handle special problems referring to the separation of liquid mixtures.

In technical thermodynamics, it has become the custom to use (in the treatment of binary mixtures by graphical methods) the entha py-concentration diagram method exclusively. In distilling and rectifying, methods which rest on the assumption of constancy of molar heat of evaporation have been adopted. Each of these methods has its advantages and drawbacks. Therefore both have been considered although the second method has been used more extensively because of its importance in dealing with mixtures of several components.

As far as possible, graphical methods have been used in preference to methods requiring involved calculations. At the same time, an attempt has been made to give, along with the fundamental theoretical data, information useful in actual engineering work. Therefore, tables and curves have been included.

EMIL KIRSCHBAUM

TABLE OF CONTENTS

A. FUNDAMENTALS

The purpose of distilling and rectifying equipment is either to separate liquid mixtures into their original components or to split them up into other mixtures having compositions other than that of the original mixture.

Distillation, in the widest sense of the word, means generating vapors from a liquid and condensing them again. When vapors are generated from a mixture of liquids, distillation will also mean altering the mixture, since the vapors condensed will have a composition different from that of the original mixture. It may also be possible to convert by distillation a mixture of liquids into new mixtures having a composition different from that of the first mixture distilled. The distillates are then collected according to their varying boiling ranges. This procedure is called fractional distillation.

If the cycle of vaporizing and liquefying is repeated, the process is called rectification. This is an application of the basic physical procedure consisting of moving liquid and vapor countercurrently, so that as a result of intimate contact of the two phases an exchange of heat and change in composition will occur.

Since in rectifying vapors are usually generated and condensed, and liquids are cooled, a *rectifying unit* comprises the rectifying column with its heat exchangers performing the distillation, condensation and cooling. Rectification will accomplish a fairly complete separation of liquid mixtures.

B. THEORY

I. Units

It is customary, in continental engineering practice, to measure quantities in kilograms, cubic meters, etc., and to use percentage by weight to indicate the composition or concentration of a mixture.

In distillation and rectification problems, however, it is desirable and sometimes even necessary to express quantities as mols, and compositions in mol per cent.

A kilogram-mol, or simply mol, is that number of kilograms which corresponds to the numerical value of the molecular weight of the substance considered. Thus, 1 mol water = 18.01 kg.

Consequently, the mol composition of a mixture is determined by its average molecular weight, μ. In a ternary mixture, let μ_x, μ_y and μ_z = the molecular weights of the individual components, and their percentages by weight = ξ, o and π. Since

$$\frac{\xi}{\mu_x} + \frac{o}{\mu_y} + \frac{\pi}{\mu_z} = \frac{100}{\mu}$$

the average molecular weight would be

$$\mu = \frac{100}{\dfrac{\xi}{\mu_x} + \dfrac{o}{\mu_y} + \dfrac{\pi}{\mu_z}} \tag{1}$$

In the case of a binary equation, with ξ per cent by weight of the low boiler having a molecular weight of μ_x, (the molecular weight of the other substance being μ_y) equation (1) becomes

$$\mu = \frac{100}{\dfrac{\xi}{\mu_x} + \dfrac{100 - \xi}{\mu_y}} \tag{1a}$$

Whenever some definite mixture is mentioned, the low boiler will always be named first.

Example. Find the average molecular weight of an ethyl alcohol-water mixture containing 10 per cent by weight of alcohol. Molecular weight of water is 18.01, of alcohol 46.05.

Solution.

$$\mu = \frac{100}{\dfrac{10}{46.05} + \dfrac{100 - 10}{18.01}} = 19.15$$

The numerical value of the mol percentage of a given substance in a mixture is determined by the number of mols of that substance in 100 mols of the mixture. Thus, if ξ, o and π = weight percentage of the substances considered, and x, y and z = mol percentage of these substances, we have:

$$\left.\begin{aligned} x &= \frac{\frac{\xi}{\mu_x}}{\frac{\xi}{\mu_x}+\frac{o}{\mu_y}+\frac{\pi}{\mu_z}} \times 100 \text{ mol per cent} \\ y &= \frac{\frac{o}{\mu_y}}{\frac{\xi}{\mu_x}+\frac{o}{\mu_y}+\frac{\pi}{\mu_z}} \times 100 \text{ mol per cent} \\ z &= \frac{\frac{\pi}{\mu_z}}{\frac{\xi}{\mu_x}+\frac{o}{\mu_y}+\frac{\pi}{\mu_z}} \times 100 \text{ mol per cent} \end{aligned}\right\} \quad (2)$$

In the case of a binary mixture, equations (2) become:

$$x = \frac{\frac{\xi}{\mu_x}}{\frac{\xi}{\mu_x}+\frac{100-\xi}{\mu_y}} \times 100 \text{ mol per cent} \quad (2a)$$

To convert the mol percentage of a binary mixture into weight percentage, the following relation derived from (2a) is employed:

$$\xi = \frac{\frac{x}{\mu_y}}{\frac{100-x}{\mu_x}+\frac{x}{\mu_y}} \times 100 \text{ per cent by weight} \quad (2b)$$

Example. If a benzol-toluol mixture has a weight composition of 30 per cent of benzol, what is the mol percentage of benzol in the mixture? Molecular weight of benzol, $\mu_x = 78.05$; of toluol, $\mu_y = 92.06$.

Solution.

$$x = \frac{\frac{30}{78.05}}{\frac{30}{78.05}+\frac{70}{92.06}} \times 100 = 33.6 \text{ mol per cent (benzol)}$$

To simplify similar conversions, the interrelation between weight percentage and mol percentage may be shown graphically; weight percentage being plotted as abscissas and mol percentage as ordinates. A conversion chart can be constructed for any given mixture. Figure 1 presents the conversion chart for the two mixtures ethanol-water, and benzol-toluol. It is seen, for instance, that 10 per cent by weight of ethanol in an ethanol-water mixture corresponds to an alcohol content of 4.25 mol percent.

The average molecular weight may also be found from the mol percentage of the individual materials. The content in low boiler in mol percentage is

$$x = \frac{\dfrac{\dfrac{\xi}{\mu_x}}{100}}{\mu} \times 100 = \frac{\xi}{\mu_x} \times \mu$$

From this it follows that:

$$\xi = \frac{\mu_x}{\mu} \times x$$

$$o = \frac{\mu_y}{\mu} \times y$$

$$\pi = \frac{\mu_z}{\mu} \times z$$

Summing up these three equations, we have:

$$\xi + o + \pi = 100 = \frac{1}{\mu}(x\mu_x + y\mu_y + z\mu_z)$$

which in turn yields:

$$\mu = \frac{1}{100}(x\mu_x + y\mu_y + z\mu_z) \qquad (3)$$

In the following text, molar units will be used wherever it is necessary or useful. However, if the usual weight units are more practical, they will be employed. For the reasons given, this apparent lack of uniformity must be tolerated.

II. Interrelation between vapor concentration and partial pressure in multicomponent mixtures

To distinguish between liquid composition and vapor composition, concentrations referring to liquids have subscript F, while those referring to vapors have subscript D. Consequently, the mol percentage of the liquid components of a ternary mixture will be designated as x_F, y_F, z_F, and the corresponding vapor components as x_D, y_D, and z_D. Furthermore, p_x, p_y, p_z refer to the absolute partial pressures of the vapors of the individual

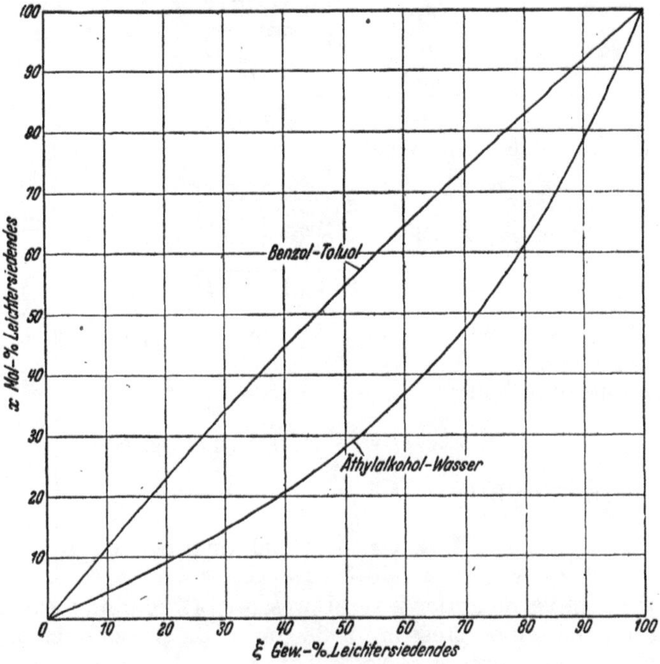

FIGURE 1

Conversion curves for ethanol-water and benzol-toluol mixtures

(The low boiler is named first.)
Mol % Leichtersiedendes—Mol per cent of low boiler
Gew. % Leichtersiedendes—Per cent by weight of low boiler
Äthylalkohol-Wasser—Ethanol-Water

components in a mixture at a given temperature, t, with the total pressure, p. Let the volume of 100 kg of a vapor mixture be V cubic meters. Then, the quantity of low boiler present is ξ_D kg. If we assume that each component as well as the vapor mixture obeys the ideal gas law, we have:

$$\frac{\xi_D}{\mu_x} = \frac{p_x \times V}{848\ (273 + t)}$$

$$\frac{o_D}{\mu_y} = \frac{p_y \times V}{848\ (273 + t)}$$

$$\frac{\pi_D}{\mu_z} = \frac{p_z \times V}{848\ (273 + t)}$$

$$\frac{100}{\mu} = \frac{p \times V}{848\ (273 + t)}$$

For the low boiler, this yields:

$$x_D = \frac{\frac{\xi_D}{\mu_x}}{\frac{100}{\mu_x}} \times 100 = \frac{\frac{p_x \times V}{848\ (273 + t)}}{\frac{p \times V}{848\ (273 + t)}} \times 100 = \frac{p_x}{p} \times 100 \text{ (mol per cent)}$$

for the medium boiler:

$$y_D = \frac{p_y}{p} \times 100 \text{ (mol per cent)} \qquad (4)$$

and for the high boiler:

$$z_D = \frac{p_z}{p} \times 100 \text{ (mol per cent)}$$

For example, if under a total pressure of 760 mm, a substance has a partial vapor pressure of 380 mm, the vapor mixture will contain 50 mol percent of the substance under consideration.

III. Equilibrium of boiling multicomponent mixtures

1. General remarks

To investigate the mechanism of distillation and rectification, some knowledge of the equilibrium conditions resulting when a multicomponent mixture boils is required. In this connection the compositions of the liquid and vapor at a given boiling temperature t, and total pressure, p, must be considered.

From a thermodynamic standpoint, the state of equilibrium is characterized by a maximum free enthalpy.[1] If we assume that the ideal gas laws hold for the case of the vapor, we have the following relationship based on the thermodynamic equilibrium conditions:[2]

$$x_F\left(\frac{\partial \ln p_x}{\partial x_F}\right)_t + y_F\left(\frac{\partial \ln p_y}{\partial y_F}\right)_t + z_F\left(\frac{\partial \ln p_z}{\partial z_F}\right)_t + \cdots = 0 \qquad (5)$$

The subscript, t, indicates that the differential quotient holds for constant temperature. Constancy of the total pressure may be disregarded, since it is assumed that the properties of the liquid are independent of pressure. In the case of a binary mixture, equation (5) becomes:

$$x_F\left(\frac{\partial \ln p_x}{\partial x_F}\right)_t = (100 - x_F)\left(\frac{\partial \ln p_y}{\partial (100 - x_F)}\right)_t \qquad (6)$$

This relation was developed by Duhem[3] and Margules.[4] The partial vapor pressure curve of the one substance as a function of liquid concentration is more easily plotted, if the corresponding curve of the other substance is known.

Equation (6) has a certain practical value since it can be used to evaluate the partial-vapor-pressure curves in some important cases. For this purpose, both sides of equation (6) are written as equal to a convenient function, say a constant C.[4] Integration then yields:

$$p_x = \left(\frac{x_F}{100}\right)^C \times P_x$$

$$p_y = \left(\frac{100 - x_F}{100}\right)^C \times P_y \qquad (7)$$

where P_x and P_y = boiling pressures of the pure components at the boiling temperature considered.

If:

$C = 0$. The liquid components are insoluble in each other; partial pressures are independent of the liquid concentration.

$C = 1$. Complete miscibility, ideal solution. Partial vapor-pressures are proportional to liquid concentrations.

$C = 0.5$, or $C = 2$ (examples, not general cases). Partial vapor-pressures vary with the liquid concentration at constant temperature, according to some curve. With a given concentration, the total pressure has a maximum (maximum vapor pressure at $C = 0.5$), or a minimum (minimum vapor pressure at $C = 2$).

These different cases will be treated in more detail; the case of partial solubility will also be discussed. However, the characteristic behavior of a liquid mixture must first be found by experiment.

With the exception of the cases of complete immiscibility and ideal solutions, the paths of the different curves must always be found experimentally. Accordingly the curves can be fitted mathematically (by some equation), with any desired degree of precision if each side of equation (6) is considered to be a power series instead of a constant, e.g.,

$$(C + C_1 \times x_F + C_2 \times x_F^2 + C_3 \times x_F^3 + \cdots)$$

2. Immiscible binary mixtures

The behavior of completely immiscible substances is best explained by the example of the immiscible binary mixture. In such a mixture, each substance behaves as if the other were

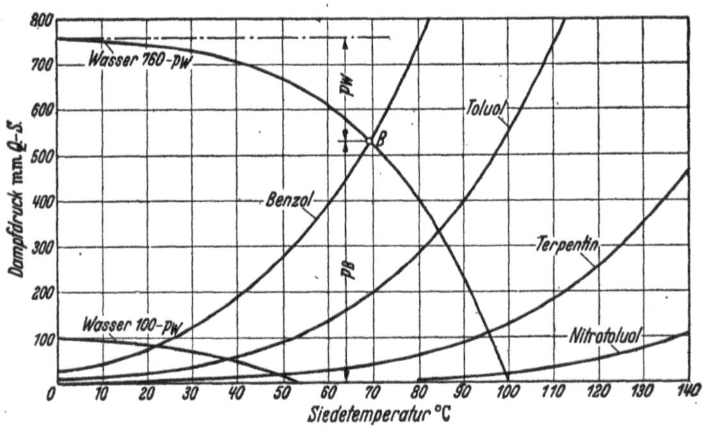

FIGURE 2

Boiling-point curves of substances insoluble in water, and determination of the boiling temperature and of the partial vapor pressures of immiscible binary systems

Dampfdruck mm Q.-S.—Vapor pressure, mm Hg
Siedetemperatur—Boiling temperature
Wasser—Water
Terpentin—Turpentine

absent. That is, at a given temperature, each substance has a partial pressure independent of the composition of the boiling liquid and equal to the boiling pressure (vapor pressure) of the pure substance at the same temperature.

In industrial practice, one of the immiscible components is usually water. For this case, Figure 2 shows the boiling temperature of the mixture and the vapor composition can be found for any total pressure. In this plot the abscissas give the boiling temperature, and the ordinates give the boiling pressure at the corresponding temperature of the substances mixed with water.

To find the conditions for a total pressure of 760 mm plot the difference $760 - P_W$ rather than P_W, the boiling pressure of the water against the boiling temperature.[4] The intersection of this $(760 - P_W)$ line with the boiling curve of a given mixture gives the boiling temperature of the mixture and the partial vapor-pressures directly. Intersection *B* in Figure 2 gives for a benzol-water mixture a boiling point of 69°C, a partial pressure of benzol $P_B = 534$ mm, and a partial vapor pressure of water $P_W = 226$ mm.

The benzol concentration in the vapor, according to equation (4), is:

$$x_D = \frac{534}{760} \times 100 = 70.2 \text{ mol per cent}$$

By a similar process, boiling temperatures and vapor compositons can be found if the total pressure is some other value, e.g., 100 mm instead of 760 mm. Then the difference $100 - p_W$ is plotted as a function of the boiling temperature instead of the boiling curve for water.

This curve is also shown in Figure 2. Again, p_W stands for the vapor pressure of pure water as a function of its boiling temperature. Under the decreased total pressure, the boiling temperature of the mixture is lowered to 21.5°C, and the benzol concentration in the vapor becomes 82 mol per cent. Thus the benzol concentration in the vapor increases as pressure decreases.

A characteristic of immiscible mixtures is that the partial vapor pressure of each component and the total pressure are independent of the concentration x_F of the low boiler; this is seen in Figure 3.

However, since completely immiscible mixtures are seldom encountered in industrial practice, the heavy lines in Figure 3 should be replaced, for very low concentrations of the different components, with the dotted lines shown.

From Figure 3 and equation (4) it follows that, in the case of immiscible mixtures, the vapor generated is of constant composition, independent of the composition of the liquid; see also Figure 4. The horizontal line indicating the benzol concentration in the vapor intersects the diagonal at *A*. The latter indicates the condition at which liquid and vapor have identical compositions. A similar point is more generally called an azeotropic point. In the case of immiscible mixtures, the vapor has the composition of the azeotropic point.

In Figure 2 it may be seen that the boiling temperature of an immiscible mixture is lower than the boiling temperatures of the individual components. For a total pressure of 760 mm, the boiling temperature of a benzol-water mixture is 69°C, although that of water is 100°C and that of benzol is 80°C.

This phenomenon is used to advantage industrially in steam distillation. It is employed in the case of liquids which have very high boiling points or which decompose at their own boiling

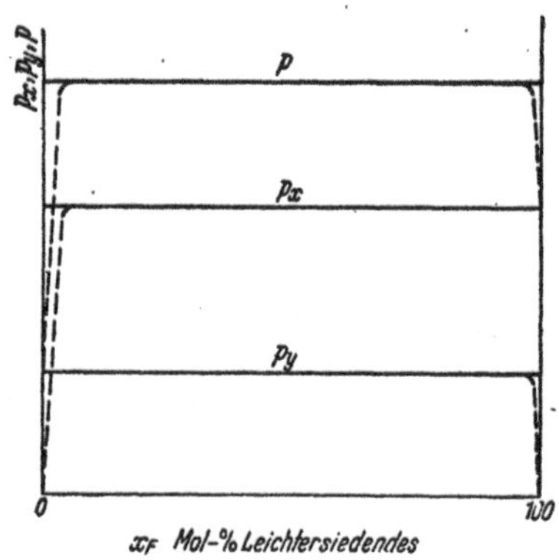

FIGURE 3

Partial vapor pressures p_x and p_y, and total pressure p, as functions of the liquid composition of an immiscible binary system at constant boiling temperature

Mol % Leichtersiedendes—Mol per cent of low boiler

temperature, e.g., fatty acids. The purpose of this procedure is to remove impurities, especially non-volatile ones. In actual practice steam is usually blown directly into the liquid. The steam should then be thoroughly dispersed, and the distillate and water must reach the evaporating surface minutely divided to attain the equilibrium state for immiscible mixtures as shown in Figure 2.

Irregular conditions and deviations from this equilibrium state will result whenever the steam passes through the liquid

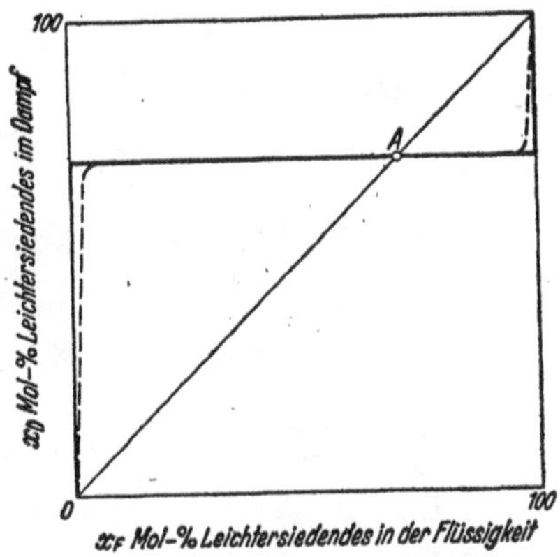

FIGURE 4

Vapor composition x_D as a function of the liquid composition x_F of an immiscible binary system for constant total pressure (p = 760 mm Hg). Benzol and water.

Mol % Leichtersiedendes im Dampf—Mol per cent of low boiler in vapor

Mol % Leichtersiedendes in der Flüssigkeit—Mol per cent of low boiler in liquid

in the form of large bubbles, and when the contact between both liquids and the evaporation surface is incomplete.

The steam will suffice to supply all of the heat required, as will be seen. However, to prevent excessive condensation in the still, the heat required for evaporation may be transmitted, partly or wholly, by heating surfaces.

Example. If steam at 1.5 atm absolute and 120°C is available, how much of it is required to produce 100 kg turpentine by atmospheric distillation? Turpentine has a heat of evaporation of 74 kcal/kg and a molecular weight, $\mu_y = 136$. Molecular weight of water, $\mu_x = 18.01$.

From Figure 2, the boiling temperature of the mixture is 96°C, the partial vapor pressure of the turpentine, $p_y = 114$ mm, and that of the water, $p_x = 646$ mm. Thus, the vapor contains

$$x_D = \frac{646}{760} \times 100 = 85 \text{ mol per cent water.}$$

According to equation (2b), this corresponds to

$$\xi_D = \frac{\dfrac{x_D}{\mu_y}}{\dfrac{100 - x_D}{\mu_x} + \dfrac{x_D}{\mu_y}} \times 100 = \frac{\dfrac{85}{136}}{\dfrac{15}{18.01} + \dfrac{85}{136}} = 42.8 \text{ per cent by weight of water.}$$

Thus, with $G_T = 100$ kg turpentine, the corresponding quantity of water is:

$$G_W = \frac{42.8}{100 - 42.8} \times 100 = 74.8 \text{ kg water.}$$

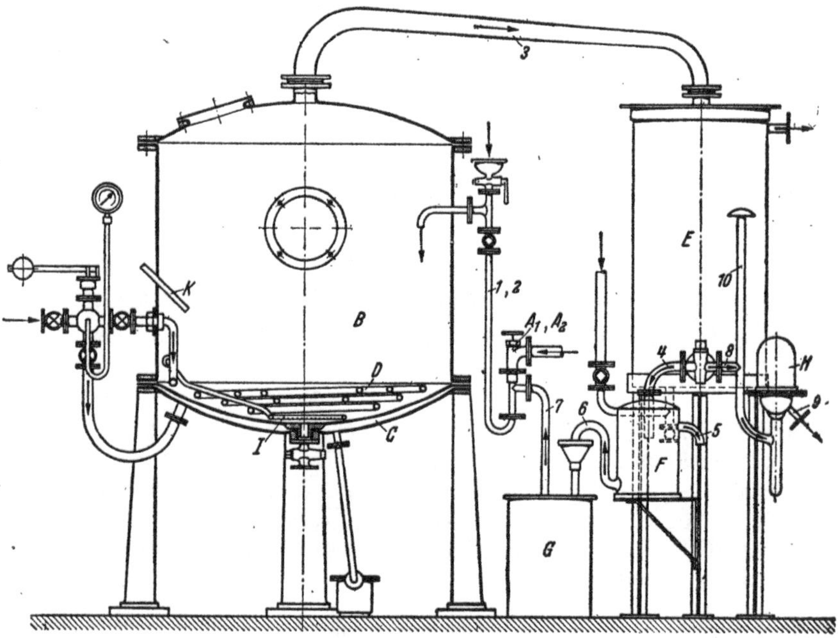

FIGURE 5
Steam distillation unit

Figure 5 shows a steam distillation unit. Two steam ejectors A_1 and A_2, arranged in parallel, feed the still B by means of parallel lines 1 and 2, with oil and water. The heat of evaporation is provided by a steam jacket C, and a steam coil D. The vapor mixture passes through line 3 and heat exchanger E, condenses and is cooled. By means of a three-way cock and line 4 it passes on to decanter F. From the latter, line 5 discharges the oil produced, while water flows through line 6 to tank *G*.

The water is transferred by ejector A_1 via line 7 to the still. If continuous distillation without refluxing is desired, the indirect heating system is disconnected, and steam is injected directly through line *J*. If no water is used in distillation, the product is recovered via the three-way cock, line 8, tailbox *H* and line 9, while noncondensible gases are removed through vent 10. *K* is a thermometer well which facilitates checking the progress of the distillation.

To evaporate the turpentine, $100 \times 74 = 7400$ kcal are required. The total heat in the steam is of 648 kcal/kg, and that of the water vapor at 96°C = 637 kcal/kg. Thus, to drive off 100 kg turpentine,

$$G_K = \frac{7400 - 74.8\ (648 - 637)}{648 - 96} = 11.9 \text{ kg}$$

This quantity of water vapor will condense in the liquid mixture. Therefore, the total amount of steam required for producing 100 kg turpentine is $74.8 + 11.9 = 86.7$ kg.

Figure 2 indicates that simple steam distillation, in which saturated steam passes into the vapor mixture, can be used only as long as the distillation temperature is below the condensing temperature of the pure steam.

In all other cases, superheated steam must be applied. It is then necessary to inject superheated steam into the liquid, and to transfer the heat of evaporation through fire or steam heated surfaces.

Strictly speaking, the same effect can be obtained by injecting air into the liquid, since the air exerts a certain partial pressure, and consequently lowers the boiling temperature of the liquid. However, since the air cannot be condensed conveniently, and would always entrain appreciable amounts of distillate, such a procedure would be wasteful. Steam is used because it can be condensed along with the product vapors and the water formed can be separated by decanting because of immiscibility.

The laws applying to binary mixtures may be extended to cover insoluble ternary and multicomponent mixtures.

3. Partially miscible binary mixtures

In contrast to the case of complete miscibility, with partial miscibility the partial pressure of the vapor of that material

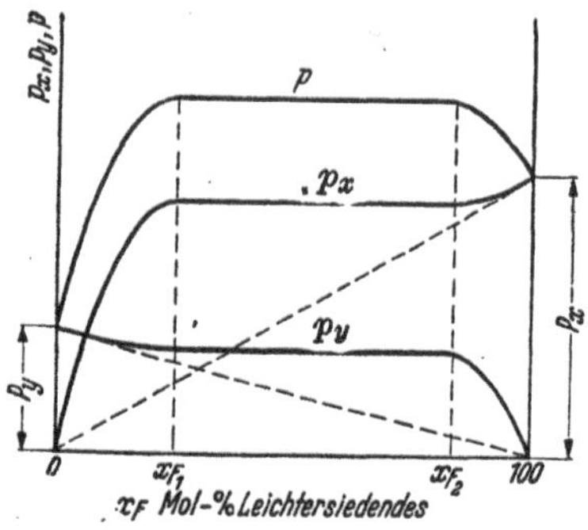

FIGURE 6

Partial vapor pressures p_x and p_y, and total pressure p, as functions of the concentration x_F mol per cent of low boiler in the liquid of a partially miscible binary system, at constant boiling temperature

Mol % Leichtersiedendes—Mol per cent of low boiler

present in the smallest amount, is decreased. This gives the interrelation between partial pressures, p_x, of the low boiler and, p_y, of the high boiler as well as of total pressure p, and the concentration x_F mol per cent of the low boiler in the liquid at constant boiling temperature, as is indicated in Figure 6.

If the percentage of low-boiler in the liquid is small, it will be dissolved completely in the mixture. Its partial pressure will start at zero at first increase as concentration increases, and will remain constant as the concentration passes into the immiscible region bounded by liquid compositions x_{F1} and x_{F2}.

For high x_F values, p_x will increase further up to value P_x, which is the partial vapor pressure of the pure low boiler at the given boiling temperature. In the case of the partial vapor pressure of the high boiler, which, in its pure state, at constant boiling temperature has a boiling pressure of P_y, the situation is similar. Beginning with P_y, the total pressure increases as the amount of low boiler in the liquid increases, up to a value which remains constant in the immiscibility region, and drops to P_x in the case of high concentrations of the low boiler. Thus, the constant p value within the immiscibility region represents a maximum value of the vapor pressure.

The following is characteristic of the immiscibility region.

Although in the case of liquid composition below x_{F1} and exceeding x_{F2} the two liquids are completely miscible, the composition within the immiscibility region will consist of liquid mixtures that are immiscible. If these are permitted to settle at the boiling temperature (Figure 6), two separate layers will result.

One layer will have a concentration x_{F1}, and the other x_{F2}. Depending on whether the average liquid concentration x_F of the mixture approaches either x_{F1} or x_{F2}, one of the layers will be denser. A liquid of concentration x_{F1} is on the verge of forming two layers. If, however, low boiler is added, the layer having the composition x_{F2} will begin to increase. If sufficient low boiler has been added to the mixture to make its average concentration equal to x_{F2}, the two layers merge into one again.

Thus, in the case of complete immiscibility and at constant boiling temperature, the partial vapor pressures of the individual components are independent of the liquid concentration and equal to the boiling pressures of the pure components at that boiling temperature. In this case,

$$p_x = P_x$$

and (8)

$$p_y = P_y$$

In the case of partial miscibility (Figure 6) within the area of immiscibility the partial vapor pressures are smaller than the boiling pressures of the pure components. Therefore, as the solubility of the liquids increases, both partial and total pressures diminish. At constant boiling pressure, however, boiling temperature increases as solubility increases. While, for example, the binary mixture, water + *n*-butanol, at a total pressure of 760 mm Hg, would within the immiscibility region possess a boiling temperature of 92.7°C,[7] in the case of complete immiscibility the procedure outlined in Figure 2 would yield a boiling temperature of only 89°C, at the same total pressure.

In order to determine whether a partially soluble mixture can be separated by distillation or rectification, it is necessary to know the interdependence of the vapor concentration x_D, mol per cent of the low boiler and the liquid concentration x_F, mol per cent at constant total pressure. This relation can be found directly by measuring the liquid and vapor concentrations when boiling at constant pressure, or from the curve of partial and total pressures at different boiling temperatures.

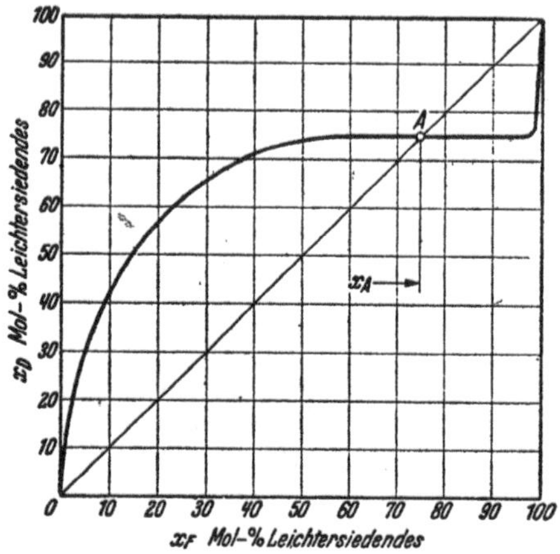

FIGURE 7

Equilibrium curve of the partially miscible system water-n-butanol at a total pressure of 760 mm Hg

(The low boiler is named first.)
Mol % Leichtersiedendes—Mol per cent of low boiler

In the second case, equation (4) is used to determine the vapor content x_D, p being the boiling pressure given, and p_x the partial vapor pressure corresponding to a specified liquid concentration x_F. The interdependence of the concentration of the low boiler in the vapor, x_D, and concentration in the liquid, x_F, in the case of the partially soluble mixture, water + n-butanol, is shown in Figure 7.

As the percentage of water in the liquid increases, its concentration in the vapor increases within the miscible region, until a constant value for the immiscibility region is reached. The latter will yield, with the diagonal in the graph, intersection A. With larger x_F percentages, it increases, within the second miscible region, up to x_D = 100 mol per cent.

The vapor composition vs. liquid composition curve is generally labelled "equilibrium curve." Within the region of immiscibility, the equilibrium curve is horizontal, the boiling temperature of the mixture remaining constant. In the case of the pair of substances mentioned, it becomes 92.7°C at a total pressure of 760 mm Hg.

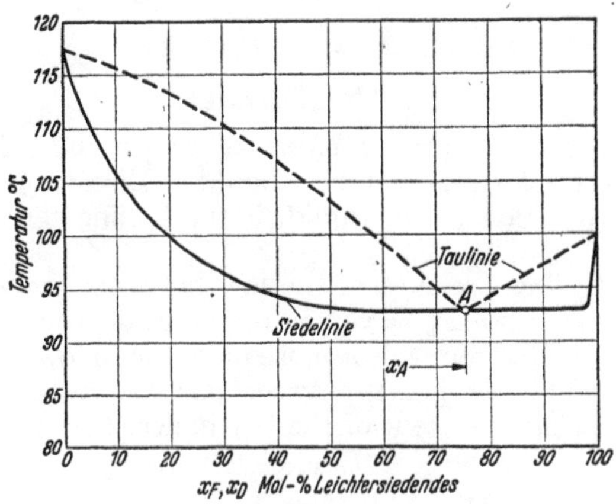

FIGURE 8

Boiling-point and dewpoint curve of the partially miscible system water-n-butanol for a pressure of 760 mm Hg

Temperatur—Temperature
Mol % Leichtersiedendes—Mol per cent of low boiler
Taulinie—Dewpoint curve
Siedelinie—Boiling-point curve

The boiling temperature increases as the water concentration, x_F, decreases, up to 117.4°C with $x_F = 0$. It also increases, outside of the immiscibility region, as the water content increases, up to 100°C at $x_F = 100$.

If the boiling temperature of a mixture is plotted against the liquid and vapor compositions, the boiling-point and the dewpoint lines are obtained. In Figure 8, they are shown for the substances given in Figure 7. Both lines exhibit a minimum boiling temperature, and intersect at A, a point corresponding to the concentration $x_A = x_D = x_F = 75.2$ mol per cent, for the azeotropic solution.

On the basis of complete immiscibility, calculation of the azeotropic point gives the concentration of water as 67.7 mol per cent. Thus, partial solubility will effect an increase of low boiler in the vapor. (See also Appendix, Equilibrium Curves and Tables.)

4. Completely miscible binary mixtures

a. Ideal mixtures

In the case of an ideal mixture of two liquids, the partial vapor pressure of each substance should be directly proportional to its concentration in the liquid, if the boiling temperature is constant.

If P_x = boiling pressure of the low boiler at a given boiling temperature t, and P_y = vapor pressure of the high boiler at the same boiling temperature, we have the following relations, at a constant boiling point, between partial pressures and concentration x_F mol per cent of the low boiler:

$$p_x = \frac{x_F}{100} \times P_x \tag{9}$$

and

$$p_y = \frac{100 - x_F}{100} \times P_y \tag{9a}$$

Thus the total pressure becomes:

$$p = p_x + p_y = \frac{x_F}{100} \times P_x + \frac{100 - x_F}{100} \times P_y \tag{10}$$

In other words, both the variation of the partial pressures and the total pressure are linear functions of the liquid composition, as indicated in Figure 9.

However, distillation and rectification occur at constant total pressure, not constant temperature. Therefore, it is the relationship between partial vapor pressure and vapor or liquid composition, at constant boiling pressure, which is important.

This relation is found as follows: If P_{x1} = boiling pressure of the pure low boiler corresponding to boiling temperature t_1, and P_{y1} = boiling pressure of the pure high boiler, the liquid composition given by equation 10, and corresponding to t_1 is

$$x_{F1} = \frac{p - P_{y1}}{P_{x1} - P_{y1}} \times 100 \tag{11}$$

From this the composition of the vapor formed at the boiling temperature, t_1, from a liquid composition x_{F1} is

$$x_{D1} = \frac{\dfrac{x_{F1}}{100} \times P_{x1}}{p} \times 100 = \frac{P_{x1}}{p} \times x_{F1} \tag{12}$$

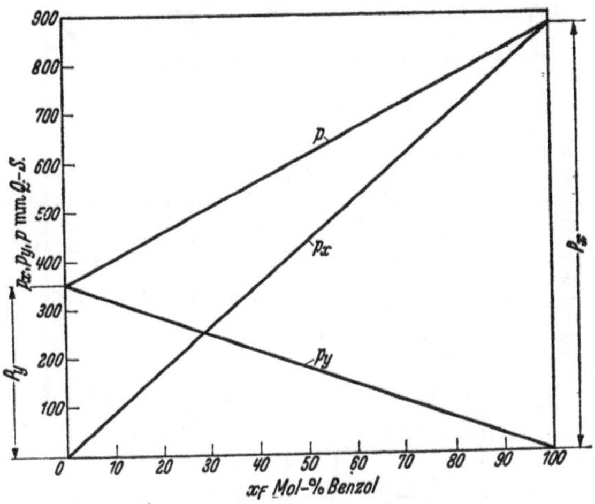

FIGURE 9

Partial vapor pressures and total pressure of the ideal binary mixture benzol-toluol as functions of the concentration of low boiler x_F mol per cent benzol in the liquid. Boiling temperature 85°C

mm Q.-S.—mm Hg

In this way, pairs of x_{F1} and x_{D1} values are found for a given t_1. Similarly, another liquid composition, x_{F2}, and another vapor composition, x_{D2}, are found to correspond to boiling temperature t_2, as well as a further x_{F3} and x_{D3} corresponding to t_3, etc.

By plotting x_{D1}, x_{D2}, x_{D3}, etc., against x_{F2}, x_{F2}, x_{F3}, etc., the equilibrium curve is obtained. Figure 10 shows the variation between vapor and liquid composition for an ideal mixture of benzol and toluol.

Throughout the entire concentration range, the curve remains above the diagonal of the graph. This is, however, not a general characteristic of all ideal solutions. Each point on the equilibrium curve corresponds to a definite boiling temperature. If this temperature is plotted against liquid concentration, the boiling curve is obtained, while the dewpoint curve gives the relation between vapor condensation temperature and vapor composition. Both lines are presented in Figure 11.

From such a plot, it is possible to determine the corresponding liquid and vapor compositions, at any given boiling temperature,

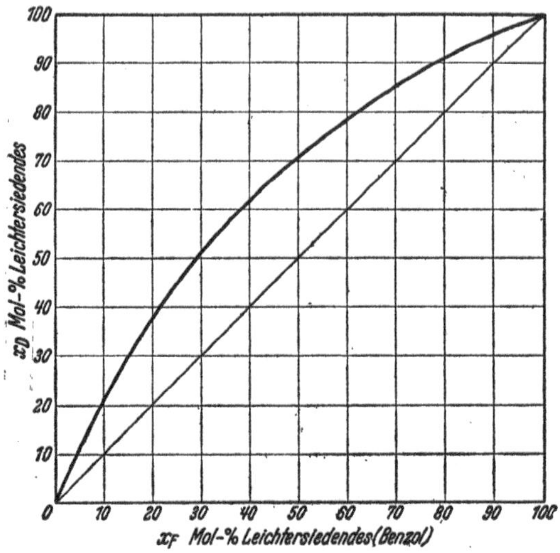

FIGURE 10

Equilibrium curve of the ideal mixture benzol-toluol at a pressure of 760 mm Hg

Mol % Leichtersiedendes—Mol per cent low boiler

as the intersection of the boiling-point and dewpoint curves with the horizontal at the given temperature level.

It is assumed that ideal mixtures result when the two components are chemically similar. Another condition is that the critical pressures of the components should be equal, and that their molecular attractions should follow some definite law.[8,9] If a_1 and a_2 = attracting forces of molecules of the specified components, and $a_{1,2}$ = force of attraction between molecules of the different components, the condition to be fulfilled is

$$a_{1,2} = \sqrt{a_1 \times a_2}$$

In mixtures of the immiscible liquids, the molecular attraction between different components is negligible as compared to the attraction between molecules of the same kind.

Partial vapor pressures and vapor composition (Equilibrium curve) can be determined, by the method explained, from the boiling pressures of the pure components, but only in the case of immiscible and of completely miscible ideal mixtures. In all

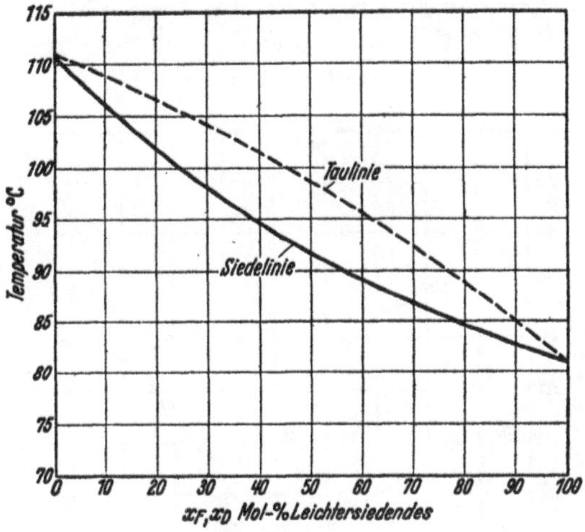

FIGURE 11

Boiling-point and dewpoint curve for the ideal mixture benzol-toluol at a pressure of 760 mm Hg

Taulinie—Dewpoint curve
Siedelinie—Boiling-point curve
Temperatur—Temperature
Mol % Leichtersiedendes—Mol per cent of low boiler

other cases, these relationships must be determined experimentally.

b. Mixtures with maximum boiling points (minimum vapor pressure)

In the case of an ideal mixture, the attractive forces between molecules of one kind are equal to the forces of attraction between unlike substances. However, if force $a_{1,2}$ is appreciably larger than the attractive forces between similar molecules, a_1 and a_2, then, when the mixture is boiled, the molecules are retained in the liquid, and pass into the vapor less easily than in the case of ideal solutions. As a consequence, the partial pressures become smaller.

This condition is shown in Figure 12, in which, for the completely miscible liquid mixture nitric acid + water,[10] the partial pressures p_x of the low boiler (nitric acid) and p_y of the high boiler are plotted against the liquid concentration. Both partial pressure curves are located below the dotted straight lines which correspond to the partial pressures of ideal solutions, in accordance with the vapor pressures of the pure components P_x and P_y, at the temperatures considered.

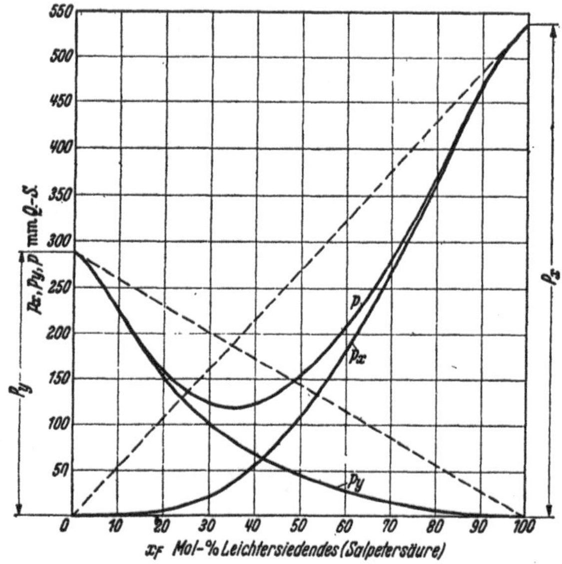

FIGURE 12

Partial pressures p_x and p_y and total pressure p of the mixture nitric acid-water at the constant boiling temperature of 75°C

(This system has a minimum vapor pressure.)
mm Q.-S.—mm Hg
Mol % Leichtersiedendes—Mol per cent of low boiler
Salpetersäure—Nitric acid

Since the partial-pressure lines are tangent to the straight connecting lines mentioned, the behavior of the liquid mixture will approach that of an ideal solution as the pure components are approached.

The sum of partial pressures which corresponds to a given liquid composition gives the total pressure curve p. At some definite liquid composition, this total pressure exhibits a minimum.

From the path of the partial pressure curves applying for different boiling temperatures and the total pressure, it is possible to determine the equilibrium curve by means of equation (4) for a given total pressure. For the mixture mentioned, the derived curve is shown in Figure 13. For low liquid concentration, the equilibrium curve is located below the diagonal of the graph, intersects it at point A and then remains above the diagonal. A miscible mixture having the composition of the point A is generally called an azeotropic mixture.

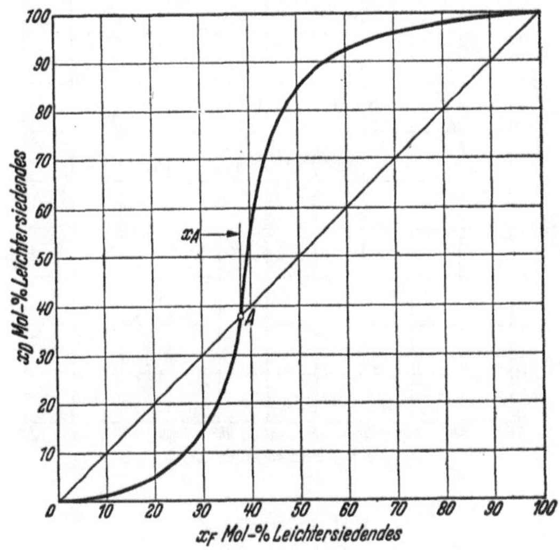

FIGURE 13

Equilibrium curve of the mixture nitric acid-water at a pressure of 760 mm Hg

(This system has a maximum boiling point.)
Mol % Leichtersiedendes—Mol per cent of low boiler

In the case shown, and for a total pressure of 760 mm Hg, the azeotropic mixture contains 38.3 mol per cent or 68.5 per cent by weight of nitric acid.

The azeotropic point hinders industrial separation of liquid mixtures as simple distillation or rectification cannot proceed beyond this composition. It thus forms a barrier to further separation.

The variation of the boiling temperature with liquid and vapor composition for the mixture, exhibiting a maximum boiling point, which has been discussed, is shown in Figure 14 which gives the boiling-point and dewpoint curves. As the nitric acid concentration increases, the boiling temperature exceeds 100°C, reaching a maximum at 122°C at point *A*. Then the boiling temperature diminishes again, and reaches that of pure nitric acid, 87°C.

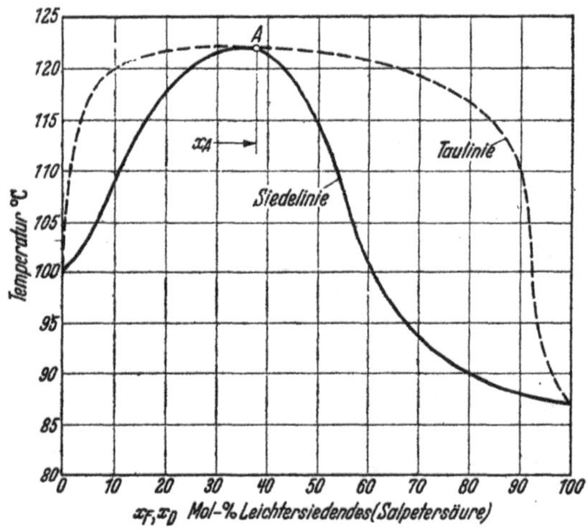

FIGURE 14

Boiling-point and dewpoint curves for the liquid mixture nitric acid-water at a pressure of 760 mm Hg

(This system has a maximum boiling point.)
Taulinie—Dewpoint curve
Siedelinie—Boiling-point curve
Temperatur—Temperature
Mol % Leichtersiedendes—Mol per cent of low boiler
Salpetersäure—Nitric acid

The boiling-point curve shows, for instance, that there are two liquid mixtures that boil at 110°C, one containing 11 mol per cent, and the other containing 53 mol per cent HNO_3, producing vapors containing respectively 1.5 mol per cent HNO_3, and 90 mol per cent of HNO_3. In both Figure 13 and Figure 14, point *A* corresponds to composition x_A of the azeotropic mixture.

In many of the known binary mixtures having a maximum boiling point one of the components is an acid. There are other cases, where liquid mixtures have a minimum vapor pressure,[11,12] although the complete details of the equilibrium curves and the vapor-pressure curves for many of these have not yet been determined.*

c. Mixtures with minimum boiling points (maximum vapor pressure)

Mixtures with maximum vapor pressure are more common than mixtures with minimum vapor pressure. The latter may result where the molecular attraction $a_{1,2}$ between molecules of different materials is very small compared to that between molecules of one kind, a_1 or a_2.

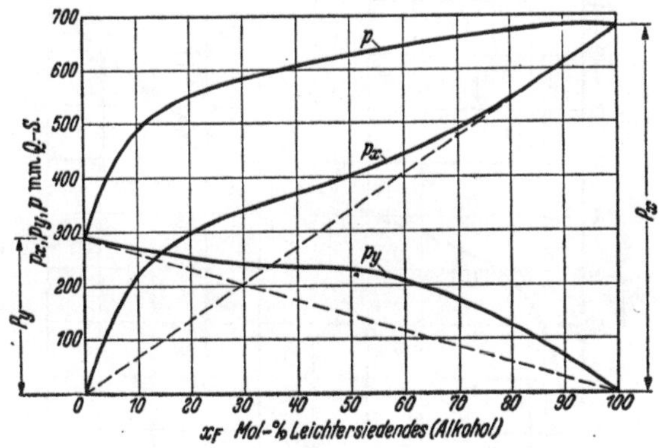

FIGURE 15

Partial pressures p_x and p_y and total pressure p of the mixture ethanol-water, at constant boiling temperature of 75°C

(This system has a maximum vapor pressure.)
mm Q.-S.—mm Hg
Mol % Leichtersiedendes—Mol per cent of low boiler
Alkohol—Alcohol

Another phenomenon may also cause a maximum vapor pressure. In the case of association of one, or both, of the liquids, in the pure state, this association may counterbalance attraction between different molecules to some extent, so that

*See also equilibrium curves and tables of industrially important binary mixtures in the Appendix.

new types of molecules are exposed which cause an increase in partial vapor pressure.

Figure 15, which refers to ethanol-water mixtures,[13] demonstrates this phenomenon in the case of the partial pressures, p_x and p_y, of alcohol and water respectively which are located above the straight dotted lines.

The fact that the partial pressure, p_x, of alcohol increases rapidly with small concentrations, x_F, seems to indicate that pure water associates considerably. The total pressure in the p curve is derived from the partial pressure. For the boiling temperature selected, 75°C, the mixture exhibits a maximum pressure at a liquid concentration of 90.3 mol per cent alcohol.

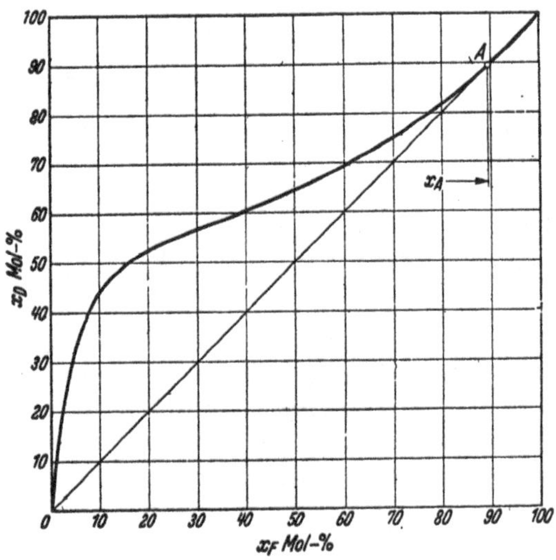

FIGURE 16

Equilibrium curve of the mixture ethanol-water at a pressure of 760 mm Hg

(This system has a minimum boiling point.)

Figure 16 shows the equilibrium curve, which is important in the development of separation methods. At small concentrations of the low boiler, the equilibrium curve is located above the diagonal, intersects at A and lies below the diagonal at high alcohol concentrations. In this case, the azeotropic point at a pressure of 760 mm corresponds to an azeotropic mixture of 89.5 mol per cent ethyl alcohol or 95.57 per cent by weight.

The variation of both the vapor and liquid concentrations with both temperatures is given by the boiling-point and dew-point curves in Figure 17. The boiling temperature decreases below 100°C as the alcohol concentration increases, until a minimum of 78.15°C, at an alcohol concentration of 89.5 mol per cent at 760 mm Hg is reached.

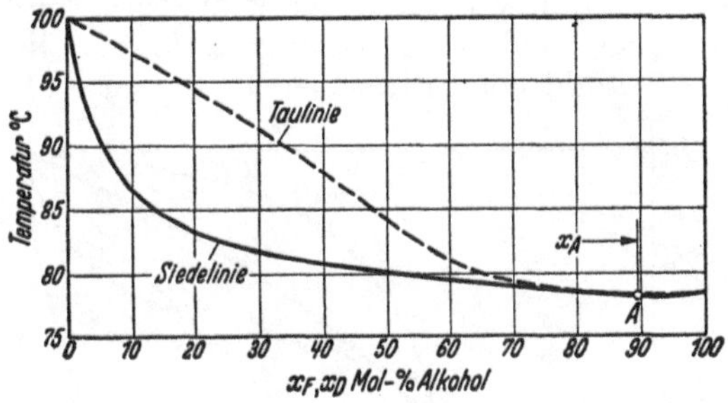

FIGURE 17

Boiling-point and dewpoint curves for the mixture ethanol-water, at a pressure of 760 mm Hg

(This system has a minimum boiling point.)
Taulinie—Dewpoint curve
Siedelinie—Boiling-point curve
Temperatur—Temperature
Mol % Alkohol—Mol per cent of alcohol

Once this minimum has been passed, the boiling temperature increases again as x_F increases, and reaches a temperature of 78.3°C for an x_F of 100 (pure alcohol). In the region between the concentration of $x_F = 89.5$ and $x_F = 100$ mol per cent the boiling point and dewpoint curves are very near each other, but do not merge. (See also data on mixtures with minimum boiling points, Appendix.) A very large number of binary mixtures with minimum boiling points is known; however, equilibrium and partial-pressure curves have so far been investigated in detail for a limited number only.

In addition to the cases mentioned, another condition under which an azeotropic point may form is when the vapor-pressure curves (vapor pressure as a function of the boiling temperature) of the two pure liquids intersect.

5. Review of typical equilibrium curves of binary mixtures

The characteristic behavior of the boiling properties of binary mixtures permits one to predict the nature of a binary mixture from the outline of an equilibrium curve.

Figure 18 indicates the different possibilities.

In the case of a completely immiscible mixture, the equilibrium curve, *a*, is represented by a straight horizontal line over the entire range of liquid concentrations; in the case of a partially

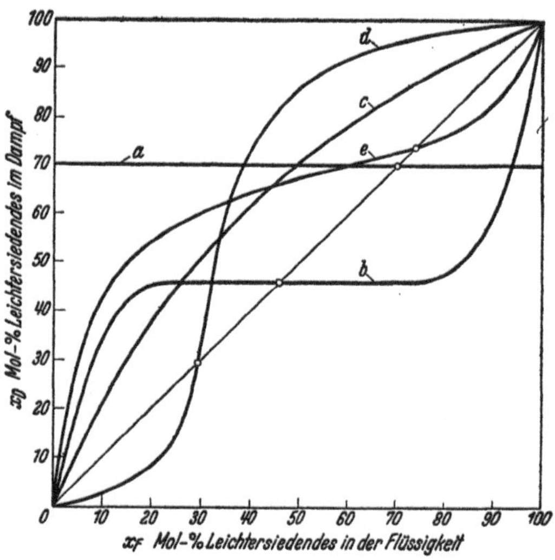

FIGURE 18

Equilibrium curves for different types of binary mixtures at constant pressure

a. Complete immiscibility
b. Partial miscibility
c. Complete miscibility; no azeotropic point (Ideal solution)
d. Maximum boiling point
e. Minimum boiling point

Mol % Leichtersiedendes im Dampf—Mol per cent of low boiler in vapor phase
Mol % Leichtersiedendes in der Flüssigkeit—Mol per cent of low boiler in liquid phase

miscible mixture, the curve, *b*, is horizontal only over a limited concentration range.

Both lines intersect the diagonal at an azeotropic point, where the compositions of the liquid and vapor are alike.

In all other cases, the equilibrium curve is a curved line. For an ideal mixture it follows path *c* always above the diagonal. For mixtures with a maximum boiling point (curve *d*) it remains below the diagonal with small percentages of low boiler, intersects it at the azeotropic point, and remains above the diagonal with high liquid concentrations.

In the case of a mixture with a minimum boiling point (curve *e*) for small concentrations of low boiler in the liquid the amounts of low boiler in the vapor exceed the former. They are equal at the azeotropic point. For liquid concentrations exceeding that of the azeotropic point, vapor concentration is below that of the liquid.

6. Ternary mixtures

a. General remarks

In establishing equilibrium conditions for ternary mixtures, the same principles hold as in the case of binary ones. Since, however, one additional component is present, a different mode of graphical representation must be selected, preferably that utilizing an equilateral triangle.

The apices of the triangle designate the three components and the sides are scaled off in mol per cent as shown in Figure 19. A general point *A* describing a ternary mixture of the components benzol, toluol and *m*-xylol corresponds, for example, to a composition of 25 mol per cent benzol, 25 mol per cent toluol and 50 mol per cent *m*-xylol. The basic geometrical principles underlying this method of representation is that the sum of the perpendicular distances h_1, h_2 and h_3 equals altitude h of the triangle.

The concentration of the low boiler in the liquid is x_F mol per cent, that of the medium boiler y_F mol per cent, that of the high boiler z_F mol per cent. Then:

$$x_F \times y_F \times z_F = 100 \qquad (13)$$

Correspondingly, for the vapor concentrations x_D, y_D and z_D mol per cent:

$$x_D \times y_D \times z_D = 100 \qquad (14)$$

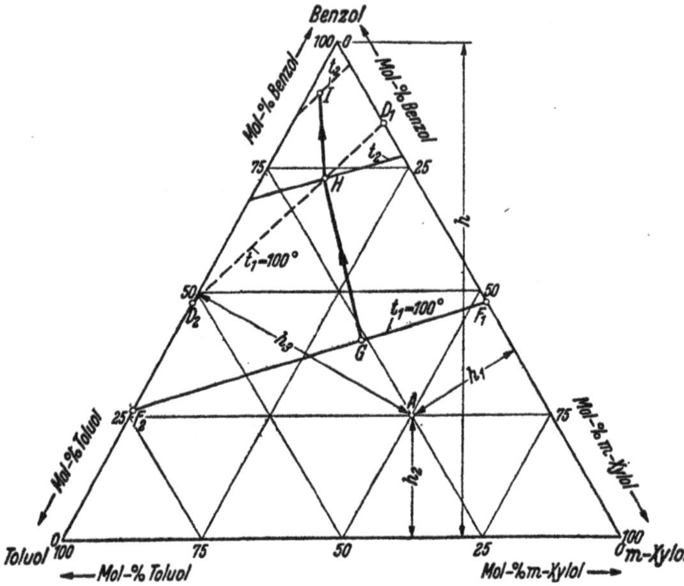

FIGURE 19

Determination of the liquid and vapor isothermals, and distillation paths

Ideal ternary mixture

If p_x = vapor pressure of the low boiler in the vapor mixture at a given boiling temperature,

p_y = vapor pressure of the medium boiler,

p_z = partial vapor pressure of the high boiler,

then the total pressure p is the sum:

$$p_x + p_y + p_z = p \tag{15}$$

As in the previous case, ternary mixtures may be subdivided into subgroups, depending on mutual solubility of the components. If all three components are insoluble in each other, each of these substances will exert a partial pressure equalling the vapor pressure of the pure component at the same boiling temperature. From the industrial standpoint, mutually insoluble ternary mixtures are unimportant. Usually the three components are miscible with each other, or individual pairs of components comprising the ternary mixture are miscible.

b. Ideal ternary mixtures (single distillation area in the triangular plot)

As in the case of an ideal binary mixture, an ideal ternary mixture is characterized by the fact that at constant boiling temperature the partial vapor pressure of each substance increases linearly with its concentration in mol per cent in the liquid.

If P_x, P_y and P_z = boiling pressures of the pure components at a given boiling temperature, subscript x referring to the low and z to the high boiler, the partial pressures of the pure components corresponding to a definite liquid concentration at a given constant boiling temperature are given by the following relations:

$$p_x = \frac{Px}{100} \times x_F \quad (16)$$

$$p_y = \frac{P_y}{100} \times y_F \quad (17)$$

$$p_z = \frac{P_z}{100} \times z_F \quad (18)$$

If values of p_z from equation (15) and z_F from equation (13) are used in equation (18):

$$p - p_x - p_y = \frac{P_z}{100} \times (100 - x_F - y_F) \quad (18a)$$

In this equation, p_x and p_y can be replaced by their corresponding values given in equations (16) and (17):

$$p - P_z = \frac{x_F}{100} \times (P_x - P_z) + \frac{y_F}{100} \times (P_y - P_z) \quad (19)$$

At a given distillation pressure and constant boiling temperature, only x_F and y_F are variable in equation (19). If one of the variables is selected, the other can be computed.

Thus, equation (19) indicates the locus for all compositions having the same boiling temperature. For this case, all points corresponding to concentrations x_F and y_F are located on the triangle on the straight line path which cuts two sides of the triangle at two points, for instance F_1 and F_2 in Figure 19. This path in figure 19 thus represents a liquid isothermal for a given temperature, e.g., t_1.

In equation (19), the variable liquid compositions x_F and y_F may be replaced by the vapor concentrations at the same boiling temperature, by replacing the partial pressures p_x and p_y by their values obtained from equations (16) and (17) in equation (4):

$$x_D = \frac{p_x}{p} \times 100 = \frac{P_x}{100} \times x_F \, 100 \frac{1}{p} = \frac{P_x}{p} \times x_F$$

Thus:

$$x_F = \frac{p}{P_x} \times x_D$$

$$y_F = \frac{p}{P_y} \times y_D \qquad (20)$$

$$z_F = \frac{p}{P_z} \times z_D$$

On substituting the values x_F and y_F of equation (20) into (19), the equation of the vapor isothermal is obtained:

$$p - P_z = \frac{p}{P_x \times 100} \times x_D \, (P_x - P_z) + \frac{p}{P_y \times 100} \times y_D \, (P_y - P_z) \qquad (21)$$

In the triangular plot, this relation again yields a straight line. On this line are located all the points representing vapors of the ternary mixture having the same condensation temperature. The straight line intersects two sides of the triangle in points D_1 and D_2, as indicated in Figure 19.

The liquid and vapor isothermals applying for a given temperature are fixed if the points of intersection F_1 and F_2 as well as D_1 and D_2 (Figure 19) are given. They may be computed from the relations holding for ideal binary mixtures, equations (11) and (12). However, a graphical method, suggested by Dr. W. Quade, permits a more rapid solution. From equations (15) to (18) it can be shown that

$$p = \frac{P_x}{100} \times x_F + \frac{P_y}{100} \times y_F + \frac{P_z}{100} \times z_F \qquad (22)$$

This equation represents a plane in a three-dimensional coordinate system where the x_F, y_F and z_F axes are mutually perpendicular. All of the variable points on the plane x_F, y_F and z_F correspond to the same total pressure p and the constant boiling temperature for which the boiling pressures P_x, P_y and

P_z of the pure components apply. If this plane intersects the plane specified by equation (13), the line of intersection of the two planes gives a straight line which is the locus of all liquids having the same boiling temperature and total pressure p.

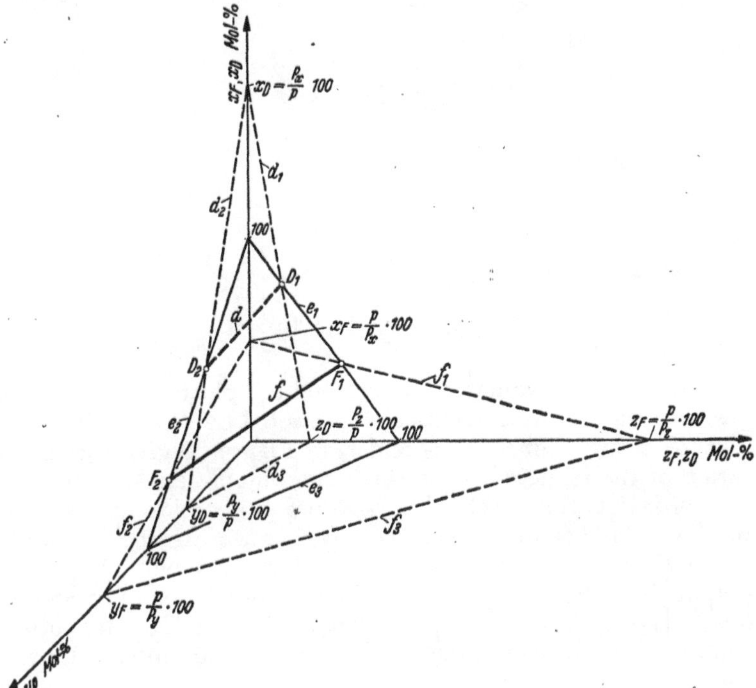

FIGURE 20

Determination of the liquid and vapor isothermals of a ternary mixture, shown with a perspective plot

The plane corresponding to equation (13) is plotted after intersections with the three coordinate planes have been found. In equation (13), when $x_F = y_F = 0$, z_F is 100. With $x_F = z_F = 0$, $y_F = 100$, and when $y_F = z_F = 0$, $x_F = 100$. Thus, if on the three coordinate axes the value of 100 mol per cent is plotted, and each pair of the points thus obtained is connected by a straight line, the intersecting line of the plane with the three coordinate planes is obtained in accordance with equation (13).

In Figure 20, these intersecting lines have been labeled e_1, e_2 and e_3. The plane given by equation (22) is drawn in a similar fashion. As a matter of fact, if $y_F = z_F = 0$, it is evident from equation (22) that

$$x_F = \frac{p}{P_x} \times 100 \tag{23a}$$

and from $x_F = z_F = 0$:

$$y_F = \frac{p}{P_y} \times 100 \tag{23b}$$

and from $x_F = y_F = 0$:

$$z_F = \frac{p}{P_z} \times 100 \tag{23c}$$

The x_F, y_F and z_F values thus obtained are plotted on the three axes, and the resulting points are connected by straight lines.

These straight lines, f_1, f_2 and f_3 in Figure 20 give the intersection of the plane with the three coordinate planes in accordance with equation (22). Lines e_1 and f_1 give intersection F_1, and lines e_2 and f_2 intersection F_2. By connecting line f between F_1 and F_2 the required liquid isothermal is obtained. It is seen that, at the temperature selected, lines e_3 and f_3 do not intersect. However, such an intersection would be obtained for other temperatures; in such cases, e_1 or e_2 would not intersect with f_1 or f_2.

The vapor isothermal for the same boiling temperature is found in the same way as the liquid isothermal. From equations (20) and (13),

$$\frac{p}{P_x} \times x_D + \frac{p}{P_y} \times y_D + \frac{p}{P_z} \times z_D = 100 \tag{24}$$

This relation represents another plane containing all points having the variable vapor concentrations x_D, y_D and z_D mol per cent, corresponding to the same total pressure p and to the same boiling temperature applying to the vapor pressures P_x, P_y and P_z of the pure components. For $y_D = z_D = 0$,

$$x_D = \frac{P_x}{p} \times 100 \tag{25a}$$

For $x_D = z_D = 0$,

$$y_D = \frac{P_y}{p} \times 100 \tag{25b}$$

and for $x_D = y_D = 0$:

$$z_D = \frac{P_z}{p} \times 100 \tag{25c}$$

If these x_D, y_D and z_D values are plotted on their appropriate axes, and if the corresponding points are connected by straight lines, intersection lines d_1, d_2 and d_3 of the isothermal plane with the three coordinate planes are obtained, in accordance with equation (24). This isothermal plane intersects the plane given by equation (14). The latter is obtained by plotting on the axes the values of 100 mol per cent and connecting the resulting points by straight lines. In Figure 20, these lines are designated again as e_1, e_2 and e_3. Points e_1 and d_1 yield intersection D_1; points e_2 and d_2 give D_2. The connecting line d between D_1 and D_2, the line intersecting the planes mentioned, represents a three-dimensional vapor isothermal. On this line are located all vapor compositions which have a constant condensation temperature at a constant pressure p.

The ideal ternary mixture benzol-toluol-*m*-xylol will be discussed as a concrete example. A boiling temperature of 100°C is selected. In accordance with Figure 21, the boiling lines of the individual components at 100°C give these vapor pressures

$$P_x = 1345 \text{ mm for benzol}$$
$$P_y = 550 \text{ mm for toluol}$$
$$P_z = 230 \text{ mm for xylol}$$

To find the characteristic plane for liquids having the same boiling temperature, the coordinate intersections are computed from equations (23a), (23b) and (23c), for a total pressure of 760 mm Hg:

$$x_F = \frac{760}{1345} \times 100 = 56.5 \text{ mol per cent}$$

$$y_F = \frac{760}{550} \times 100 = 138 \text{ mol per cent}$$

$$z_F = \frac{760}{230} \times 100 = 330 \text{ mol per cent}$$

These values are plotted on the appropriate axes, which are drawn 120 degrees apart, as shown in Figure 22. The intersections are connected by the dotted lines. Two sides of the triangle are intersected by the dotted lines. The straight line connecting these intersections yields the liquid isothermal for 100°C which is shown in Figure 22 as a solid line.

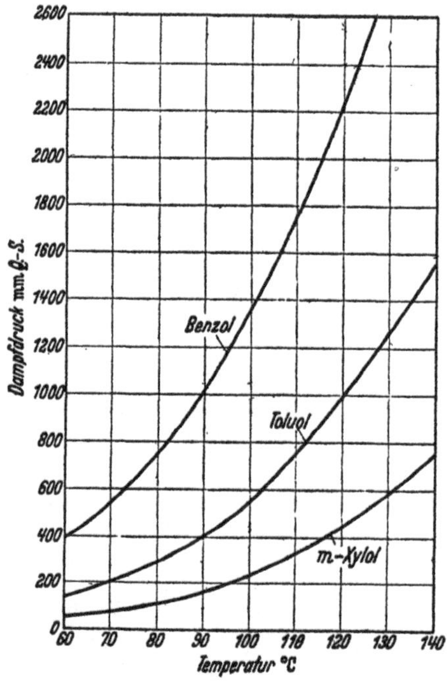

FIGURE 21

Vapor-pressure curves of benzol, toluol and m-xylol

Temperatur—Temperature
Dampfdruck mm Q.-S.—Vapor pressure, mm Hg

The vapor isothermal is obtained in the same fashion. To locate the plane containing all vapor components of constant temperature, the following ordinate intersections are computed from equations (25a), (25b) and (25c):

$$x_D = \frac{1345}{760} \times 100 = 177.0 \text{ mol per cent benzol}$$

$$y_D = \frac{550}{760} \times 100 = 72.5 \text{ mol per cent toluol}$$

$$z_D = \frac{230}{760} \times 100 = 30.2 \text{ mol per cent xylol.}$$

These ordinate values are plotted on the three axes, and the intersections obtained are connected by the dotted lines in Figure 22. Two of these straight lines intersect with two sides of the triangle. The connecting line between these two intersections yields the vapor isothermal for 100°C, shown as a heavy dotted line in Figure 22.

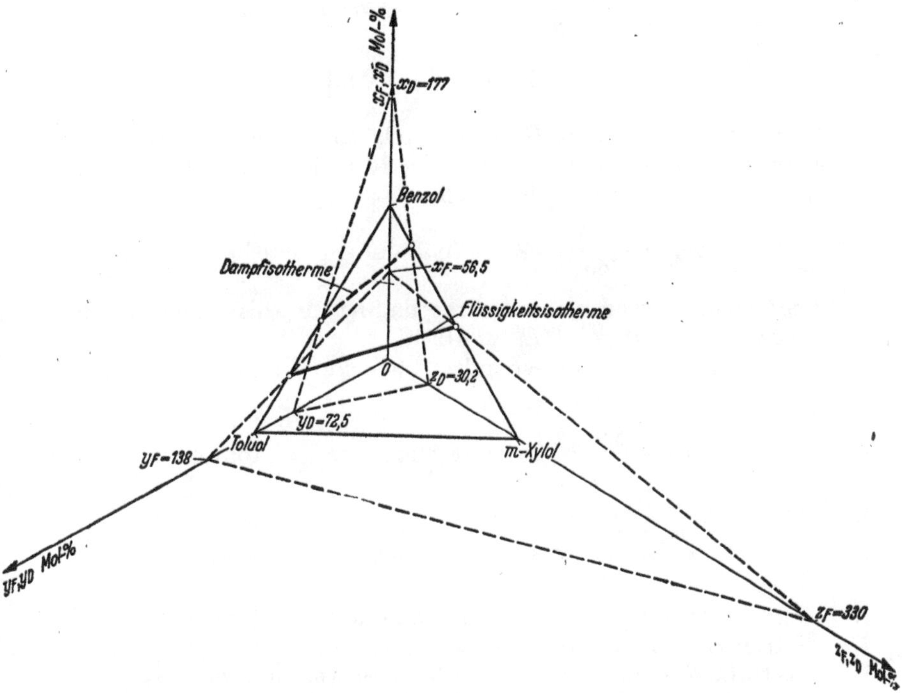

FIGURE 22

Determination of the liquid and vapor isothermals of the ideal ternary mixture benzol-toluol-m-xylol, for a temperature of 100°C and a total pressure of 760 mm Hg

Flüssigkeitsisotherme—Liquid isothermal
Dampfisotherme—Vapor isothermal

If the point representing a liquid at a specified temperature is given, and hence is located on a specified liquid isothermal on the triangle, such as point *G* for 100°C in Figure 19, the point representing the vapor arising from liquid *G* must be located on the vapor isothermal for 100°C. Furthermore, according to equation (20) the following relations exist between vapor concentrations x_D, y_D and z_D and liquid concentrations x_F, y_F and z_F:

$$x_D = \frac{P_x}{p} \times x_F \text{ mol per cent benzol}$$

$$y_D = \frac{P_y}{p} \times y_F \text{ mol per cent toluol}$$

$$z_D = \frac{P_z}{p} \times z_F \text{ mol per cent xylol}$$

If, for instance, point *G* corresponds to a liquid composition, $x_F = 40$ mol per cent benzol, $y_F = 26$ mol per cent toluol, and $z_F = 34$ mol per cent xylol, then

$$x_D = \frac{1345}{760} \times 40 = 70.9 \text{ mol per cent benzol}$$

The vapor concentration corresponding to this liquid is the vapor isothermal point *H* for 100°C.
This point *H* also corresponds to a composition for the other components

$$y_D = \frac{550}{760} \times 26 = 18.8 \text{ mol per cent toluol,}$$

and

$$z_D = \frac{230}{760} \times 34 = 10.3 \text{ mol per cent xylol.}$$

Isothermals for any desired temperature can be plotted on the triangle. Not only does the vapor isothermal for $t_1 = 100°C$ pass through point *H* but also a liquid isothermal for a temperature t_2. Therefore, *H* may also be considered as a possible condition for the liquid state. Similarly, a point *J* located on the vapor isothermal for t_2 may be determined, which indicates the composition of the vapor which results from a boiling liquid represented by point *H*. *H* may be found from *G*, and *J* may be found from *H*. In the same manner, another point representing a vapor can be found from *J*. Also, *G* may be con-

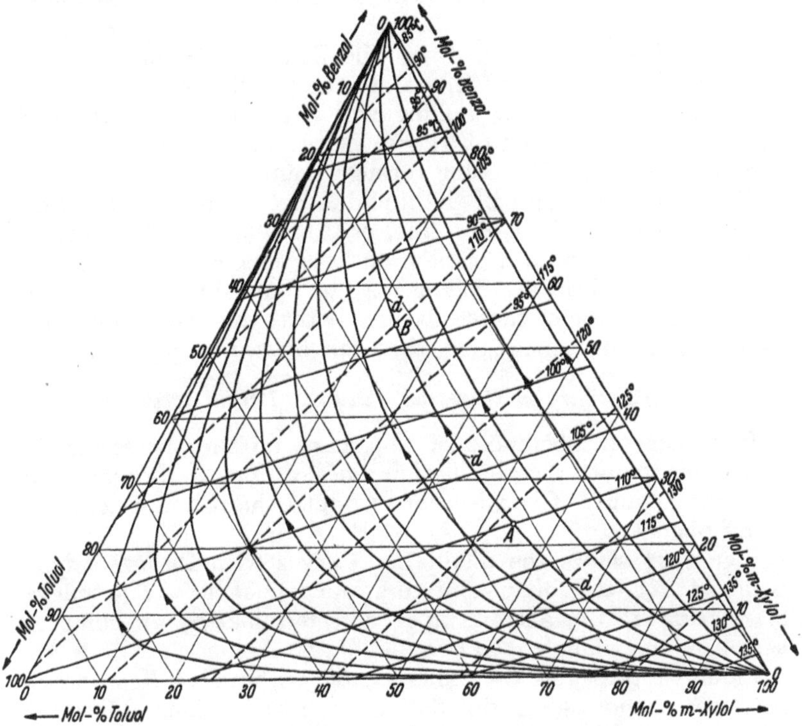

FIGURE 23

Liquid and vapor isothermals and distillation paths for the mixture benzol-toluol-m-xylol

——— Liquid isothermals
- - - - Vapor isothermals

sidered as a vapor point, obtained from a corresponding liquid point.

If all such points are connected with each other, a distillation path is obtained. This curve and the arrows shown on it indicate the direction taken by the vapor concentration during boiling, depending upon the liquid concentration.

If condensation takes place the process is reversed. Any desired number of such curves may be plotted in a triangle, as shown in Figure 23, for the ideal ternary mixture benzol-toluol-*m*-xylol.

If, for instance, point *A* for the liquid is specified on the liquid isothermal for 110°C, and on the distillation curve *d*, the vapor generated during boiling will have the composition *B* (in Figure 23,) located on the 110°C vapor isothermal and on the distillation line *d* passing through point *A*.

If the starting point of the liquid is not situated directly on an isothermal or distillation line, or if it is not the intersection of an isothermal with a distillation line, the composition of the vapor generated during boiling may be found by interpolation.

A characteristic of these mixtures is that the distillation lines pass only through one region, the triangle used in the plot.

c. Ternary mixtures with two distillation areas

If a pair of substances of a ternary mixture possesses an azeotropic point, the distillation lines will pass through two separate regions. Consider, for example, the mixture acetone-water-phenol.

Figure 24 shows the distillation curves with the azeotropic point *A* for the binary mixture water-phenol.[14] The curve originating at the acetone apex and terminating at point *A* separates the triangle into two distillation regions.

Liquids represented by points located within area *I* can never produce phenol upon distillation or rectification. Mixtures whose compositions lie within region *II* can never yield water. The distillation lines in region *I* originate at the acetone apex and terminate at that for water, while the distillation curves in region *II* originate at the same apex of the triangle and meet at the phenol apex. The curve separating the acetone apex and point *A* is not intersected by any distillation curve.

A three component mixture in which one component is insoluble in the other two and in which the other two components are completely miscible, forming an ideal binary mixture, will also yield a triangular plot with two distillation areas.[15] This is true, for example, for the system benzol-water-toluol.

For a given boiling temperature, let:

P_B = pressure of the pure component benzol
P_T = pressure of the pure component toluol
P_W = pressure of the pure component water

Then, the concentration of benzol in the liquid, x_F mol per cent, corresponding to this boiling temperature, may be found from the following relation ($x_F + y_F = 100$)

$$\frac{x_F}{100} \times P_B \times \frac{P_T}{100} \times (100 - x_F) = p - P_W$$

The vapor composition corresponding to the x_F value is then computed from equation (12).

d. Ternary mixtures with three distillation areas

If each of the three pairs of components of a ternary mixture possesses an azeotropic point (a similar point may also be present due to immiscibility), the concentration change occurring

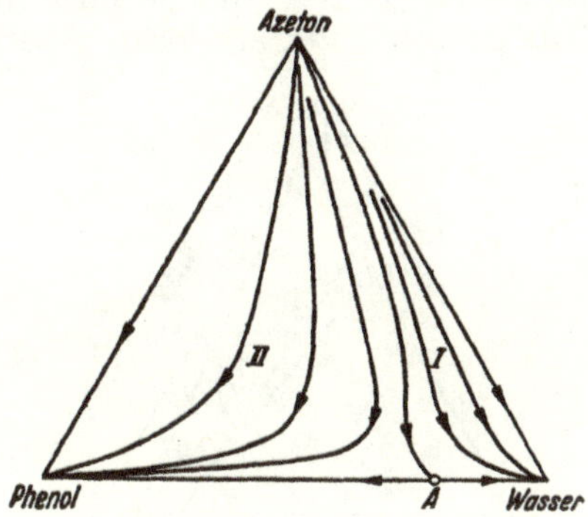

FIGURE 24

Distillation paths for a ternary mixture with two distillation areas in the triangular plot

(Acetone-water-phenol)
Azeton—Acetone
Wasser—Water

upon vaporization is characterized by three distillation regions in the triangular plot.

While there are no known ternary mixtures having maximum boiling points, mixtures with minimum boiling points are of commercial importance.

An example, of such a case occurs in the system ethyl aclohol-benzol-water. In the triangular plot shown in Figure 25, *D* is the minimum boiling point. Point *A* represents the minimum boiling point for the system ethanol-benzol at 68.25° C. Point *B* represents the minimum boiling point for the mixture ethanol-water at 78.15°C. *C* is the minimum boiling point of the mixture water-benzol at 69°C.

The minimum boiling point *D* of the ternary mixture for a total pressure of 760 mm Hg, corresponds to a temperature of 64.85°C, and to a composition of 7.5 mol per cent water, 18.5 mol per cent ethyl alcohol and 74 mol per cent benzol. It is situated within the region enclosed by curve *k* and the straight line connecting the water and benzol apices.

All liquids having compositions within this area are emulsions which separate into two layers.

Curves *a*, *b* and *c* divide the triangle into three distillation regions. The distillation curves are located within these three

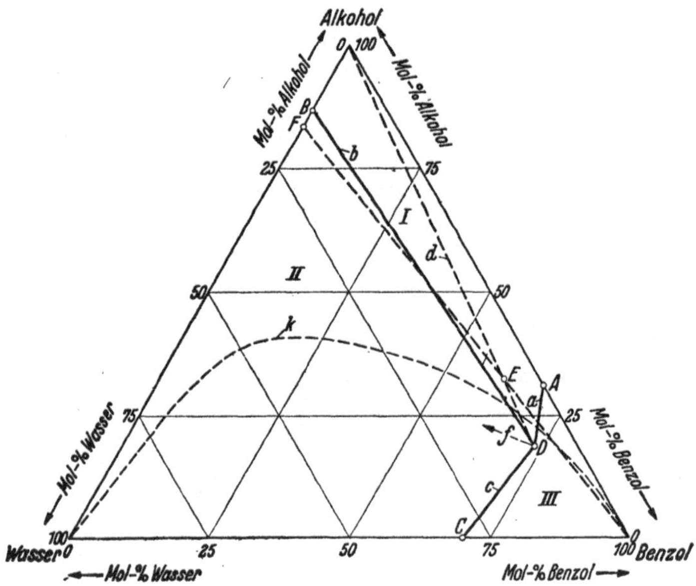

FIGURE 25

Triangular plot of a ternary mixture with a minimum boiling point and three distillation areas

(Ethanol-benzol-water)
Wasser—Water
Alkohol—Alcohol

regions and never intersect the boundary curves *a*, *b* and *c*. Mixtures represented by points located in area *I*, can therefore be converted by distilling or rectifying into other mixtures whose compositions must also lie in region *I*.

Since the distillation curves in region *I* terminate in the azeotropic point *D* of the ternary mixture, the composition representing this point can be attained by rectification. If, for instance, a mixture represented by a point lying in region *I* is to be split into pure alcohol and a mixture having the composition of the azeotropic point of the ternary mixture, the point representing this initial mixture must be located on the connecting line *d* between point *D* and the alcohol apex; as shown in Figure 25.

Such a mixture may be obtained from a binary mixture of ethyl alcohol-water by adding benzol. If, for instance, the composition of the original mixture of ethyl alcohol and water is given by *F*, it will be necessary to add·benzol until a ternary mixture is formed such that the composition is located on the line connecting intersection of *F* and the benzol apex, and line *d*.

This point is labelled *E* on Figure 25. A ternary mixture of this composition can be split into pure alcohol and a ternary mixture having the azeotropic composition *D*.

This procedure was developed by Young,[16] who added to an ethyl alcohol-water mixture (such as point *F* in Figure 25) an amount of benzol sufficient to produce that ternary mixture having points lying in region *I* (such as point *E*).

This method of azeotropic dehydration which originated as a batch process, was later converted into a more efficient continuous process.

Since an azeotropic ternary mixture does not exhibit complete miscibility, but forms two layers upon settling, the azeotropic point is also called a eutectic point, and a mixture of azeotropic composition is also called a eutectic.

7. Influence of pressure on equilibrium conditions

a. Binary mixtures

Industrial processes employing pressures above or below atmospheric in separating liquid mixtures are frequent. Thus the question arises whether the equilibrium conditions for any given pressure can be determined if data are known for a given pressure, for example, 1 atmosphere absolute pressure. Much research has been devoted to this problem, but no general law has been found thus far.

Wrewsky[11] states, "When the temperature of solutions, whose vapor-pressure curves exhibit maxima, is increased, the relative concentration of the component having the greater heat of vaporization in the constant boiling mixture is increased."

This rule is followed in the case of ethyl alcohol-water mixtures.[17] Table 1 which gives the variation of minimum boiling-point temperatures and water concentrations as a function of the absolute pressure, shows that the azeotropic point disappears at an absolute pressure of 70 mm Hg and a temperature of 27.96°C.

TABLE 1

Variation of Water Concentration and Minimum Boiling Point with Absolute Pressure for an Alcohol-Water Mixture

Absolute Pressure, mm Hg	Water Content, per cent	Minimum Boiling Point, °C
70	0.0	27.96
94	0.5	33.35
129	1.3	39.20
298	2.7	47.63
404	3.75	64.04
760	4.4	78.15
1075	4.65	87.12
1451	4.75	95.35

Figure 26 and 27 show the variation of the boiling-point and dewpoint curves for this mixture at pressures below and above atmospheric.[18] The curves for sub-atmospheric pressure indicate that Wrewsky's rule, which holds at the azeotropic point, does not apply over the entire concentration range. Figure 28 shows that the equilibrium curves for different pressures may even overlap.

The difference in the relative concentrations of vapor and liquid which arises during distillation will increase with the difference in boiling temperatures of the individual pure components at any given temperature. This relationship, however, is not always valid. This is shown in the case of the boiling-

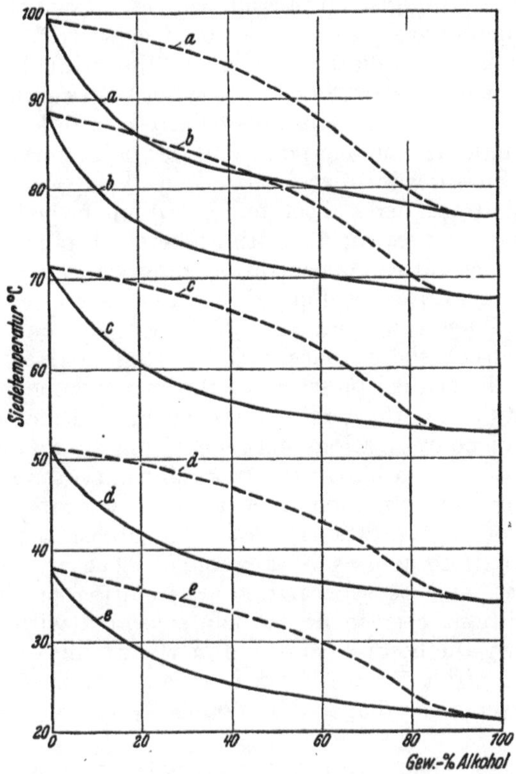

FIGURE 26

Bubble-point curves ——— *and dewpoint curves* - - - - *for various low pressures*

(Ethanol-water)
a. 740 mm Hg
b. 500 mm Hg
c. 250 mm Hg
d. 100 mm Hg
e. 50 mm Hg
Siedetemperatur—Bubble point
Gew. % Alkohol—Per cent by weight of alcohol

point and dewpoint at different pressures for mixtures of nitrogen-oxygen [7] and ammonia-water [2]; curves shown in Figures 29 and 30 for usual industrial pressure ranges. For both mixtures, the enrichments are greater at lowered pressure.*

At very high distillation pressures, the critical pressure of a component may be exceeded, or that of both components may be exceeded. Then the mixtures behave in a different way.

For example, the boiling-point and dewpoint curves of carbon dioxide-sulfur dioxide mixtures are considered for pressures of 41.3 and 93 atmospheres absolute.[21,22] Although the boiling-point and dewpoint curves for 41.3 atmospheres absolute are normal and cover the entire concentration range, the two curves for the higher pressure are limited within the concentrations of $x_1 = 33$ mol per cent and $x_2 = 72.5$ mol per cent. The two curves intersect these concentrations, the intersections being limited by the critical pressures of the two components.

All points below the boiling-point curve and included within the range of concentrations between x_1 and x_2 represent liquid mixtures. The area above the dewpoint curve and within the range of the concentrations mentioned represents superheated vapors. For concentrations below x_1 or above x_2 simultaneous existence of both phases is impossible. The two phases can only coexist, and be separated by distillation or rectification, when conditions correspond to points located within the area confirmed by the boiling-point and dewpoint curves.

At the limiting concentrations x_1 and x_2, the equilibrium curve intersects the diagonal at points A_1 and A_2, as shown in Figure 32.

Two intersections with the diagonal result whenever the distillation pressure exceeds 77 atmospheres absolute. At low pressures, the two intersection points disappear simultaneously, and the equilibrium curve takes on the normal appearance.

As the pressure is increased above 77 atmospheres absolute, points A_1 and A_2 approach each other and finally coincide, in this case at 98.5 atmospheres absolute. The area normally confined within the boiling-point and dewpoint curve shrinks to a point.

At a sufficiently high pressure, the equilibrium curve also may intersect the diagonal at only one point; the intersection is again determined by the critical pressure. An example of this behavior is shown in the equilibrium curve for the mixture nitrogen-oxygen at a pressure of 41.3 atmospheres absolute, at which the

*See also Equilibrium curves, Appendix.

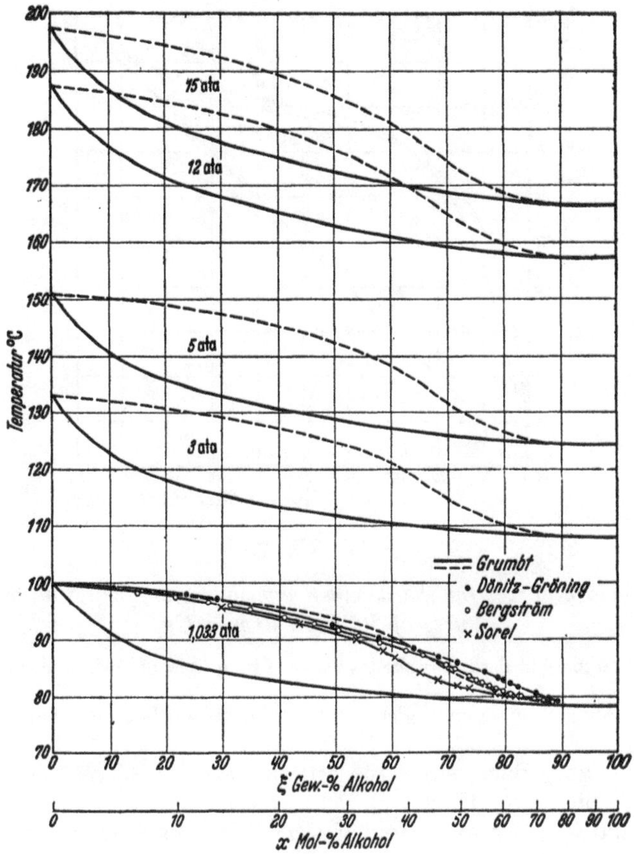

FIGURE 27

Bubble-point and dewpoint curves for the mixture ethanol-water

Pressures up to 15 kg/cm² absolute
Gew. % Alkohol—Per cent by weight of alcohol
Mol % Alkohol—Mol per cent alcohol
Temperatur—Temperature

diagonal is intersected for a concentration of 64 mol per cent nitrogen, while, at 10.16 atmospheres absolute the equilibrium curve no longer intersects the diagonal, as shown in Figure 33.

For this system, the equilibrium curve always intersects the

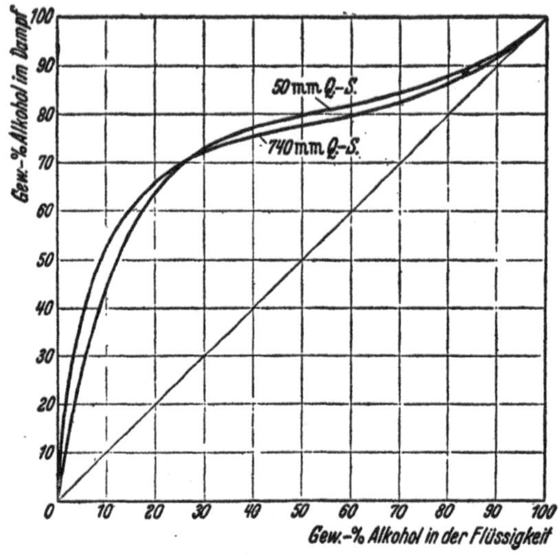

FIGURE 28

Equilibrium curves for the mixture ethanol-water, for absolute pressures of 50 and 740 mm Hg

Gew. % Alkohol in der Flüssigkeit—Per cent by weight of alcohol in the liquid phase
Gew. % Alkohol im Dampf—Per cent by weight of alcohol in the vapor phase

diagonal only once, since the critical pressure of oxygen is the pressure of the mixture.

An industrial application of very high distillation pressures is found in the separation of gas mixtures which are non-condensable at low pressures, but which at high pressures yield liquid mixtures which can be concentrated by distillation and rectification.

b. Mixtures of more than two components

As in the case of binary mixtures, for ternary mixtures equilibrium conditions at different pressures can be calculated for ideal solutions only. In all other cases, either estimates or tests must form the basis for design.

In the case of the industrially important mixture ethyl alcohol-benzol-water, the shift of the azeotropic point with pressure has been subjected to detailed study. When the distilling

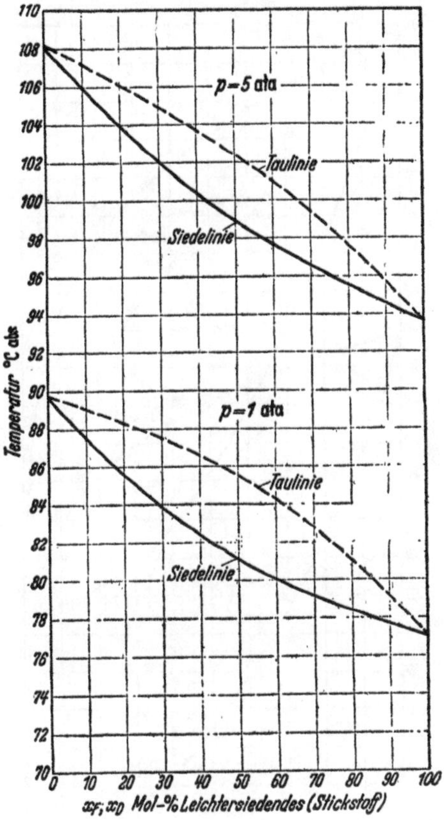

FIGURE 29

Bubble-point and dewpoint curves for the mixture nitrogen-oxygen at pressures of 1 and 5 kg/cm² absolute

Mol % Leichtersiedendes (Stickstoff)—Mol per cent of low boiler (nitrogen)
Temperatur, °C abs.—Temperature, degrees Kelvin
Taulinie—Dewpoint curve
Siedelinie—Bubble-point curve
ata—Absolute atmospheres

pressure is increased, the water concentration is increased in both the benzol-water mixture and the alcohol-water mixture.

This might lead one to expect that increasing the distillation

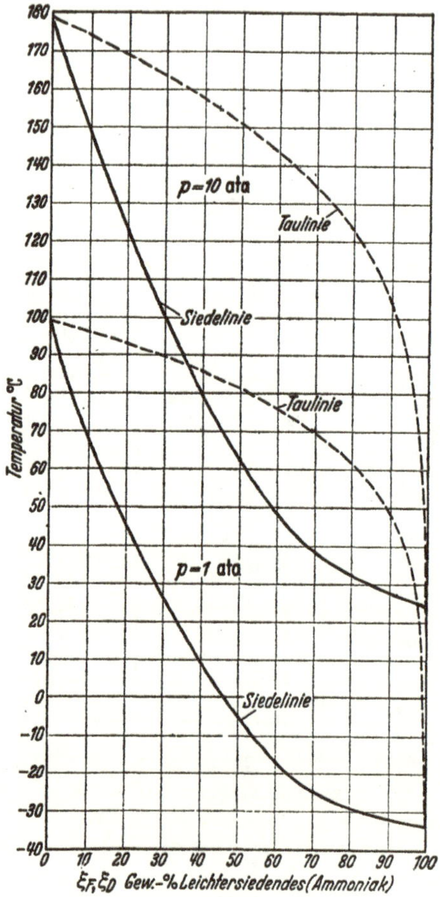

FIGURE 30

Bubble-point and dewpoint curves for absolute pressures of 1 and 10 kg/cm² absolute

Ammonia-water

Gew. % Leichtersiedendes (Ammoniak)— Per cent by weight of low boiler (ammonia)

Temperatur—Temperature
Taulinie—Dewpoint curve
Siedelinie—Bubble-point curve

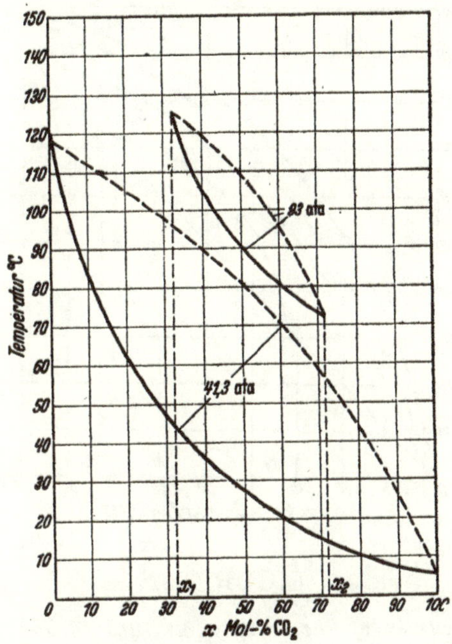

FIGURE 31

Bubble-point and dewpoint curves for the binary mixture carbon dioxide-sulfur dioxide for pressures of 41.3 and 93 atm. absolute atmospheres

pressure will yield an azeotropic mixture richer in water and leaner in benzol. Tests have confirmed this.[23] It has been concluded that, by increasing the pressure distillation for binary ethyl alcohol-water mixtures upon the addition of benzol, the dehydration could be completed in a more economical fashion than at ordinary pressure. If the pressure increases, the azeotropic point will shift in the direction of the arrow *f* in Figure 25.

IV. Partial condensation of mixtures

1. Fundamentals

Concentration of liquid mixtures within certain limits may be carried out not only by vaporization (distilling), but also by

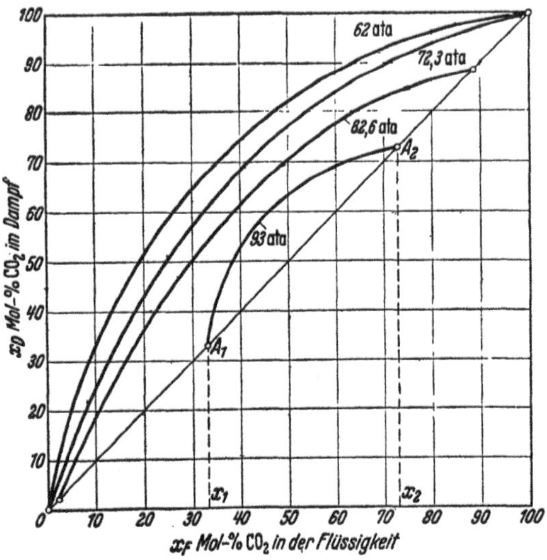

FIGURE 32

Equilibrium curves of the binary mixture carbon dioxide-sulfur dioxide for various pressures

Mol % in der Flüssigkeit—Mol per cent in liquid
Mol % im Dampf—Mol per cent in vapor
ata—Absolute pressure, kilograms per square centimeter

partial condensation of the vapor of a multicomponent mixture. The residual vapor has a composition different from that of the original vapor. These conditions are discussed for the case of a binary mixture. Where there are no particular reasons for using mol units, the usual weight units are employed.

If a vapor is condensed on a cooling surface, a layer of condensate forms on this surface. The heat of condensation of the vapor must pass through this layer. A characteristic of the condensation process is that at a given temperature sufficient heat is withdrawn from the vapor to liquefy it. Therefore, vapor and liquid are in a state of phase equilibrium over their common area of contact.[24]

As the distance from this area of contact increases, the concentration of the low boiler in the vapor decreases, provided that the equilibrium curve is located above the diagonal of the

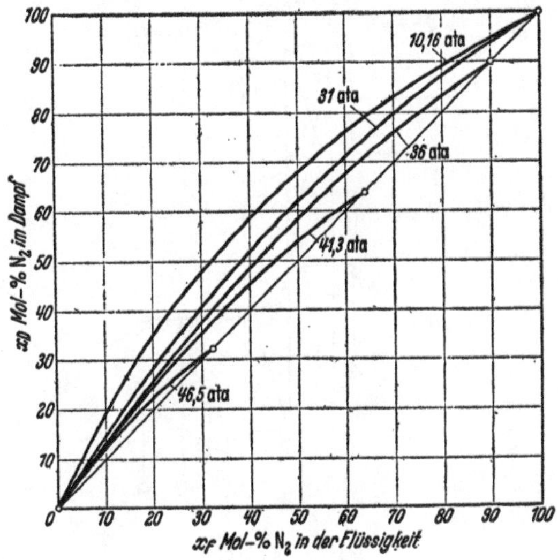

FIGURE 33

Equilibrium curves for various pressures
(Nitrogen-oxygen)

Mol % in der Flüssigkeit—Mol per cent in liquid
Mol % im Dampf—Mol per cent in vapor
ata—Absolute pressure, kilograms per square centimeter

ξ_D, ξ_F plot shown in Figure 35. This change in concentration in the vapor occurs in the case of turbulent flow in a boundary layer adjacent to the surface of contact between vapor and liquid.

The increased concentration in low boiler in the direction toward the surface of contact is due to the fact that the condensate from the vapor has a composition different from that of the vapor itself. The change of concentration in the boundary layer causes an interchange of material between the surface of contact and the bulk of the vapor. According to Figure 34, the vapor concentration of the low boiler, which is constant in a turbulent vapor flow, increases in the boundary layer from ξ_D to ξ_{D0}. Consequently, vapor temperatures within the boundary layer decrease from t_D to a value t_0 at the surface. The concentration of low boiler in the liquid increases from the value ξ_{F0}

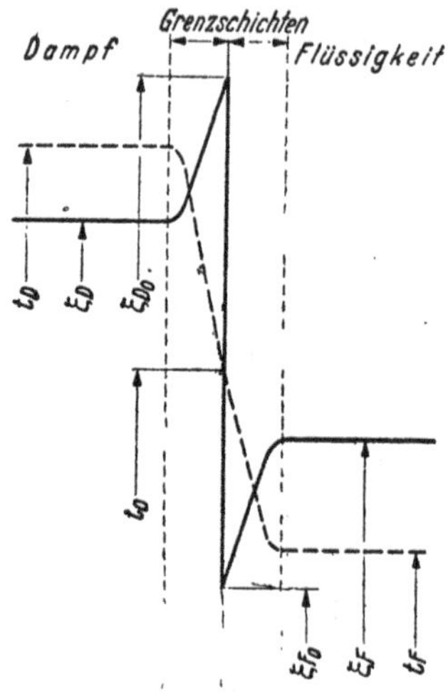

FIGURE 34

Variations of concentration and temperature in vapor and liquid mixtures

Dampf—Vapor
Flüssigkeit—Liquid
Grenzschichten—Boundary layers

at the surface to the value ξ_F in the bulk of the liquid, while the liquid temperature decreases from t_0 to t_F.

If the mass of vapor D, containing ξ_D per cent by weight of low boiler, passes across an infinitely small cooling area, an infinitely small amount of condensate dC is condensed, containing ξ_{F0} of the low boiler. An amount of vapor $(D - dC)$ containing $(\xi_D + d\xi_D)$ weight per cent of low boiler remains. $d\xi_D$ is the enrichment in vapor surviving. Thus,

$$D \times \left(\frac{\xi_D}{100}\right) = dC \times \frac{\xi_{F0}}{100} + (D - dC)\left(\frac{\xi_D + d\xi_D}{100}\right) \qquad (26)$$

and, neglecting second order differentials, the enrichment of low boiler in the vapor becomes:

$$d\xi_D = \frac{dC}{D}(\xi_D - \xi_{F0}) \qquad (27)$$

Thus, enrichment is increased when the weights of the vapor mass passing across the cooling surface are small as compared to the mass of condensate, and when the difference between ξ_D and ξ_{F0} is large.

If D_e = total mass of vapor entering a condenser in kg/hr, and D_a = residual mass of vapor leaving the condenser, in kg/hr, the mass of condensate becomes

$$C_g = D_e - D_a \tag{28}$$

To integrate equation (27), a relation between the vapor mass flowing to the condenser D_e and the mass of condensate C formed between condenser inlet and a certain point of the cooling surface must be available. Thus,

$$D = D_e - C \tag{29}$$

Inserting the D value from equation (29) into equation (27) gives the differential equation for the enrichment produced by partial condensation

$$\frac{dC}{D_e - C} = \frac{d\xi_D}{\xi_D - \xi_{F0}} \tag{30}$$

At the inlet of the cooling surface $C = 0$ and the concentration of the vapor entering is ξ_{De} weight per cent, while at the outlet of the cooling surface the residual vapor leaves with a low-boiler concentration of ξ_{Da} weight per cent. The latter concentration is found by integrating equation (30) within the limits $C = 0$ and $C = C_g$:

$$\int_0^{C_g} \frac{dC}{D_e - C} = \int_{\xi_{De}}^{\xi_{Da}} \frac{d\xi_D}{\xi_D - \xi_{F0}} \tag{31}$$

The solution of the integral on the left-hand side of equation (31) is

$$\int_0^{C_g} \frac{dC}{D_e - C} = \ln \frac{D_e}{D_e - C_g} \tag{32}$$

This relation takes on a simple appearance if the reflux ratio of the partial condensation is introduced, defined as the ratio of the total condensate formed and the residual vapor leaving the cooling surface:

$$v = \frac{C_g}{D_a} \tag{33}$$

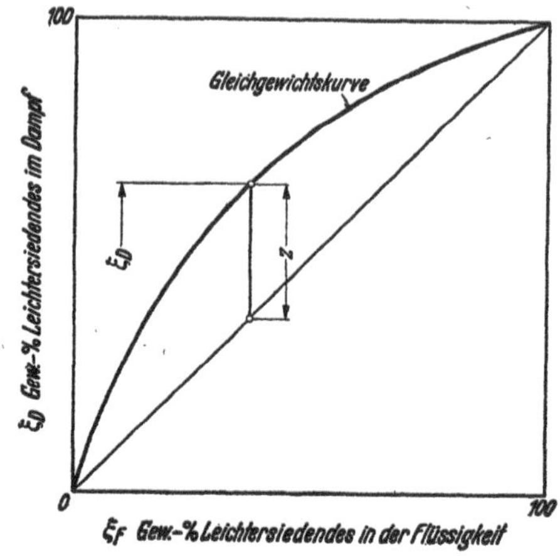

FIGURE 35

Vapor concentration vs. liquid concentration during condensation

Gew. % Leichtersiedendes in der Flüssigkeit —Per cent by weight of low boiler in liquid

Gew. % Leichtersiedendes im Dampf—Per cent by weight of low boiler in vapor

Then,

$$\frac{D_e}{D_e - C_g} = \frac{D_a + C_g}{D_a} = (v + 1) \tag{34}$$

Equation (31) becomes

$$\ln (v + 1) = \int_{\xi_{De}}^{\xi_{Da}} \frac{d\xi_D}{\xi_D - \xi_{F0}} \tag{35}$$

This basic equation for enrichment by partial condensation states that, for a given initial concentration of a mixture, the enrichment obtained (by partial condensation in a given condenser) in the concentration of the low boiler is determined exclusively by the reflux ratio.

The integral on the right side of equation (35) can be solved from the relationship between ξ_D and ξ_{F0}, which is derivable from the equilibrium conditions of the condensation process.

The condensation which causes the accumulation of a substance in the vapor results in an exchange of substances and a corresponding heat exchange. It is even possible to derive a relationship between the heat-exchange and substance-exchange characteristics from the basic phenomena.[25]

The heat of condensation given up by the vapor must be passed on to the liquid layer, and from the latter to the cooling surface.

When

α_D = film coefficient between the bulk of the vapor and the contact surface of the liquid, kcal/m² × hr × °C, and

α_F = film coefficient between the bulk of the liquid and the contact surface (see Figure 34), the following condition applies:

$$\frac{\alpha_D}{\alpha_F} = \frac{t_0 - t_F}{t_D - t_0} \tag{36}$$

If α_D is very large in comparison with α_F, the temperature difference $t_D - t_0$ becomes small both in magnitude and in comparison to $t_0 - t_F$. In the extreme case it may become zero. Then,

$$t_D = t_0 \tag{37}$$

Experiments prove that such an extreme case exists for ethanol-water mixtures. From equations (36) and (37),

$$\xi_D = \xi_{D0} \tag{38}$$

so that for the difference, $\xi_D - \xi_{F0}$, in equation (35) it may be considered that the vapor with concentration ξ_D is in phase equilibrium with a liquid of composition ξ_{F0}. This condition is stated symbolically as

$$\xi_D / / / \xi_{F0} \tag{39}$$

Thus, the difference, $\xi_D - \xi_{F0}$ for a given ξ_D corresponds to the ordinate difference

$$z = \xi_D - \xi_{F0} \tag{40}$$

between the equilibrium curve and the diagonal in the ξ_D versus ξ_F plot, Figure 35.

2. Use of the Basic Equations for Rectifying by Partial Condensation; Example: Ethanol and Water

It should be possible, with the help of equation (35), to determine the concentration of alcohol in the residual vapor leaving the condenser, for a given reflux ratio and for a given initial concentration of the ethanol-water vapor entering the condenser. These correlations are shown in a plot.

To evaluate equation (35), integral values, J_e, corresponding to the initial concentration, ξ_{De}, and J_a, corresponding to the final concentration, ξ_{Da}, are employed, so that equation (39) gives

$$\ln (v + 1) = \int_{\xi_{De}}^{\xi_{Da}} \frac{d\xi_D}{\xi_D - \xi_{F0}} = J_a - J_e \tag{41}$$

The integral in this equation can be solved graphically as shown in Figure 36. The ordinate differences between the equilibrium curve and the diagonal represent the differences $\xi_D - \xi_{F0}$. Their reciprocals are plotted over ξ_D, giving a curve for the expression $1/(\xi_D - \xi_{F0})$. Figure 36 gives, for $\xi_D = 75$ per cent by weight alcohol, both the curve required for evaluating the $1/(\xi_D - \xi_{F0})$ curve as well as the $1/(\xi_D - \xi_{F0})$ curve itself. Then the shaded area, f, evaluates the value of the integral in equation (41) between the limits $\xi_{De} = 18$ and $\xi_{Da} = 20$ (weight per cent):

$$f = \int_{18}^{20} \frac{d\xi_D}{\xi_D - \xi_{F0}} \tag{42}$$

Since differences of integral values are expressed in equation (41), for any value of ξ_{De} a numerical value of J_e may be assumed. In Figure 36, for instance, with $\xi_{De} = 4$ an integral value $J_e = 3$ has been selected.

If the value corresponding to the area below the $1/(\xi_D - \xi_{F0})$ curve between a ξ_{De} and a ξ_{Da} is added to J_e, the sum gives the integral value, J_a. This is plotted versus ξ_{Da}. In this way, for any assumed ξ_{De}, any number of J_a values may be computed; these yield an integral curve

$$\int \frac{d\xi_D}{\xi_D - \xi_{F0}} \quad \text{(See Figure 36)}$$

From the latter, an integral value, J_e, corresponding to a concentration, ξ_{De}, and an integral value, J_a, corresponding to a ξ_{Da}, can be obtained.

By means of equation (41), the required concentration ξ_{Da} can now be found for a given reflux ratio v and a given ξ_{De} value.

Numerical example. Given: reflux ratio 5, concentration of the vapor entering the condenser $= \xi_{De} = 40$ weight per cent. What is the concentration of the final vapor leaving the condenser, ξ_{Da}?

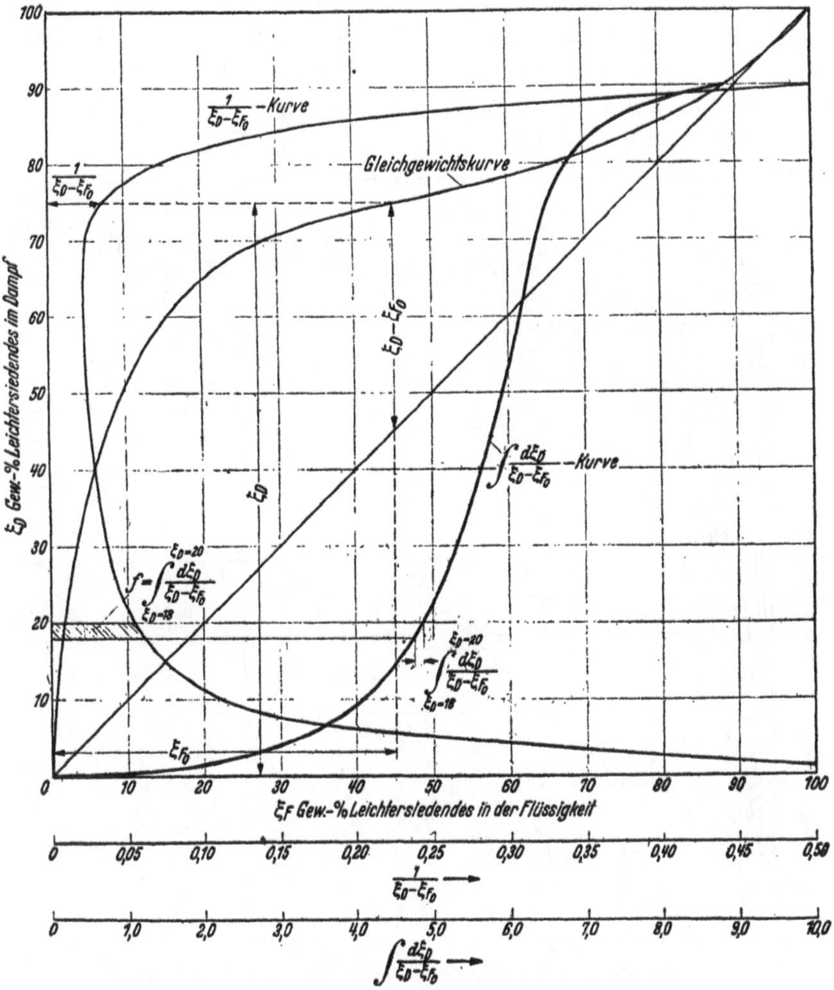

FIGURE 36

Development of the curves for computing the enrichment of the mixture ethanol-water

Gew. % Leichtersiedendes in der Flüssigkeit—Weight per cent of low boiler in liquid
Gew. % Leichtersiedendes im Dampf—Weight per cent of low boiler in vapor
Kurve—Curve
Gleichgewichtskurve—Equilibrium curve

The integral curve in Figure 36 gives a J_e value of 5.65 for $\xi_{De} = 40$. From equation (41) the integral value J_a is computed

$$J_a = 5.65 + \ln(5 + 1) = 5.65 + 1.79 = 7.44$$

For $J_a = 7.44$ the integral curve in Figure 36 gives the required ξ_{Da} as 85.7 weight per cent of alcohol. The vapor entering

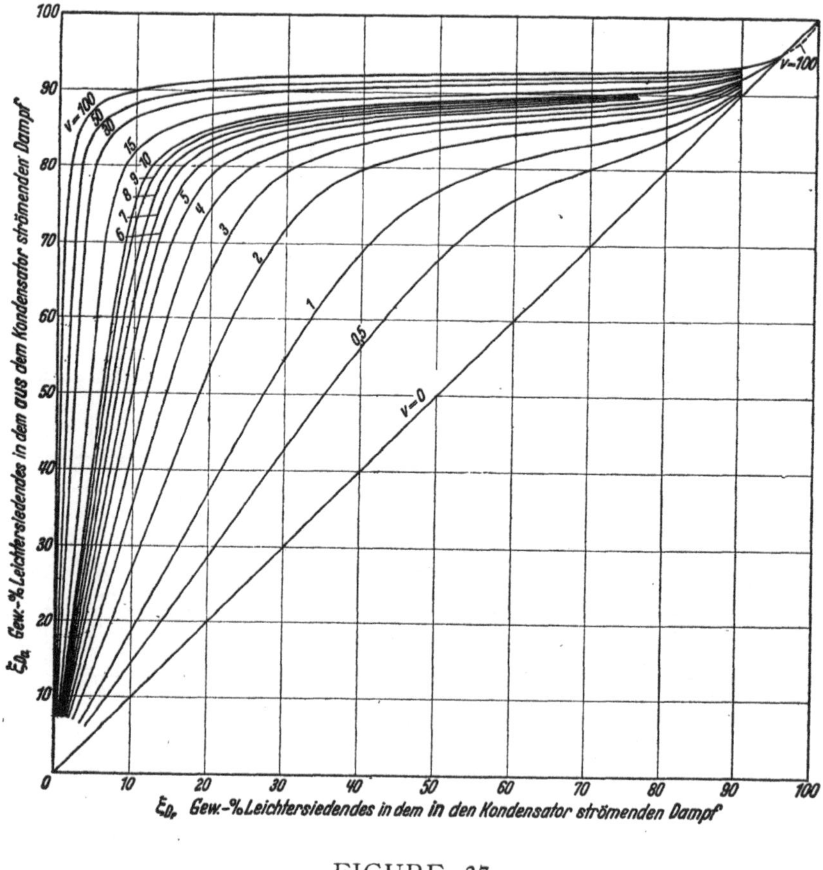

FIGURE 37

Enrichment curves for the mixture ethanol-water

Gew. % Leichtersiedendes in dem in den Kondensator strömenden Dampf—Per cent by weight of low boiler in vapor entering condenser

Gew. % Leichtersiedendes in dem aus dem Kondensator strömenden Dampf—Per cent by weight of low boiler in vapor leaving condenser the condenser with $\xi_{De} = 40$ weight per cent is concentrated this much merely by partial condensation.

In a similar manner, for any reflux ratio selected, a corresponding ξ_{Da} may be found for each ξ_{De}. If the ξ_{Da} values are plotted against corresponding ξ_{De} values for a given reflux ratio, a curve results which facilitates finding the composition ξ_{Da} up to which the residual vapor is concentrated, starting with an initial concentration ξ_{De}. For any reflux ratio, ξ_{De} versus ξ_{Da} curves can be plotted in a similar manner.

Representative enrichment curves for alcohol-water mixtures are shown in Figure 37. If mol units, instead of weight units, are used in calculations, enrichment curves are obtained in which the abscissa indicates the alcohol concentration in mol per cent in the vapor entering the condenser, while the ordinate indicates the alcohol mol per cent in the vapor leaving the condenser. Such enrichment curves are shown in Figure 38.

Comparison of the individual enrichment curves shows that the greater the reflux ratio, the greater is the enrichment for a given initial vapor concentration. The enrichment in low boiler can be increased only by increasing the reflux ratio. Increasing the reflux ratio by a fixed amount has moreover a larger effect for small v-values than for large v-values.

Numerical example. Find the alcohol concentration in the vapor leaving a condenser, if the vapor entering the condenser contains 20 weight per cent of alcohol, and if the ratio of vapor condensed in the exchanger to residual vapor is 6.

From Figure 37, for $\xi_{De} = 20$, with $v = 6$, the alcohol concentration in the residual vapor is $\xi_{Da} = 81.1$ weight per cent.

Numerous tests have shown that the enrichment curves given in Figures 37 and 38 are correct for a mixture of ethyl alcohol and water. In these tests a glass heat exchanger of the type shown in Figure 39 was used. The vapor was introduced at the lower end, and the condensate was also drained off at the lower end. The residual vapor escaped at the upper end.

When the tests are made in this fashion, the vapor and the condensate in the exchanger move countercurrently. The reflux ratio was varied between the limits 0.75 and 10, with the initial vapor concentrations between 2 and 90 weight per cent. The average error of the results of approximately fifty tests as compared with the theoretical values was only 0.04 per cent, while the maximum deviations were 8.3 and −8.5 per cent,

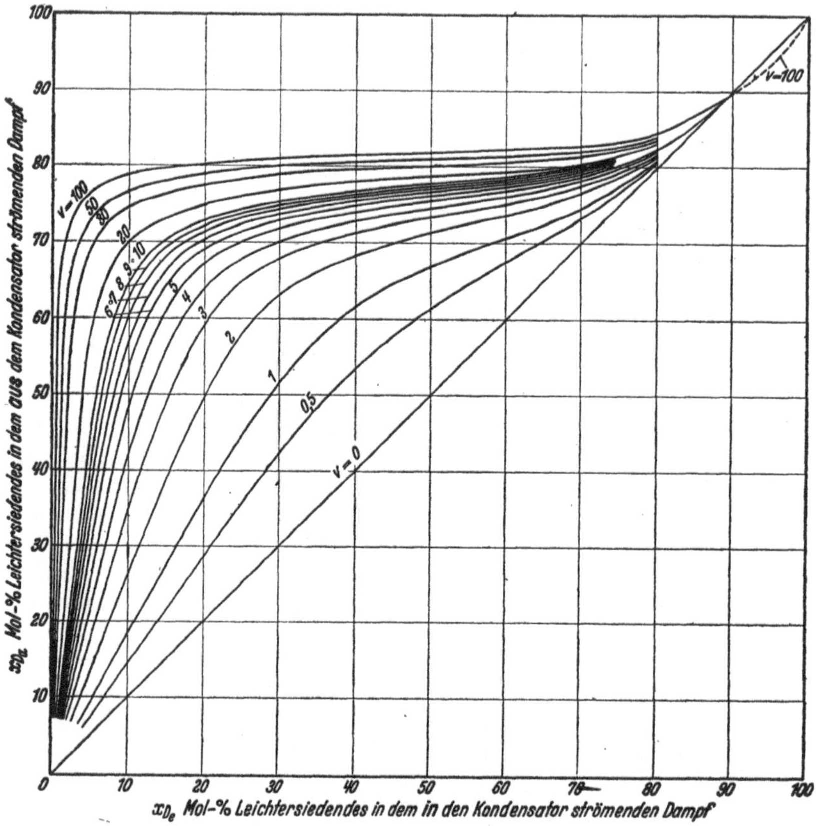

FIGURE 38

Enrichment curves for the mixture ethanol-water

Mol % Leichtersiedendes in dem in den Kondensator strömenden Dampf—Mol per cent low boiler in vapor entering condenser

Mol % Leichtersiedendes in dem aus dem Kondensator strömenden Dampf—Mol per cent low boiler in vapor leaving condenser

Thirty per cent of the results were in perfect agreement with theoretically computed results.

However, when the vapor was introduced through the top connection of the condenser, resulting in parallel flow of the vapor and condensate, and when the residual vapor was withdrawn at the same end as the condensate, deviations of the test results from the computed values ranged up to 31 per cent.

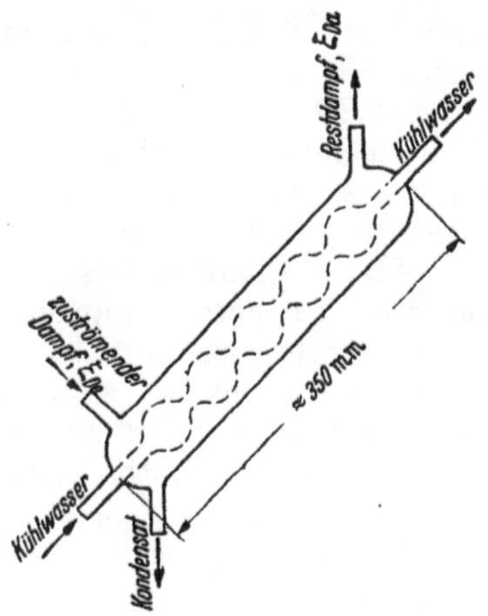

FIGURE 39

Type of condenser used in the experiments on enrichment through partial condensation

Kühlwasser—Cooling water
Kondensat—Condensate
Zuströmender Dampf—Vapor entering
Restdampf—Residual vapor

Even then, a large number of test results agreed with the computed values.

Entrainment of liquid by the residual vapor can probably account for these discrepancies. In fact, in all cases in which deviations occurred, the enrichment was less than would have been expected on the basis of the theoretical computations.

3. Rectification of a benzol-toluol mixture by partial condensation

The film coefficients between a vapor and a heat exchanging surface, and a liquid and a heat exchanging surface, depend on the physical characteristics of the materials, velocities, and the dimensions of the cross-sectional areas. If their numerical values are of the same order, the concentration at the contact surface between the vapor and liquid is different from that in the main mass of the vapor. In this case, enrichment by partial condensation is lower than that occurring under the limiting conditions where the vapor concentration at the contact surface is the same as that in the mass of the vapor.

Operating data for determining the condition of the contact

surface between vapor and liquid can be used in the equations on hand for computing the enrichment by partial condensation and for obtaining the enrichment curves for a given reflux ratio.

Table 2 presents data obtained with a mixture of benzol and toluol in a condenser of commercial size, equipped with 48 vertical metal tubes of a 26.7 mm inner diameter[26] and having an available surface of 1.56 square meters. The vapor mixture in the tubes and the condensate moved countercurrently. Table 2 gives values for different reflux ratios v, and the corresponding benzol concentrations as per cent by weight, ξ_{De} being the initial benzol concentration at the inlet, while $\xi_{D\Delta}$ is the final benzol concentration of the residual vapor at the outlet.

TABLE 2

Experimental results on a toluol-benzol mixture; enrichment by partial condensation

v	ξ_{De}	ξ_{Da}	v	ξ_{De}	ξ_{Da}
2.3	63.2	72.7	3.8	84.7	92.0
2.9	52.3	64.0	5.6	59.2	70.5
2.4	34.6	44.8	8.8	24.5	34.3
2.3	20.8	31.3	10.4	90.3	93.8
			8.3	85.6	91.7
3.9	82.0	87.5	9.2	78.0	82.2
3.8	70.5	78.7	10.0	89.3	93.0
3.9	22.3	33.7	10.0	84.4	90.0
3.5	9.7	20.0	10.0	77.0	84.0
			10.0	6.5	11.5

The ξ_{De} values for constant reflux ratios are plotted against numerical figures for the integral J_e. By means of the relation derived from equation (41):

$$J_a = J_e + \ln (v + 1) \tag{43}$$

the J_a corresponding to a given J_e can be computed and plotted over the ξ_{Da} range obtained during the same test. When all of the test data are correlated and used, the integral curve shown in Figure 40, can be drawn. From this, by means of equation (35) the $1/(\xi_D - \xi_{F0})$ curve shown can be found; from this curve the difference $\xi_D - \xi_{F0}$ for any ξ_D value may be obtained; this in turn yields the $\xi_D - \xi_{F0}$ curve, also shown in Figure 40. This curve indicates that the benzol concentration in the vapor is considerably higher at the liquid contact surface that it is in the mass of vapor, and that the film and mass transfer coefficient for the vapor side is smaller than it is for the liquid film in contrast to the ethyl alcohol-water mixtures.

Using the integral curve and equation (43), corresponding values of ξ_{De} and ξ_{Da} can be found for any desired reflux ratio; employ the same procedure as shown for ethyl alcohol and water.

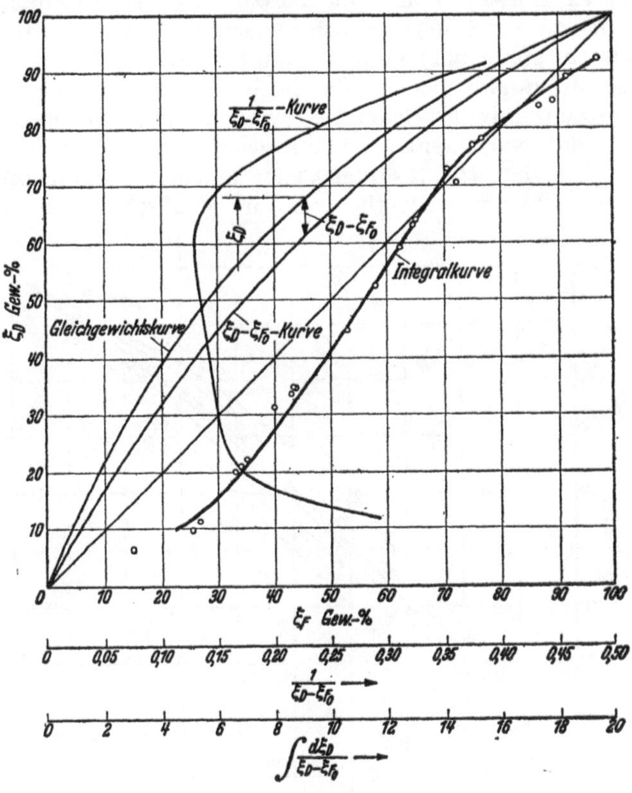

FIGURE 40

Auxiliary graph for determining the enrichment curves for the benzol-toluol mixture

Gew. %—Per cent by weight
Kurve—Curve
Gleichgewichtskurve—Equilibrium curve
Integralkurve—Integral curve

Thus, enrichment curves for a mixture of benzol and toluol result, as shown in Figure 41.

It is seen, for example, that a vapor entering the condenser with $\xi_{D_e} = 50$ weight per cent of benzol is enriched on partial condensation with a reflux ratio of 5 up to 64 weight per cent of benzol.

In comparison to the ethyl alcohol-water system, average enrichment is less. This is due not only to differences in physical constants, film and mass transfer coefficients, but also to differences in the shape of the equilibrium curves.

The enrichment curves for both ethyl alcohol-water and benzol-toluol mixtures can be found by both theoretical deduction and experiment. They show that the frequently made assumption that the residual vapor leaving a condenser is in phase equilibrium with the total mass of the condensate is incorrect.[27]

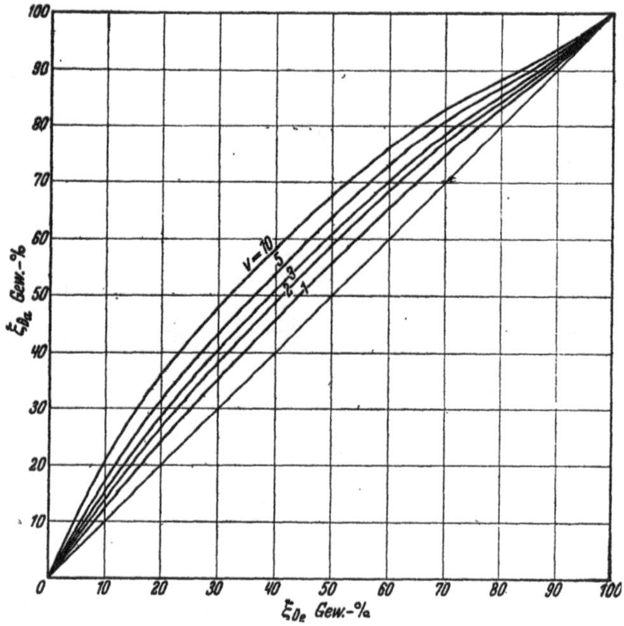

FIGURE 41

Enrichment curves for benzol-toluol mixtures

Gew. % — Per cent by weight

V. Heat of evaporation of mixtures

In distillation and rectification the processes encountered are evaporation of liquid mixtures and condensation of vapor mixtures. In order to make calculations for a process and to determine the consumption figures, it is necessary to know the heats of evaporation and condensation of mixtures.

1. Heat of evaporation per unit of weight

As an example the method used to compute the heat of evaporation of a binary mixture is demonstrated. Let

ρ_1 = Heat of evaporation of the low boiler, kcal/kg
ρ_2 = Heat of evaporation of the other substance
ρ = Heat of evaporation of the mixture

The concentrations of the low boiler in the liquid and in the vapor are ξ_F and ξ_D (in weight per cent), respectively.

a. Immiscible mixtures

Each component vaporizes at the temperature corresponding to the boiling point of the mixture which in turn is related to the partial vapor pressure of the components. If the saturation heats of vaporization of the components are taken for this boiling temperature of the mixture, the saturation heat of vaporization is

$$\rho = \frac{\xi_D}{100} \times \rho_1 + \frac{100 - \xi_D}{100} \times \rho_2 \qquad (44)$$

If condensate is formed from a mixture of vapors, the heat of condensation is computed from the composition of the liquid formed:

$$\rho = \frac{\xi_F}{100} \times \rho_1 + \frac{100 - \xi_F}{100} \times \rho_2 \qquad (44a)$$

b. Miscible mixtures

In contrast to immiscible mixtures, superheating of the vapor of each component of a miscible mixture in the vapor mixture occurs. Thus, for the boiling temperature under consideration, the partial vapor pressure of each substance is smaller than the saturation pressure of the component at the same temperature.

In contrast to the evaporation of a pure liquid, vaporization of a miscible mixture produces superheated vapors. Tests with different liquid mixtures indicate that the evaporation of a miscible mixture requires heats of vaporization which can be determined by equation (44) from the composition of the vapor generated. Then ρ_1 and ρ_2 represent the heats of vaporization of the components at the total distillation pressure (rather than at the partial pressure) and at the boiling temperature corresponding to this distillation pressure.

Since the compositions of vapor and liquid differ under equilibrium conditions, heats of solution may become important during evaporation. In comparison to the heats of vaporization, the latter are, however, sufficiently small (or liquid and vapor concentration are so nearly alike), that heats of solution can be neglected.[28] Thus, equation (44) yields values which are sufficiently exact. The heat of condensation may also be found from equation (44a).

Numerical example. What is the heat of vaporization of a vapor mixture of ethanol and benzol containing 40 per cent by weight of ethanol at 760 mm Hg, if the heats of vaporization of benzol and ethanol respectively, at 760 mm Hg, are 94 and 207 kilocalories per kilogram?

According to equation (44)

$$\rho = \frac{40}{100} \times 207 + \frac{60}{100} \times 94 = 82.8 + 56.4$$
$$= 139.2 \text{ kilocalories per kilogram}$$

Experimental results indicate that $\rho = 137.4$ kilocalories per kilogram.

2. Molar heat of evaporation

The heat of evaporation per mol of mixture, r, in kilocalories per mol, is determined from the same basic equation used for finding the heat of evaporation per kilogram. Let

r_1 = heat of evaporation per mol of the low boiler, in kilocalories per kilogram

r_2 = molar heat of evaporation of the other component

x_F and x_D = concentrations of the low boiler in the liquid and vapor respectively, in mol per cent

If evaporation generates a vapor mixture containing x_D mol per cent of low boiler, the heat

$$r = \frac{x_D}{100} \times r_1 + \frac{100 - x_D}{100} \times r_2 \tag{45}$$

is required per mol of vapor mixture. Correspondingly, if a liquid with x_F mol per cent is generated by condensing a vapor, the heat of condensation to be removed is

$$r = \frac{x_F}{100} \times r_1 + \frac{100 - x_F}{100} \times r_2 \qquad (45a)$$

TABLE 3

Values of Trouton's Constant

Substance	Molar Heat of Vaporization kcal/mol	Boiling Temperature °K	Constant C, Experimental Value
Nitrogen	1340	77.3	17.3
Oxygen	1630	90.1	18.1
Hydrochloric Acid	3860	188.1	20.5
Ammonia	5560	239.8	23.2
Pentane	6130	309	19.85
Carbon Disulfide	6400	319.3	20
Ethanol	9550	351.5	27
Water	9710	373.1	26
Formic Acid	5430	374	14.5
Acetic Acid	5800	391.2	14.85
Aniline	10000	457	21.9
Acetone	7250	329	22
Methanol	8420	340	24.8
Isopropyl Alcohol	9550	356	26.8
Amyl Alcohol	10560	402	26.3
Benzol	7370	353	20.9
Toluol	8000	384	20.8
m-Xylol	8200	413	19.9

There is a definite relation between the heat of evaporation per mol and the absolute boiling temperature T. Pictet discovered, and Trouton stated[29] that the molar heat of vaporization is roughly proportional to the absolute boiling temperature:

$$r_1 = C \times T \qquad (46)$$

The C in equation (46) has an average value of 21. Table 3 shows that a number of substances actually behave in conformity with this rule.

Where substances exhibit a behavior differing from what would be expected from equation (46), such differences may be due to association in the vapor and liquid state. Differences are often encountered with commercially important materials. Both ethyl alcohol and water deviate in the same direction, and have almost the same Trouton constant.

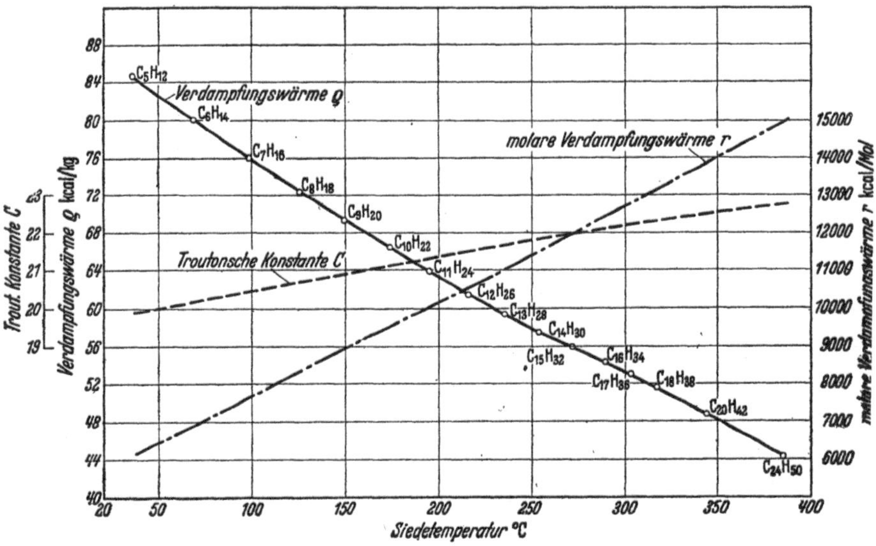

FIGURE 42

Heat of vaporization and Trouton's constant for hydrocarbons of the paraffin series as a function of the boiling point at 760 mm Hg

Verdampfungswärme—Heat of vaporization
Troutonsche Konstante—Trouton's constant
Siedetemperatur—Boiling point
Molare—molar

As a further check on the validity of Trouton's rule, the heat of evaporation of hydrocarbons of the paraffin series has been plotted against the boiling temperature in Figure 42.[30] The formulae of the different compounds have been designated at the corresponding points. Furthermore, the molar heat of evaporation, r, (kilocalories per mol) has been given as a function of the boiling temperature, so that Trouton's constant can be computed. It will be seen from Figure 42 that its average value is 21, with variations lower for lower boiling temperatures, and variations higher for higher boiling temperatures. Thus the Trouton constant for heptane (C_7H_{16}) is 20.5. Other saturated hydrocarbons also obey Trouton's rule. For methyl cyclohexane (C_7H_{14}) with a heat of evaporation of $\rho = 76.9$ kilocalories/kg, a boiling temperature, $t = 99.9°C$ and a molar heat of evaporation, $r = 7550$ kilocalories/mol, Trouton's constant under the same pressure of 760 millimeters amounts to 20.4.

In the separation of industrially important mixtures, many cases are found which follow Trouton's rule, as stated in equation (46). Considering that the absolute boiling temperatures of mixtures handled in distilling and rectifying equipment are usually close together and practically equal, it follows that the molar heats of evaporation of the individual mixtures can be considered to be equal. It further follows that the molar heat of evaporation of each of the two components equals the heat of evaporation of the mixture, and that it is independent of the concentration

$$r = r_1 = r_2 \tag{47}$$

This condition is used in estimating the capacity of distillation equipment. Of course, proper allowance may be made for any deviations from equation (47).

C. SEPARATION OF LIQUIDS BY SIMPLE DISTILLATION; THE SIMPLE POT STILL

I. Data for computation

Except in the case of azeotropic mixtures, a vapor formed from a liquid mixture will have a composition different from that of the liquid itself. This circumstance makes it possible to split up liquids into mixtures having compositions different from the original mixture.

As shown in Figure 43, the mixture to be separated is vaporized in a still, and the vapors formed are condensed in a heat exchanger and pass to a receiver. The distillate formed may also be recovered as a series of fractions, as in the case of fractional distillation.

The equations applying for simple distillation are derived for a binary mixture. Let

B = amount of liquid in the still, kilograms
B_1 = amount of liquid at the start of the distillation process

B_2 = amount of liquid at the end of the distillation process

$\xi_F, \xi_{F1}, \xi_{F2}$ = corresponding concentrations of the low boiler in the liquid, per cent by weight

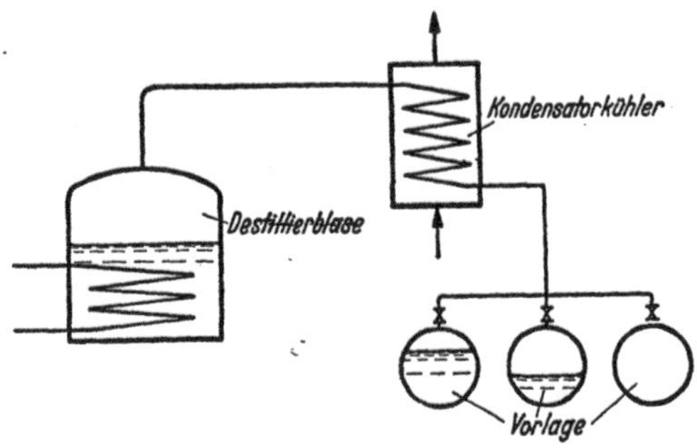

FIGURE 43

Distillation unit

Destillierblase—Still pot
Kondensatorkühler—Condenser and cooler
Vorlage—Receivers

For an infinitesimal amount of vaporization of B, with ξ_F per cent by weight of low boiler, the quantity dB containing ξ_D weight per cent of low boiler will vaporize. This will alter the liquid composition by $d\xi_F$. The following equation then applies.

$$B \times \frac{\xi_F}{100} = dB \times \frac{\xi_D}{100} + (B - dB) \times \frac{\xi_F - d\xi_F}{100} \tag{48}$$

Neglecting the infinitesimals of the second order, the differential equation becomes:

$$\frac{dB}{B} = \frac{d\xi_F}{\xi_D - \xi_F} \tag{49}$$

Integrating this equation between the limits given by the initial batch B_1 and the residue B_2[31] yields:

$$\int_{B_2}^{B_1} \frac{dB}{B} = \int_{\xi_{F2}}^{\xi_{F1}} \frac{d\xi_F}{\xi_D - \xi_F} \tag{50}$$

The integral on the left-hand can be computed, while the right-hand one is given by the difference of integral values J_1 and J_2 coordinated to concentrations ξ_{F1} and ξ_{F2}:

$$\ln \frac{B_1}{B_2} = \int_{\xi_{F2}}^{\xi_{F1}} \frac{d\xi_F}{\xi_D - \xi_F} = J_1 - J_2 \tag{51}$$

Figure 44 shows the graphical solution of equation (51), for an ethanol-water mixture. Any given concentration of the liquid $(\xi_D - \xi_F)$, corresponds to the vertical distance between the equilibrium curve and the 45° diagonal line. Reciprocals of this plotted against ξ_F give a $1/(\xi_D - \xi_F)$ curve. An area below this curve, of width $d\xi_F$, represents the value

$$\frac{d\xi_F}{\xi_D - \xi_F}$$

Since equation (51) involves the difference of two integral values, for any ξ_{F2} a numerical value of J_2 may be selected. In Figure 44, the value $J_2 = 1$ has been selected for $\xi_{F2} = 26$.

If, to this value the numerical value of the area below the $1/(\xi_D - \xi_F)$ curve is added, starting at ξ_{F2} and ending at any desired ξ_{F1} (J_2 having been assumed), the value of J_1 can be computed for different ξ_{F1} values. The $\int \frac{d\xi_F}{\xi_D - \xi_F}$ curve is evaluated in this manner; see Figure 44.

Using this integral curve, the curve applying to the batch still for a desired charge may be found by the following method: A value for the mass of the final batch B_2 is assumed for a ξ_{F2} and a corresponding J_2. In Figure 44, for instance, B_2 has been assumed as 51 for the curve *b* with $\xi_{F2} = 26$. The corresponding J_1 value is found for a ξ_{F1} value, and the batch B_1 corresponding to ξ_{F1} is computed by means of equation (51). This value is plotted against F_1, which yields another point on curve *b*. In this way, any number of B_1 values corresponding to different values of ξ_{F1} can be determined. Upon connecting these points, a batch-capacity curve *b* is obtained.

If the ordinates of such curves are multiplied or divided by the same factor, a number of still-capacity curves are obtained, labelled *a*, *b* and *c* in Figure 44. Depending on the concentration of the feed, the proper curve is employed to eomputc the still capacity.

Corresponding computation and capacity curves for an acetic acid-water mixture are presented in Figure 45.

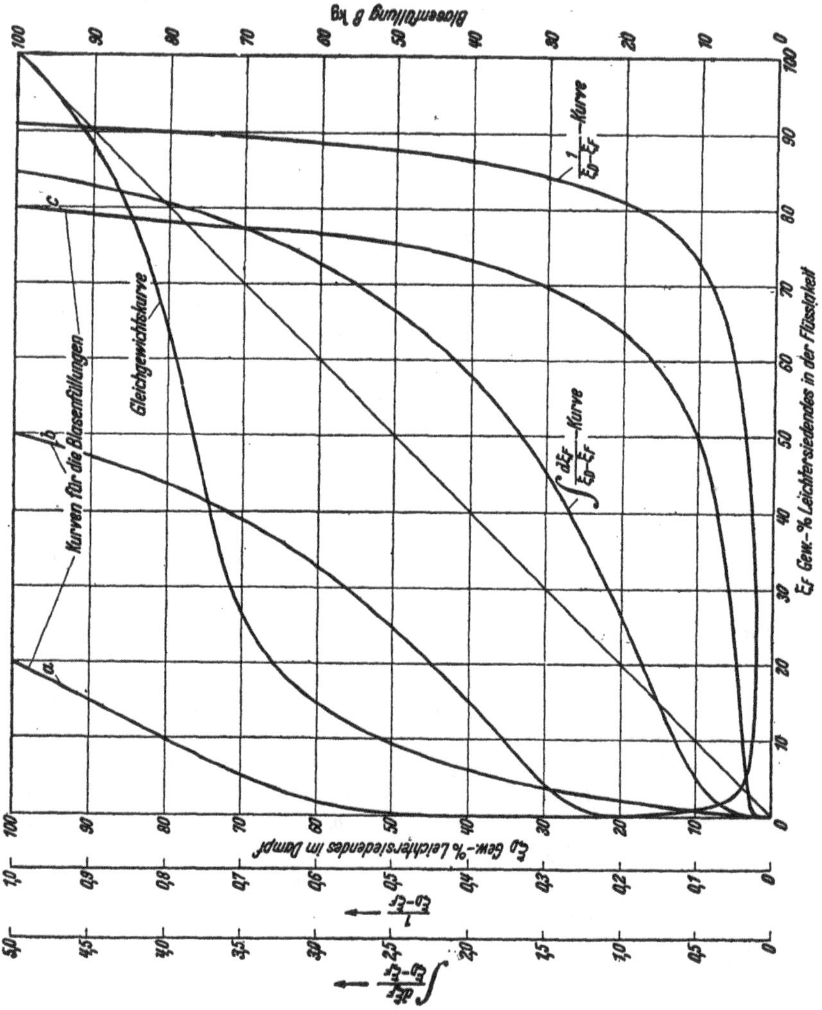

FIGURE 44

Curves for computation of simple batch stills
(Ethanol and water)

Gew. % Leichtersiedendes im Dampf—Per cent by weight of low boiler in vapor
Gew. % Leichtersiedendes in der Flüssigkeit—Per cent by weight of low boiler in liquid
Blasenfüllung—Batch charge
Kurven für die Blasenfüllungen—Curves for various batch charges
Gleichgewichtskurve—Equilibrium curve
Kurve—Curve

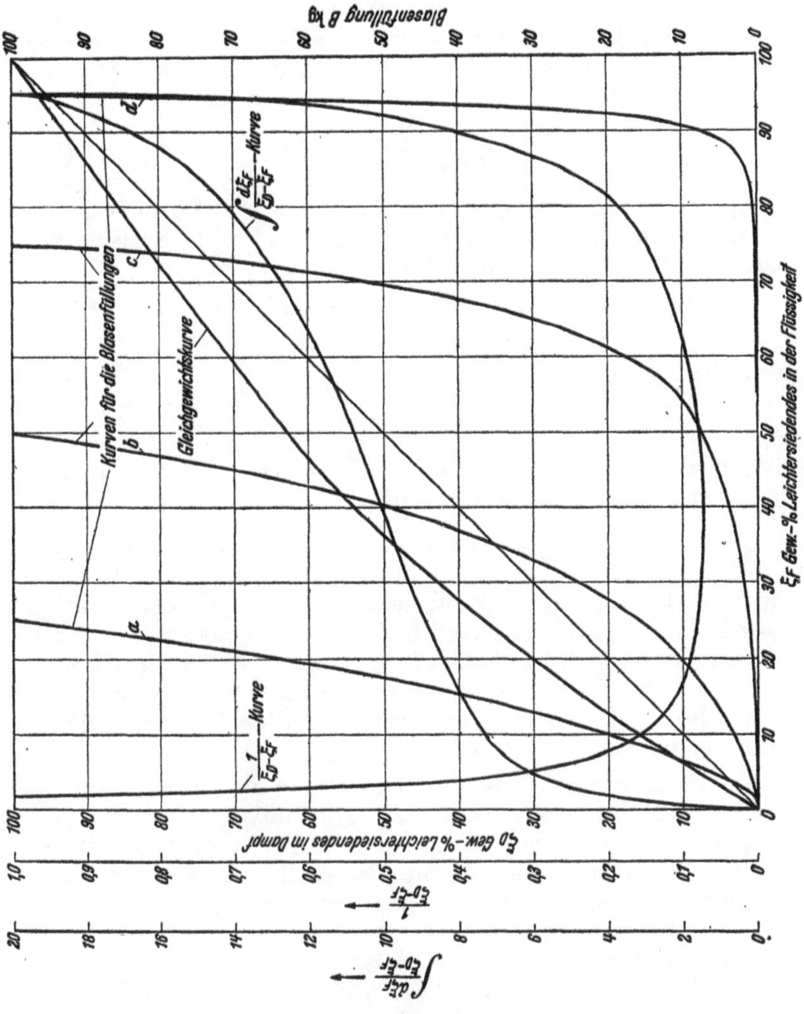

FIGURE 45

Curves for computation of simple batch stills
(Acetic acid and water)

Kurve—Curve
Kurven für die Blasenfüllungen—Curves for various batch charges
Gleichgewichtskurve—Equilibrium curve
Blasenfüllung—Batch charge
Gew. % Leichtersiedendes im Dampf—Per cent by weight of low boiler in vapor
Gew. % Leichtersiedendes in der Flüssigkeit—Per cent by weight of low boiler in liquid

If E (kilograms) is the amount of distillate obtained upon reducing the batch concentration from ξ_{F1} to ξ_{F2}, the following relation holds for the residue in the still, starting with B^* (kilograms) of feed.

$$\frac{E}{B^*} = \frac{B_1 - B_2}{B_1} \tag{52}$$

B_1 and B_2 values, corresponding to given ξ_{F1} and ξ_{F2} values, may be taken from plots as shown in Figure 44 and 45, using the same batch-capacity curve. Then from equation (52)

$$E = \frac{B_1 - B_2}{B_2} \times B^* \tag{53}$$

and

$$B^* = \frac{B_1}{B_1 - B_2} \times E \tag{54}$$

The distillate produced (fraction) has a concentration in low boiler of ξ_E weight per cent. Then,

$$\xi_E = \frac{B_1 \times \xi_{F1} - B_2 \times \xi_{F2}}{B_1 - B_2} \tag{55}$$

At the end of the run, a residue having a low-boiler concentration ξ_{F2} remains. This is removed from the still before a new batch is charged. As the residue always contains some low boiler, and since the concentration of the remainder varies continuously during distillation, complete separation by simple distillation is impossible.

Numerical example. It is desired to obtain, in one run, 100 liters of distillate from an ethanol-water mixture containing 14 weight per cent alcohol. What will be the composition of the distillate, and how much of the raw product must be charged if the residue is to contain a maximum alcohol content of 0.5 per cent by weight?

Refer to Figure 44. Curve *b* gives, for a concentration of $\xi_{F1} = 14$, a value of B_1 of 39 kg, and for $\xi_{F2} = 0.5$ a B_2 value of 24 kg. From equation (55) the alcohol concentration of the distillate is

$$\xi_E = \frac{39 \times 14 - 24 \times 0.5}{39 - 24} = 35.6 \text{ weight per cent}$$

For this distillate concentration, the density[32,33] at 15°C γ is 0.9472 kg per liter, so that the amount of distillate is

$$E = 100 \times 0.9472 = 94.72 \text{ kg}$$

According to equation (54), the initial charge must be

$$B^* = 94.72 \times \frac{39}{39 - 24} = 247 \text{ kg}$$

or 252 liter of mixture.

II. Design of kettle stills

The design of kettle stills is determined primarily by the method of heating and the nature of the crude material. Figure 46 shows a steam-heated cast iron still; (see also Figure 90, cast iron still heated by direct fire). Figure 47 shows a carbon steel kettle heated by direct fire and Figure 48, a still heated by a steam coil, built up of rolled metal sheet.

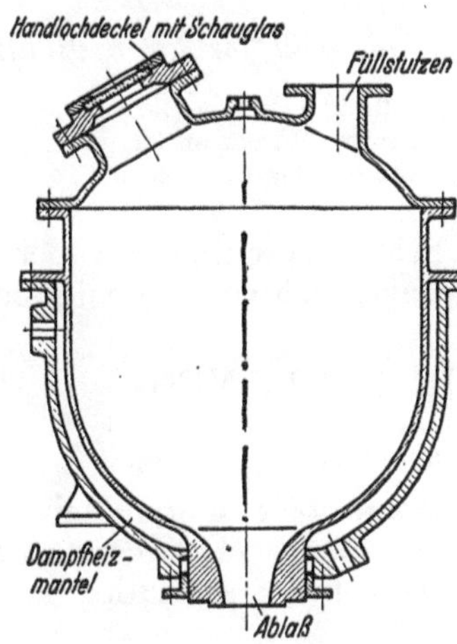

FIGURE 46

Cast-iron still

Dampfheizmantel—Steam jacket
Handlochdeckel mit Schauglas—Handhole cover with inspection glass
Füllstutzen—Feed connection
Ablass—Drain

III. Separation by distillation and partial condensation

Separation of liquids by distillation combined with enrichment by partial condensation can be carried out in one single unit, as shown in Figure 49. Part of the vapors produced in the kettle are condensed in the condenser. The condensate

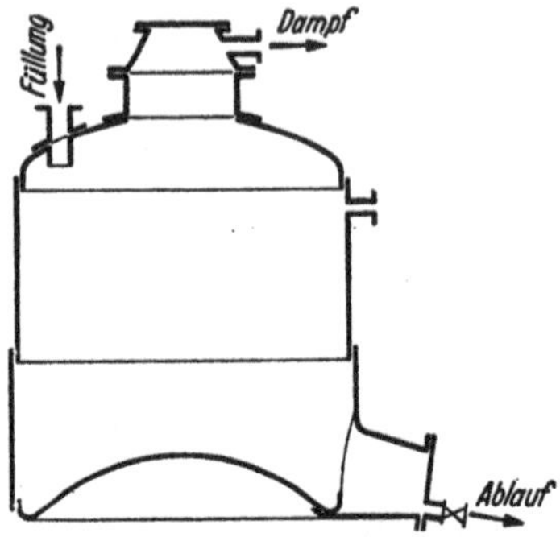

FIGURE 47

Direct-fired still for tar distillation

Füllung—Feed connection
Dampf—Vapor outlet
Ablauf—Drain

returns to the still, while the enriched vapor is liquefied in an additional heat exchanger to give the final product.

D. THE RECTIFYING COLUMN

I. Effect of rectifying plates

The best possible separation of a liquid mixture can be effected by subjecting it as frequently as necessary to the two basic processes of enrichment by vaporization and enrichment by partial condensation.

To accomplish this, vapor arising from a liquid mixture is reintroduced into the same liquid mixture. Because of the temperature difference between vapor and liquid, part of the vapor is condensed, and the heat of condensation liberates fresh vapor from the liquid. This process takes place on interchanger or rectifying plates, on which the liquid mixture accumulates, and which the vapor is added in such a way that a large area of contact becomes available between vapor and liquid, allowing better interchange of heat and materials.

Since the vapor leaving a plate should have a larger concentration in low boiler than the vapor fed to that plate, it is necessary to feed the plate continuously with fresh liquid having a low-boiler concentration which exceeds that of the liquid leaving

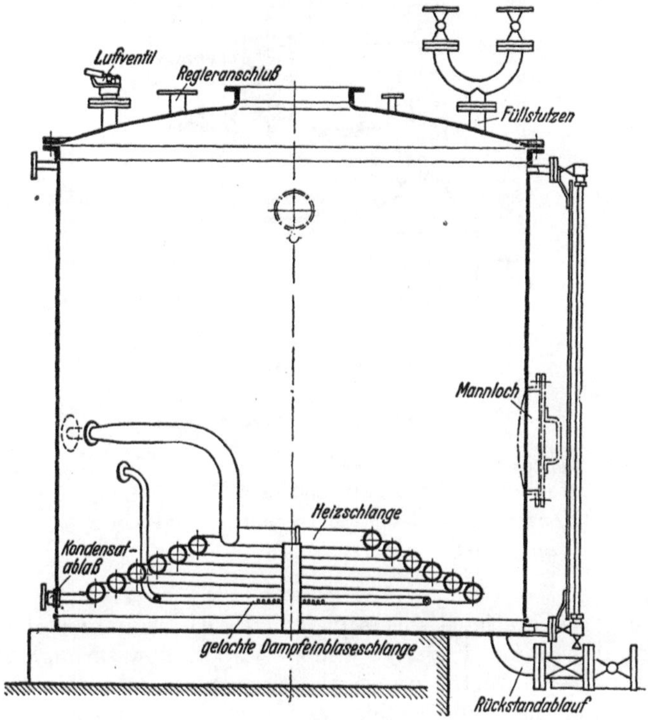

FIGURE 48

Vertical kettle-still with heating coil

Heizschlange—Heating coil
Kondensatablass—Condensate drain
Gelochte Dampfeinblaseschlange—Perforated steam injection ring (Sparger)
Luftventil—Relief valve
Regleranschluss—Controller connection
Füllstutzen—Feed connection
Mannloch—Manhole
Rückstandsablauf—Residue draw-off connection

that plate. This liquid is called the reflux. If there is no reflux, an interchanger plate cannot change the concentration of the vapor.

The commoner kinds of devices used in rectification are sieve plates and bubble-cap type plates, and their modifications. Their mode of operation is shown in Figures 50 and 51.

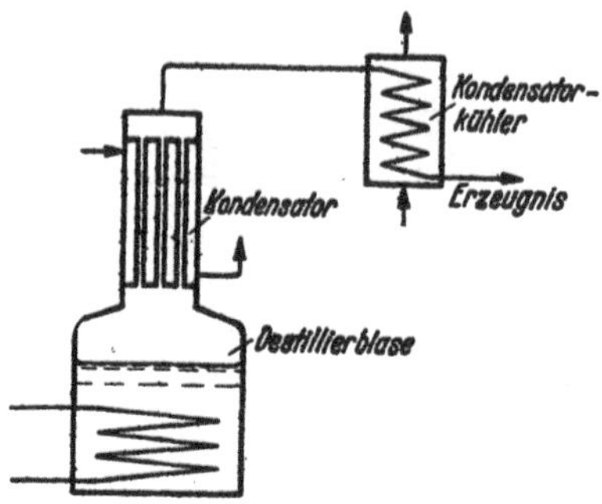

FIGURE 49

Still with reflux condenser

Destillierblase—Still pot
Kondensator—(Partial) condenser
Kondensatorkühler—(Final) condenser and cooler
Erzeugnis—Product

Sieve plates are level sheets provided with a number of holes and liquid feed and discharge weirs. Vapors flow through these holes and pass into the liquid phase, forming bubbles. In the bubble zone, the liquid is clear and relatively quiescent. At a certain distance above the plate, a layer of foam forms. In the case of non-foaming mixtures, a turbulent layer forms. It is in this zone that most of the heat and material interchange takes place.

Cap-type plates are also equipped with liquid feed and dis-

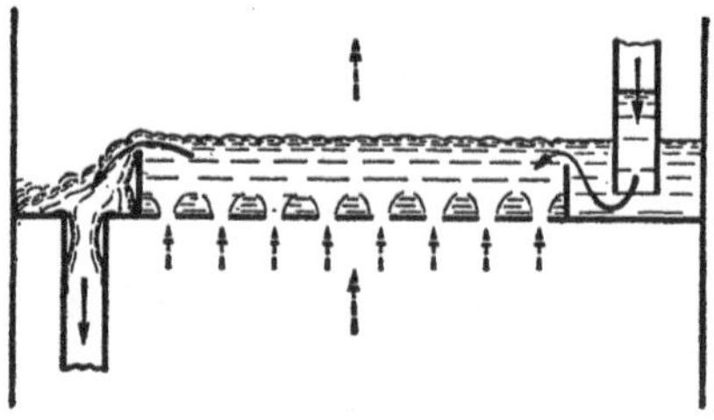

FIGURE 50

Sieve plate

charge weirs. The plate is fitted with one or several nozzles, covered with caps which dip into the liquid on the plate, so that the vapors are forced to pass through the liquid. Below the channels, there is a clear liquid layer, and above them is located the foam or turbulent layer.

In industrial practice, bubble-cap types are preferred. This is because of the wider load range possible with the cap type and its relative freedom from plugging and scale deposition due to the larger vapor passages available.

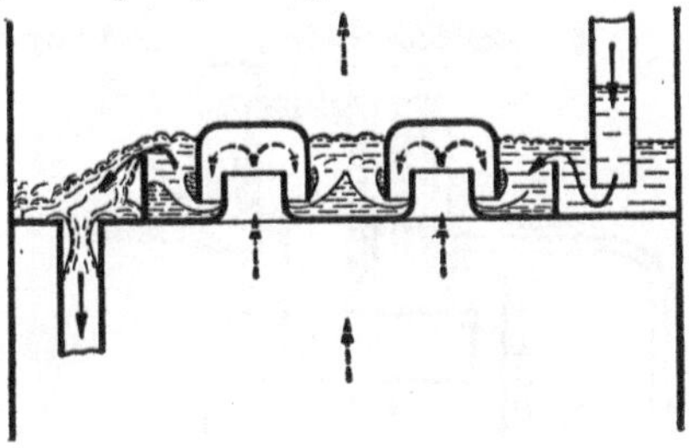

FIGURE 51

Bubble-cap type plate

Thus, the effect obtained on a plate may be repeated any desired number of times, and when a large number of plates is used the vapor may be greatly enriched in low boiler. The vapors move from any plate to the one above, and the liquid flows downward to the next plate below. Although in the usual design the vapor and liquid move in a cross-current over the plate, their relative motion, considered over the entire height of the column, is that of counter-current flow. Therefore, rectification really is a thermal separation method, by which liquid mixtures are broken up by making the vapor and liquid move countercurrently to each other, and providing an area of mutual contact between the streams which is as large as feasible.

II. Hookup of reflux condensers

The reflux which flows downward from plate to plate in a fractionating column is generated by liquefying the vapor leaving the top plate of the column in a condenser. Part of the condensate is refluxed to the top plate of the rectifying column, and the balance constitutes the product. There are two different types of reflux condenser hookups, each having its special characteristics.

In the hookup shown in Figure 52, all of the vapor leaving the column goes to a condenser and is totally condensed. The liquid leaving the heat exchanger is split up into two streams. One stream forms the reflux which goes back to the top plate of the column, while the balance passes on to a cooler and is recovered as a product.

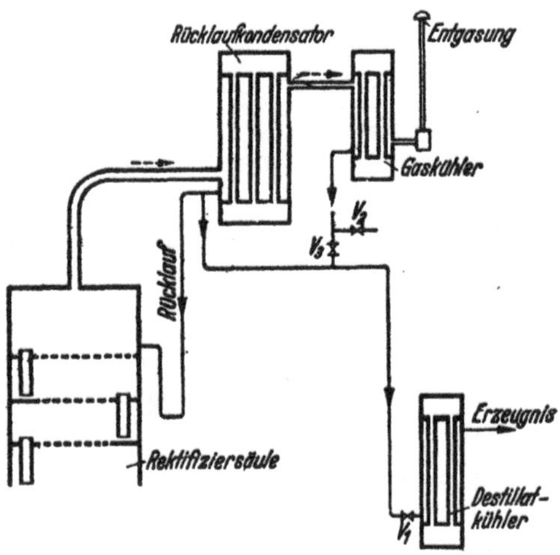

FIGURE 52

Reflux condenser hookup involving complete condensation

Rektifiziersäule—Rectifying column
Rücklauf—Reflux
Rücklaufkondensator—Reflux condenser
Gaskühler—Vent condenser
Entgasung—Vent
Destillatkühler—Distillate cooler
Erzeugnis—Product

The reflux condenser thus treats both the reflux and the product. Since the distillate vapors always contain gases consisting mainly of air, a gas cooler or vent condenser is inserted after the reflux condenser. In the vent condenser, condensible vapors are liquefied, which then usually flow to the product cooler. In addition, valves such as V_2 and V_3 in Figure

52 may be adjusted in such a way that the comparatively small amount of liquid formed in the vent condenser is drained through V_2. An advantage of this scheme is that by setting valve V_1 it is possible to adjust both the amount of product and of reflux accurately and conveniently. The gases are separated at the temperature of the liquid in the reflux condenser. This temperature exceeds that of the product flowing from the cooler. Since as a rule the solubility of gases is greater at lower temperatures, gas separation is most effective with this hookup.

Alternatively, it is possible to abstract heat from the vapors to an extent just sufficient to supply the required reflux. The reflux then passes from the condenser directly to the top plate of the column as shown in Figure 53.

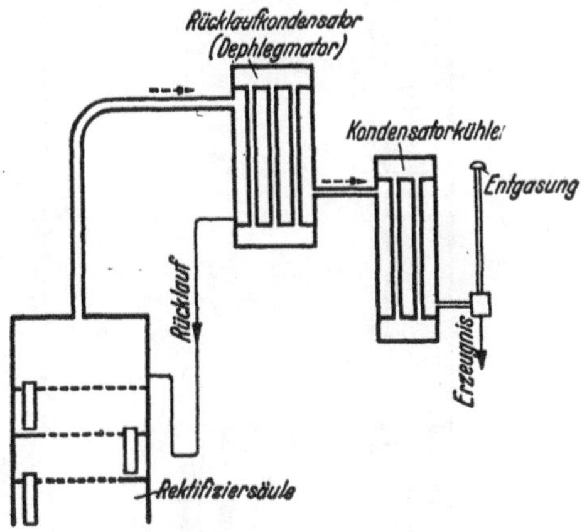

FIGURE 53

Dephlegmator hookup

Rektifiziersäule—Rectifying column
Rücklaufcondensator—Reflux condenser (Dephlegmator)
Rücklauf—Reflux
Kondensatorkühler—Condenser and cooler
Entgasung—Vent
Erzeugnis—Product

The uncondensed vapors then flow into an additional heat exchanger which is usually referred to as a condenser-cooler. In this unit not only does the distillate vapor condense, but

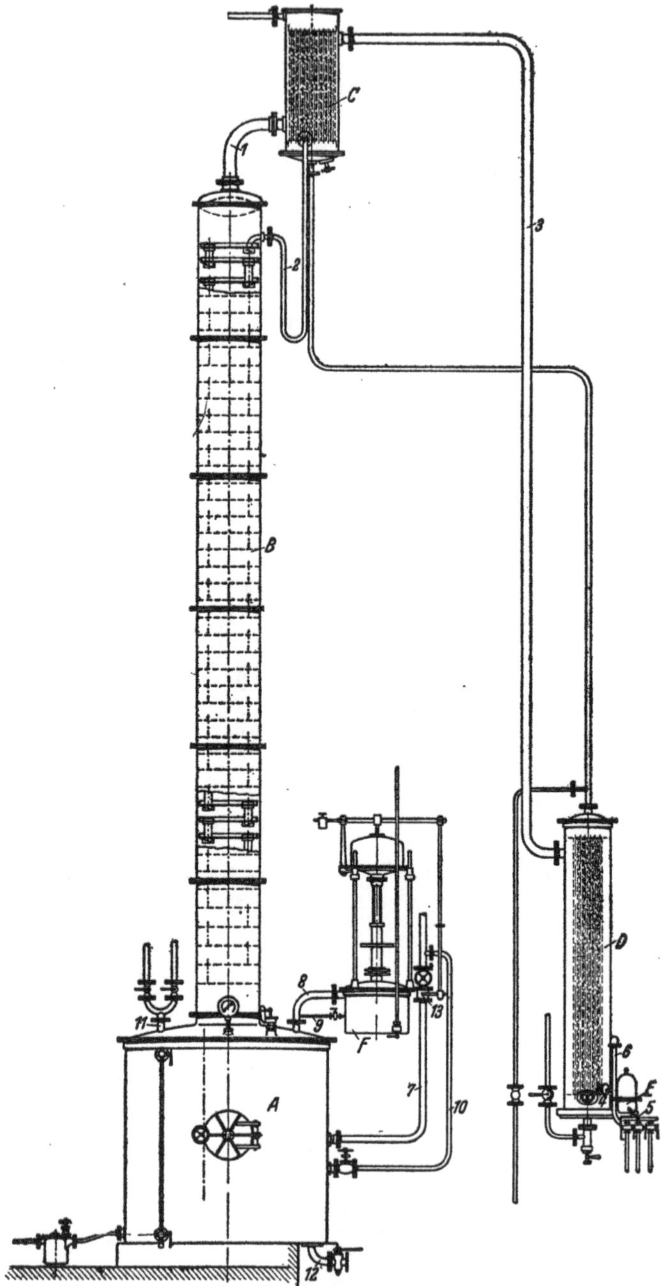

FIGURE 54

Batch-type rectifier

The vapors generated in still *A* are enriched in low boiler in rectifying column *B*, and pass through line 1 to the reflux condenser (dephlegmator) *C*. The liquid condensed is refluxed through line 2 to the top plate of the rectifying column, while the remaining vapors richer in alcohol pass through line 3 into final cooler *D* where they condense.

The product flows through connection 4 to tailbox *E*, and through line 5 to a fusel oil separator. Gases escape through vent 6. Line 7 supplies heating steam to insure constant output. Steam control valve 13 is actuated by regulator *F*. Line 8 transfers the distillation pressure impulse to the regulator. Excess liquid passes through line 9 from the regulator back to the still. Through line 10, steam may be passed to a perforated coil delivering direct steam which strips the remaining alcohol in the residue. The residue is drained, and fresh raw material introduced through connection 11. The cooling water passes first through product cooler *D*, then through dephlegmator *C*.

---

the liquid formed is also cooled below the condensation temperature and recovered as product.

With this installation, the vapor leaving the reflux condenser is richer in low boiler than the bulk of the vapor originally entering the condenser, because enrichment in low boiler by partial condensation takes place. A reflux condenser connected in this way is also called a dephlegmator.

The advantage of this hookup is that the dephlegmator gives an additional enriching effect, not only generating the reflux, but also replacing a limited number of rectifying plates. In this case, condensation of the vapors occurs in the final cooler (condenser-cooler).

III. Layout of a batch-type distillation unit

The combination of the devices described, by means of which liquid mixtures can be separated, results in a batch distillation unit. The layout of such a unit is shown in the example given in Figure 54 which refers to a setup used for separating alcohol and water.

Separating effects are exerted by the still, rectifying column, and reflux condenser when it acts as a dephlegmator. The separating effect of the column is by far the most important, while the main duties of the still and reflux condenser are to generate the vapor and reflux, respectively.

IV. Computation of the number of plates for a batch-type distillation unit

1. Theoretical number of plates

To simplify the computation for the number of plates in a rectifying unit, equation (47) is assumed to be valid. Accordingly, the heat of vaporization of a mixture is independent of its composition. It then follows that the number of mols of vapor condensed by the liquid on any plate equals that which is newly formed. Therefore, the quantities of vapor entering and leaving a plate are equal and constant over the entire column. This same condition holds for the liquid phase.

Referring to Figure 55, let

D = mols per hour of vapor of a binary mixture passing upward between two plates,
F = mols per hour of liquid flowing downward through the downcomer, in the same horizontal section,
x_D = mol per cent low boiler in the vapor,
x_F = mol per cent low boiler in the liquid,
x_E = mol per cent low boiler in the product.

Then,[34,35,36,37]

$$D = F + E \tag{56}$$

and

$$D \times x_D = F \times x_F + E \times x_E \tag{57}$$

On substituting D from equation (56) into equation (57):

$$x_D = \frac{F}{E + F} \times x_F + \frac{E}{E + F} \times x_E \tag{58}$$

Dividing the numerator and denominator on the right side of this equation by F yields the ratio

$$v = \frac{F}{E} \tag{59}$$

which is an important operating variable of a rectifying unit since it determines not only the steam and coolant consumption, but also the number of rectifying plates needed. It indicates the amount of product per unit of reflux, and is termed the *reflux ratio*.

Introducing reflux ratio v into equation (58) gives another equation indicating the relation between vapor concentration and liquid concentration at any time:

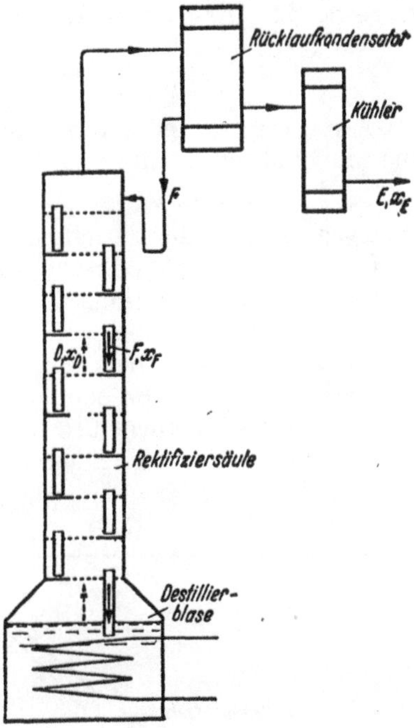

FIGURE 55
Batch-type rectifier

Destillierblase—Still pot
Rektifiziersäule—Rectifying column
Rücklaufkondensator—Reflux condenser
Kühler—Cooler

$$x_D = \frac{v}{v+1} \times x_F + \frac{x_E}{v+1} \tag{60}$$

Since only x_D and x_F are variable, this expression represents the equation of a straight line termed the *enrichment line* (line of interchange of a fractionating column). When x_F is plotted against x_D (Figure 56) this line intersects the ordinate at the point

$$o = \frac{x_E}{v+1} \tag{61}$$

and its slope with respect to the x_F axis is:

$$\tan \alpha = \frac{v}{v + 1} \qquad (62)$$

In equation (60), if $x_D = x_F$ the intersection of the enrichment line with the diagonal for the abscissa is

$$x_F = x_E$$

The two points x_E and o on the enrichment line have been labelled A and B in Figure 56. A point of this line indicates the relation between the concentrations in vapor and liquid for any two contiguous rectifying plates.

When no distillate is withdrawn from the column ($E = 0$), it works with total reflux. All of the condensate formed from the vapors leaving the column is returned to the column. Then, $v = \infty$ and the enrichment line coincides with the diagonal in the plot.

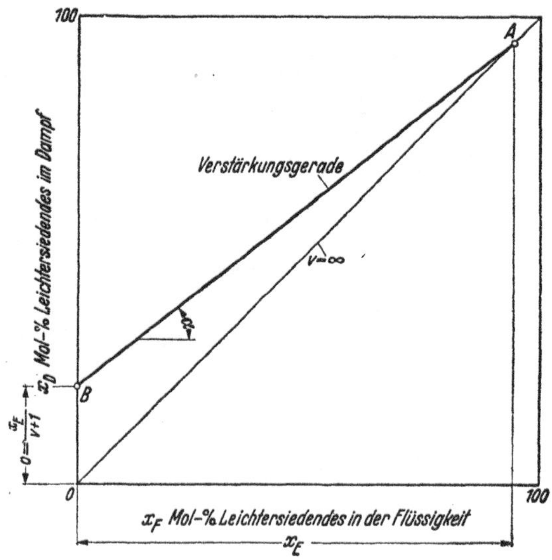

FIGURE 56

Determination of the enrichment line

Verstärkungsgerade—Enrichment line
Mol % Leichtersiedendes im Dampf—Mol per cent low boiler in vapor
Mol % Leichtersiedendes in der Flüssigkeit—Mol per cent low boiler in liquid

The composition of the vapor arising from a plate can be determined by means of the enrichment line, provided the composition of the liquid overflowing to that plate is known. Furthermore, if the concentrations of the vapor rising from a plate and of the liquid leaving that plate are known, it becomes possible to ascertain, step by step, the variations of the compositions of vapor and liquid through the entire column.

The latter relation is found by assuming that the mixing of the liquid on a plate and the interchanges of heat and materials are complete, so that phase equilibrium exists between the vapor and the liquid.

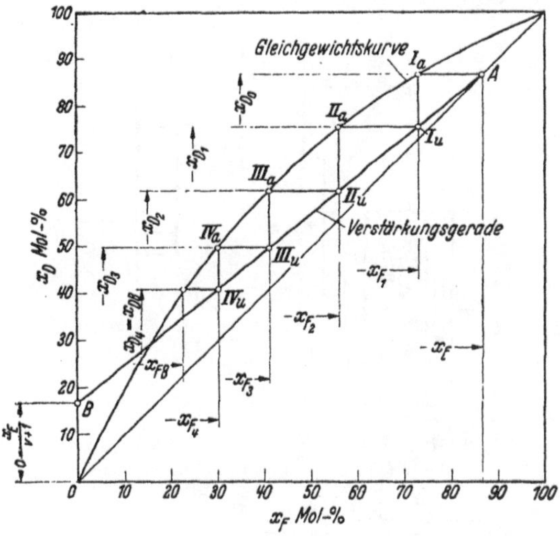

FIGURE 57

Determination of the number of plates in an enriching column

Gleichgewichtskurve—Equilibrium curve
Verstärkungsgerade—Enrichment line

This means that the vapor rising from a plate is in phase equilibrium with the liquid leaving the same plate. Thus, the relation between the concentrations of these two phases may be represented by a point on the equilibrium curve.

If these assumptions are made, the number of theoretical plates can be determined as in Figures 57 and 58. It is assumed that all of the vapor flowing to the reflux condenser is liquefied, so that there is no enrichment by partial condensation.

In this case, the vapor rising from the top plate, the reflux fed back to the top plate and the product all have the same composition, x_E mol per cent. This composition determines intersection A of the enrichment line with the diagonal (Figure 57). If the reflux ratio, v, is known, another point, B, of the enrichment line is given.

The vapor rising from the top plate of a rectifying column has a composition $x_{D0} = x_E$, represented by the ordinate value of point A. The parallel with the abscissa passing through A intersects the equilibrium curve at I_a.

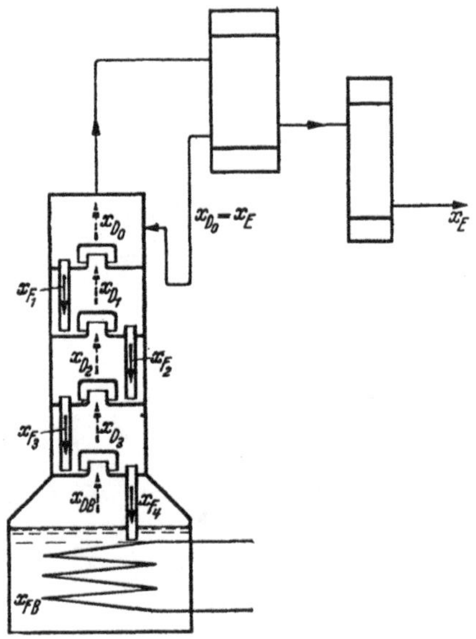

FIGURE 58

Batch-type rectifier

Location of the liquid and vapor concentrations determined in Figure 57

Point I_a indicates the relation between the concentrations of the vapor rising from the top plate and the liquid leaving the same plate. Thus, the abscissa of I_a establishes the concentration x_{F1} of the liquid leaving the top plate. It further specifies point I_u on the enrichment line, the ordinate value of which yields concentration x_{D1} of the vapor passing from the second plate to the top plate. Since this vapor is in phase equilibrium with the liquid x_{F2} leaving the second plate, the relation between x_{D1} and x_{F2} is represented by intersection II_a of the parallel to the abscissa through I_u with the equilibrium curve.

The liquid composition x_{F2} specifies point II_u on the enrichment line which establishes the relation between vapor and liquid concentrations below the second plate. Point III_a with the ordinate value x_{D2} represents a point of equilibrium for the two phases on the third plate, while point III_u indicates the relation of concentrations below the third plate, IV_a applies to conditions above the fourth plate, and IV_u to those below the fourth plate.

If the column is equipped with four perfect plates, the liquid leaving the bottom plate returns to the still with a concentration x_{F4}, and x_{D4} is the concentration of low boiler in the vapor rising from the still. This vapor with a concentration $x_{DB} = x_{D4}$ is in phase equilibrium with the liquid in the still containing x_{FB} mol per cent of low boiler.

Thus, the reflux fed back to the still has a concentration of low boiler which lies between that of the still charge and that of the vapor generated in the still.

In Figure 57, points I_a, II_a, III_a, IV_a indicate the relation of the equilibrium concentrations above the plates, and points I_u, II_u, III_u, IV_u refer to the liquid and vapor concentrations in a horizontal section below these plates.

The theoretical number of enriching plates, n_t, is thus represented by the number of points I_u, II_u, III_u, IV_u etc. on the enrichment line, which result on tracing the steps between the equilibrium curve and the enrichment line, beginning at A, up to the still concentration, x_{FB}.

The vapor and liquid concentrations discussed in Figure 57 have been indicated in Figure 58 to show their relative positions in the column.

2. The enrichment ratio

Assuming that the vapor arising from a plate is in phase equilibrium with the liquid leaving the same plate, the variation in the low-boiler concentration in the vapor which can be accomplished by means of an interchanger plate can be computed.

If the concentration of low boiler in the liquid leaving a plate is x_{Fu} (Figure 59), the concentration x_{Du}, of the vapor passing into the same plate is determined by point C located on the

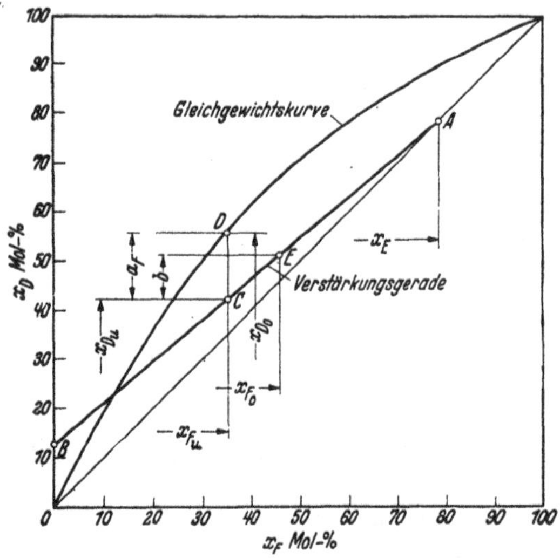

FIGURE 59

Theoretical and actual enrichment of low boiler in vapor for one plate

Mol %—Mol per cent
Gleichgewichtskurve—Equilibrium curve
Verstärkungsgerade—Enrichment line

enriching line. The vapor arising from the plate under consideration is in phase equilibrium with the liquid leaving, so that its concentration, x_{D0}, is represented by point D on the equilibrium curve, which in turn corresponds to liquid concentration, x_{Fu}.

The difference between the ordinate on the equilibrium curve and on the enrichment line for the abscissa value x_{Fu} (distance between points D and C) thus represents the theoretical enrichment in low boiler obtainable with complete mixing of the liquid and perfect interchange, i.e., $a_F = x_{D0} - x_{Du}$.

If the actual concentration of the liquid flowing to the plate under consideration is x_{F0}, intersection E (on the enrichment line) obtained by drawing a vertical line through x_{F0} (Figure 59) yields the actual concentration of low boiler in the vapor arising from the plate under consideration. Actual enrichment attains only the value b (Figure 59). The ratio of actual to theoretical enrichment

$$s = \frac{b}{a_F} \tag{63}$$

is called the enrichment ratio or interchange ratio.

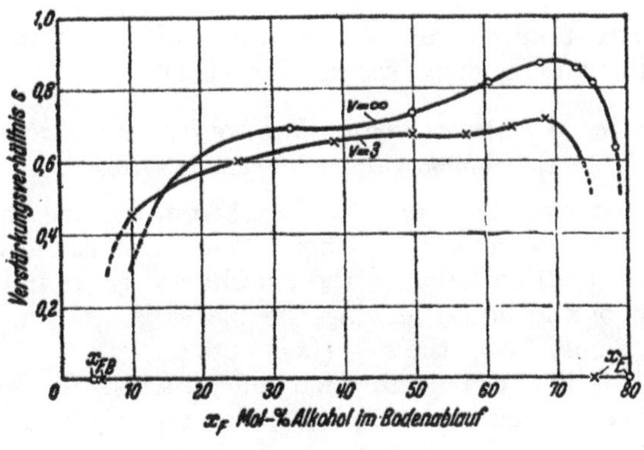

FIGURE 60

Actual enrichment ratio as a function of the final plate concentration, from tests run on a column separating ethanol and water

Mol % Alkohol im Bodenablauf—Mol per cent alcohol in liquid leaving plate
Verstärkungsverhältnis—Enrichment ratio

Experiments point out that,[38] other things being equal, the interchange ratio in a column varies from plate to plate with the concentration. According to Figure 60, the enrichment ratio increases with the concentration in low boiler up to a maximum, and then decreases again. The points plotted indicate the enrichment ratios corresponding to the concentrations in the liquids leaving the plates.

The experimentally determined $s = f(x_F)$ curves appear to intersect the abscissa at values depending on distillate concentration, x_E, and on the concentration x_{FB} of low boiler in the still.

3. Determination of the actual number of plates

a. *Use of the average enrichment ratio*

Generally, an average enrichment ratio, s_M, assumed to apply to all of the plates in a rectifying column, may be used in the computation. It usually varies between 0.6–0.8, but may have other values, depending on conditions.

The number of plates required, n, is derived from the theoretical number of plates n_t by employing the mean enrichment ratio

$$n = \frac{n_t}{s_M} \tag{64}$$

Since s may take on values exceeding unity, it should not be described as an efficiency; see Section H IV.

b. *Number of plates required, allowing for the variation of the enrichment ratio with concentration*[39]

The number of plates may be found from the equilibrium and enrichment lines directly, without using a mean enrichment ratio, if the relation between the enrichment ratio and the concentration of the liquid leaving the plates is known, as represented in Figure 61 by the $s = f(x_F)$ curve.

A vertical line drawn through the abscissa for any given liquid concentration from a plate gives the point representing the enrichment ratio on the $s = f(x_F)$ curve, and intersects the enrichment line and equilibrium curve in points III_u' and III_a, respectively. The ordinate difference a_F, between the two points represents the theoretical enrichment of low boiler in the vapor effected by the plate for the composition of liquid overflow selected.

Actually, the concentration of low boiler in the vapor is only increased by the amount $b = a_F \times s$, since there is neither complete interchange of materials nor perfect mixing of the liquid.

If the numerical value of the product $a_F \times s$ is plotted against liquid concentration, using points III_u' and III_a', the ordinates indicate the true concentration of low boiler in the vapor arising from the plate considered.

For a given concentration of the liquid leaving a plate, it has been explained how s, a_F, and the concentration of the vapor arising from the plate with the same liquid concentration are found from point III_a'. Analogously, for any concentration

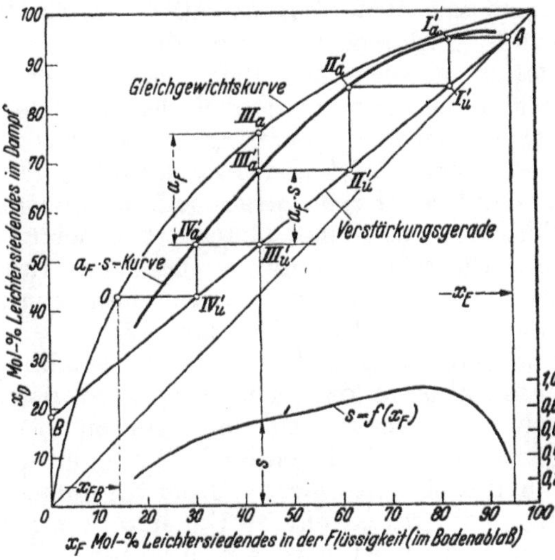

FIGURE 61

Curve for determining the actual number of rectifying plates

Mol % Leichtersiedendes in der Flüssigkeit (im Bodenablass)—Mol per cent of low boiler in liquid (at bottom drain)
Mol % Leichtersiedendes im Dampf—Mol per cent low boiler in vapor
Kurve—Curve
Gleichgewichtskurve—Equilibrium curve
Verstärkungsgerade—Enrichment line

of liquid leaving a plate, an s and an a_F value can be determined, and the product $a_F \times s$ can be plotted, yielding an $a_F \times s$ curve which begins at a point corresponding to the final plate concentration at the enrichment line. Any point on this line gives the relation between the concentration of the reflux flowing from a plate and the true concentration of low boiler in the vapor arising from the same plate.

If it is assumed that no enriching occurs in the reflux condenser, point A in Figure 61 indicates the concentration of the low boiler in the vapor rising from the top plate of the column and that of the reflux to the same plate. A horizontal line through point A intersects the $a_F \times s$ curve at I_a', the abscissa of which is the concentration of the liquid flowing from the top or first plate. It gives point I_u' which indicates the vapor concentration corresponding to said liquid concentration. Similarly, a horizontal through I_u' yields point II_a' on the $a_F \times s$ curve which gives the relation between the vapor concentration above the second plate and the concentration of the liquid leaving that plate. If, for instance, a separating column has four interchanger plates, points III_u' and IV_u' establish the relations between liquid and vapor concentrations in a horizontal section below the third and the fourth plates, while points III_a' and IV_a' determine the relation between the concentration in the vapor above the third or fourth plate, and the concentrations of the liquid flowing away from these plates.

A horizontal through IV_u' intersects the equilibrium curve at O, the concentration of low boiler in the vapor rising up from the still, which determines the still concentration, x_{FB}.

According to the method outlined, the number of plates actually required is obtained directly from the number of points I_u', II_u', III_u', etc., traced out by the step sequence between the enrichment curve and the $a_F \times s$ curve beginning at A and terminating at the still concentration. However, the horizontal through O will seldom coincide with the last step, as shown in Figure 61. Accordingly, the number of plates corresponding to the number of points in the step sequence will have to be increased by one.

4. Determinating the number of plates when enrichment by partial condensation occurs in the reflux condenser

It has been assumed in the previous discussion that the reflux condenser has no separating effect. This assumption is valid in the case of flow in accord with Figure 52, where all of the vapor entering the reflux condenser is condensed. This vapor and the liquid leaving the condenser then have identical compositions.

If, however, the reflux condenser supplies the reflux only, and if the balance of the vapor is condensed and cooled in a special exchanger, the vapor residue forming the product has a composition different from that of the vapor leaving the column. The difference in low boiler concentrations between the residual vapor and the vapor passing to the condenser results from enrichment due to partial condensation. When this occurs, the dephlegmator not only yields the reflux, but itself substitutes for a certain number of plates. As an example, Figure 62 shows the condition holding for a mixture of ethanol and water.

Let the product contain ξ_E = 90 per cent by weight or x_E = 78 mol per cent of alcohol, and assume the molar reflux ratio to be $v = 5$. The concentration of the distillate determines point A, and x_E and v fix point B. On connecting points A and B the enrichment line is obtained.

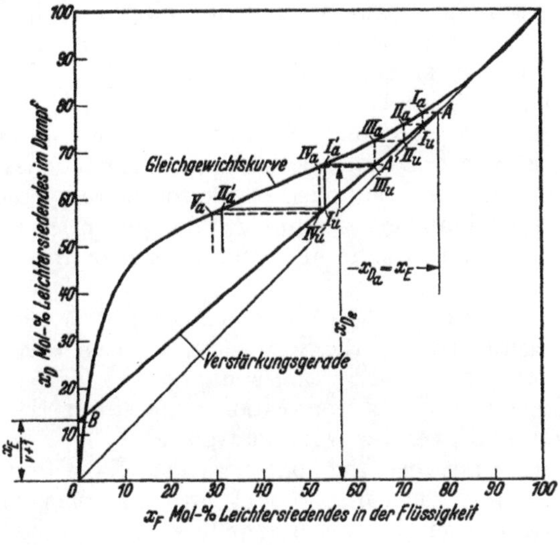

FIGURE 62

Curve for computing the number of plates, taking into account the separating effect of the reflux condenser (Ethanol-water)

Mol % Leichtersiedendes in der Flüssigkeit—Mol per cent low boiler in liquid
Mol % Leichtersiedendes im Dampf—Mol per cent low boiler in vapor
Gleichgewichtskurve—Equilibrium curve
Verstärkungsgerade—Enrichment line

In the precise determination of the theoretical number of plates, complete mixing of the liquid and interchange on the rectifying plates were assumed, and the enriching action of the reflux condenser was omitted. Then, the theoretical number of plates follows directly from the number of intersections I_u, II_u, III_u, etc.; see dotted sequence of steps in Figure 62 between equilibrium curve and diagonal. The composition of the overflow from the third plate is represented by the abscissa of point III_u.

If the reflux condenser is operated so that it enriches, the enrichment line has the same significance as it does in the case of complete condensation of the vapor entering the reflux condenser. Therefore, distillate concentration, x_E and reflux ratio v again determine points A and B on the enrichment line.

The given reflux ratio applies not only for the rectifying column, but also for the partial condensation in the reflux condenser. For the given alcohol concentration of the vapor supplied by the reflux condenser $x_E = 78$ mol per cent, and for $v = 5$, Figure 38 indicates a concentration in low boiler, x_{De}, of 67 mol per cent for the vapor entering the condenser. Vapor of this concentration is supplied to the top plate by the dephlegmator. Its ordinate gives point I_a' on the enrichment curve, and a horizontal through the latter intersects the enrichment line at A'.

The abscissa of point A' indicates 64.8 mol per cent, the alcohol concentration of the liquid flowing from the reflux condenser to the top rectifier plate, while point I_a' gives the equilibrium concentration of the vapor rising from the top plate and in the liquid overflow from that plate.

The relation between the concentration of the liquid leaving the top plate and the vapor flowing into that plate is given by point I_u' located on the enrichment line and obtained from the abscissa of point I_a'.

Thus, the theoretical number of plates, when considering the enrichment by the dephlegmator, is the number of points on the enrichment line obtained by tracing the sequence of steps I'_u, II'_u, etc., between enrichment line and equilibrium curve, starting at A' and terminating at the still concentration considered.

Figure 62 shows that under the conditions selected the dephlegmator will cause vapor enrichment equivalent to that of about three perfect plates. Assuming an average enrichment ratio of 0.75, the reflux condenser thus replaces four rectifying plates.

The number of interchanger plates which can be replaced by a dephlegmator depends primarily on the nature of the mixture to be separated. In the case of the binary mixture benzol-toluol, a distillate concentration, x_E, of 80 mol per cent benzol with a reflux ratio $v = 5$ supplies an enriching action for the dephlegmator equivalent to but one theoretical rectifying plate.

The intermediary condenser P in Figure 123 also saves rectifying plates. In the following chapters, it will be assumed for the sake of simplicity that the hookup of the reflux condenser is such that it does not have any rectifying effect.

V. Minimum reflux ratio and actual reflux ratio for batch-type rectifiers

The reflux ratio which is the basis of the computation of a rectifying unit arises from the minimum reflux ratio, v_M. This minimum reflux ratio applies when the number of plates is infinite. This number becomes finite when the minimum reflux ratio is increased by a very small amount.

Using this interpretation, the minimum reflux ratio will now be determined for an ideal binary mixture having no azeotropic point, as shown in Figure 63. It is presumed that

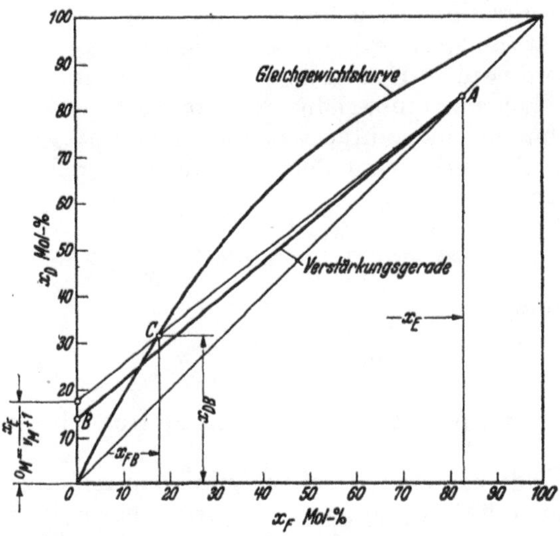

FIGURE 63

Determination of the minimum reflux ratio for a binary mixture having no azeotropic point

Gleichgewichtskurve—Equilibrium curve
Verstärkungsgerade—Enrichment line

distillate composition, x_E, and batch liquid concentration, x_{FB}, are known.

The former specifies the intersection A of the enrichment line with the diagonal of the plot, and the abscissa of the latter gives intersection C on the equilibrium curve. If A and C are connected by a straight line, the number of plates becomes infinite at point C, since by tracing a step sequence between the equilibrium curve and line AC, beginning at A, an infinite number of steps is required to reach C. Line AC is the ordinate intersection

$$o_M = \frac{x_E}{v_M + 1}$$

for which the numerical value can be read from the scale, and which yields the minimum reflux ratio:

$$v_M = \frac{x_E}{o_M} - 1 \tag{65}$$

For a concentration of low boiler equal to $x_E = 83$ mol per cent (Figure 63), and the still concentration of $x_{FB} = 18$ mol per cent, an ordinate section of $o_M = 17.5$ mol per cent is found, which yields a minimum reflux ratio of $v_M = 3.74$.

Since equilibrium point C, corresponding to x_{FB} and x_{DB} is located on the enrichment line, the minimum reflux ratio may also be found from

$$x_{DB} = \frac{v_M}{v_M + 1} \times x_{FB} + \frac{x_E}{v_M + 1}$$

corresponding to

$$v_M = \frac{x_E - x_{DB}}{x_{DB} - x_{FB}} \tag{66}$$

Concentration x_{DB} of the vapor rising up from the still is read at point C.

The second characteristic of the minimum reflux ratio is also demonstrated by the equilibrium curve shown in Figure 63. Enrichment line AC, to which the minimum reflux ratio corresponds, shifts into a new location when the ratio is increased by any small amount, as for instance that given by line AB in Figure 63. Using this enrichment ratio, a finite number of stages is required.

This characteristic, however, is not realized when the equilibrium curve possesses a point of inflection, as in the case of

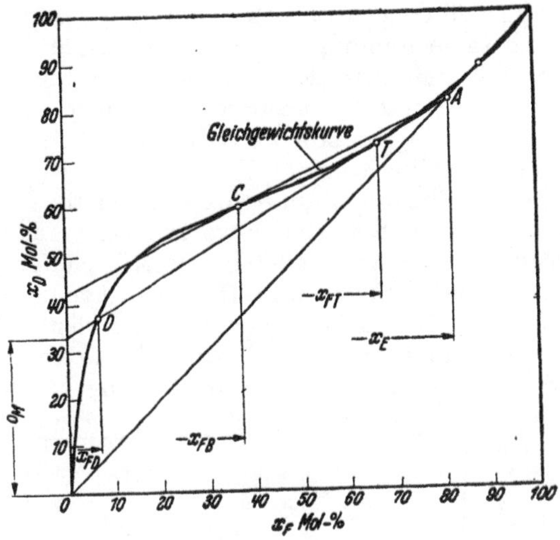

FIGURE 64

Determination of the minimum reflux ratio for a binary mixture whose equilibrium curve exhibits a point of inflection

Gleichgewichtskurve—Equilibrium curve

ethanol-water mixtures, unless a special procedure is used. In this case again, point *A* of the enrichment line is fixed by the intersection of the vertical drawn through the abscissa of distillate concentration, x_E, with the diagonal. The still concentration, x_{FB}, gives, in accordance with Figure 64, point *C* on the equilibrium curve. The connecting line *AC*, however, intersects the equilibrium curve at three points. This means that the number of steps remains infinite even if the reflux ratio corresponding to line *AC* is increased by a small amount.

The second criterion for the minimum reflux ratio is met only if a tangent to the equilibrium curve is drawn through *A* which contacts it at *T*, intersects it at *D* and yields value o_M on the ordinate axis.

From this construction, the minimum reflux ratio can be determined by means of equation (65). If this is increased by a small amount, the enrichment line through *A* becomes steeper than line *AD* and intersects the equilibrium curve at one point only.

Thus, for all still concentrations in the range between points D and T, the same minimum reflux ratio results. The ratio increases for still concentrations smaller than the liquid concentration, x_{FD}, of point D, and becomes smaller for concentrations of low boiler in the still exceeding concentration x_{FT} of low boiler corresponding to point T.

Figure 65 shows the minimum reflux ratio for the ideal benzol-toluol mixture and for the ethanol-water mixture as a function of the concentration of low boiler in the still charge. For the second mixture, two distillate concentrations have been assumed.

The minimum reflux ratio is constant for a distillate concentration, x_E, of 86 mol per cent of low boiler in the still, being represented by an enrichment line drawn tangent to the equi-

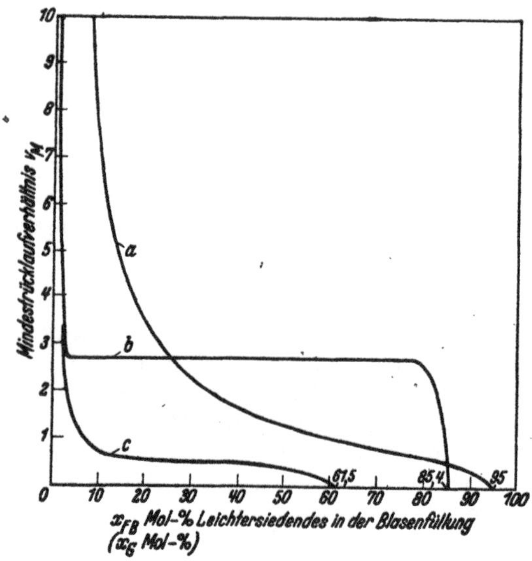

FIGURE 65

Minimum reflux ratio as a function of the concentration of low boiler in the still charge

a. Benzol-toluol; x_E = 98 mol per cent benzol in distillate
b. Ethanol-water; x_E = 86 mol per cent alcohol in distillate
c. Ethanol-water; x_E = 70 mol per cent alcohol in distillate.

Mol % Leichtersiedendes in der Blasenfüllung—Mol per cent low boiler in still charge

Mindestrücklaufverhältnis—Minimum reflux ratio

librium curve. With distillate concentrations smaller than 70 mol per cent of low boiler, the minimum reflux ratio decreases steadily, as shown for the case of an ethanol-water mixture.

In selecting an actual reflux ratio from the minimum reflux ratio determined, the influence of the reflux ratio on both the operating cost and the investment cost must be considered. As the reflux ratio increases, the requirements in heat and coolant increase steadily, while the number of plates required to separate a given mixture decreases steadily.

For this reason, the first cost of a column decreases as the reflux ratio increases. However, the amount of material needed for the heat exchangers varies in an inverse manner. Thus, the sum of the first costs of column and heat exchanger gives a minimum for a definite value of the reflux ratio.

For actual operation, that reflux ratio is selected which yields the lowest total cost, considering the annual expense for materials consumed, amortization and interest. The numerical value of the optimum reflux ratio, v, is a multiple of the minimum reflux ratio, v_M:

$$v = z \times v_M \tag{67}$$

Usual z values are between 1.3 and 5, up to 10. In alcohol distilleries, z is customarily about 1.6. See also Figure 80 which shows the cost as a function of the reflux ratio.

VI. The rectification mechanism on interchanger plates

To judge the performance of rectifying columns, a knowledge of the mechanism of exchange occurring on plates is required. Since the vapor entering a plate always has a higher temperature than the liquid leaving it, the basic phenomenon is a heat exchange associated with an interchange of materials.

Typical relationships are discussed employing a binary mixture as an example (Figure 66). The lower part of the figure shows the equilibrium curve and enrichment line AB, the upper part, the boiling-point and dewpoint curves, as well as a diagram indicating where the concentrations employed occur in the column. Complete mixing of liquid and perfect interchange are assumed to occur. Concentration x_{D1} of the vapor entering the plate under consideration, and concentration x_{F1} of the liquid leaving the same plate are fixed by point C. In turn, they determine points D and E on the boiling-point and dewpoint curves and their corresponding temperatures.

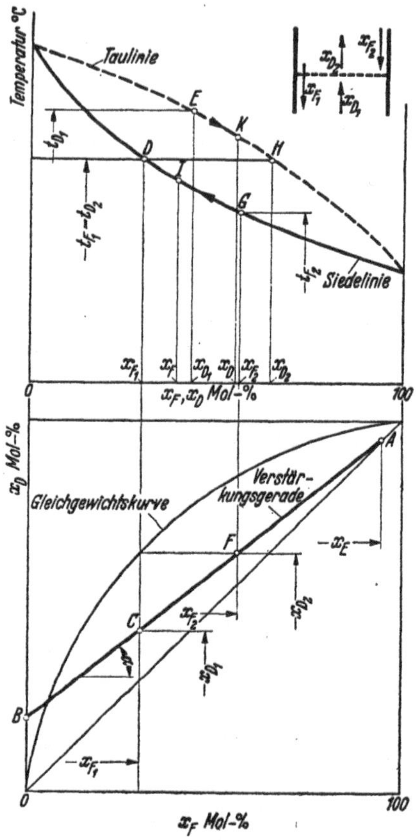

FIGURE 66

Conditions encountered in rectification by an interchanger plate

Taulinie—Dewpoint curve
Siedelinie—Bubble-point curve
Gleichgewichtskurve—Equilibrium curve
Verstärkungsgerade—Enrichment curve

The relation between the concentration x_{D2} of the vapor rising from the plate and x_{F2} of the liquid overflowing into the plate is given by point F; the temperatures of the phases are indicated by G and H.

The beginning of the interchange is indicated by the contact of the vapor entering the plate (point E) with the liquid (point G).

Assume that the materials introduced into the plate are collected for a certain time in a container, impermeable to heat, where it would be possible for them to separate. Within the container, assume a vertical rotating shaft carrying two agitating vanes. One of these vanes moves the mass of the liquid, the other the mass of the vapor. As a consequence of this turbulent motion in the two phases (in both the bulk of the liquid and vapor) temperature and concentration do not undergo any local changes. All of the temperature and concentration drops required for the heat and mass interchange then take place in two boundary layers, which are adjacent to the surface of contact of liquid and vapor. For viscous flow, the phenomena in both phases would be identical in principle.

At some intermediate point during this interchange movement, assume the liquid to contain x_F mol per cent and the vapor to contain x_D mol per cent of low boiler. The final point of the interchange movement is at the ideal conditions given by D and H where there is constancy of temperature and equilibrium between phases. Assume further, that the molar heat of evaporation is independent of the composition of the mixture. The mols condensed in the liquid from the vapor give rise to an equivalent amount of vapor from liquid. Then,

$$D \times x_{D1} + F \times x_{F2} = D \times x_D + F \times x_F = D \times x_{D2} + F \times x_{F1}$$

which yields

$$\frac{F}{D} = \frac{x_D - x_{D1}}{x_{F2} - x_F} = \frac{x_{D2} - x_{D1}}{x_{F2} - x_{F1}} = \frac{x_{D2} - x_D}{x_F - x_{F1}} = \tan \alpha = \frac{v}{v+1} \qquad (68)$$

where angle α determines the slope of the enriching line with the horizontal, and v = reflux ratio. With $v = 4$, tan α becomes 0.8.

If for some intermediate condition the concentration x_F of low boiler in the liquid is given, x_D can be computed from equation (68).

For $x_F = 40$ mol per cent (see point J in Figure 66) and $v = 4$, the initial and final conditions of Figure 66 will yield, with $x_{D1} = 43$, $x_{F1} = 30$, $x_{D2} = 63$ and $x_{F2} = 55$ mol per cent, an x_D value of 55 mol per cent; see point K. During the interchange period, the point representing the liquid shifts on the bubble-point curve from G via J to D, while the point representing the vapor moves from E via K to H.

The heat interchange between vapor and liquid causes condensation of the vapor into the liquid, and some evaporation of the latter. In this connection, the heat moves from the contact area into the bulk of the liquid up to the point where the boiling temperature corresponding to the given concentration exists. Then, evaporation begins. The heat of evaporation of the mixture under such conditions plays an important part. This has been corroborated by experiments.[40] These phenomena are encountered on every plate. The vapor with x_{D1} low boiler (Figure 66) can transmit heat only to the liquid surrounding it on a plate, with simultaneous condensation, until vapor and liquid have attained the same temperature. Constancy of temperature results on a plate in which complete mixing of liquid and perfect interchange occur as soon as the noncondensed portion of the vapor entering attains a low-boiler concentration, x_{D2}, due to enrichment by partial condensation. The additional vapor produced and rising from the plate has the same concentration, and is in phase equilibrium with the liquid leaving the plate.

Let

F_B = Liquid condensed on the plate, mols per hour
D = Total amount of vapor entering the plate, mols per hour

then, the plate reflux ratio is

$$v_B = \frac{F_B}{D - F_B} \tag{69a}$$

This plate reflux ratio bears a definite relation to the concentration, x_{D1}, of the vapor entering and the concentration, x_{D2}, of the vapor leaving. This is evident from the relationships holding for enrichment by partial condensation as shown for ethanol-water mixtures in Figure 38.

If it is assumed that the conditions applying to Figure 38 also hold for the interchange on a rectifying plate, the relationship between x_{D1}, x_{D2} and v_B can be found. Thus, if a plate of a rectifying column handling a mixture of ethanol and water is fed a vapor with x_{D1} = 54 mol per cent alcohol, the vapor generated on the plate and the vapor remaining uncondensed will contain x_{D2} = 364.2 mol per cent of low boiler (see Figure 67).

For this enrichment, the enrichment curves in Figure 38 indicate a plate reflux ratio, v_B, of 0.65. This means that, per mol vapor entering, 0.394 mol must condense, while the residual 0.606 mol with a concentration of x_{D2} leave the plate uncon-

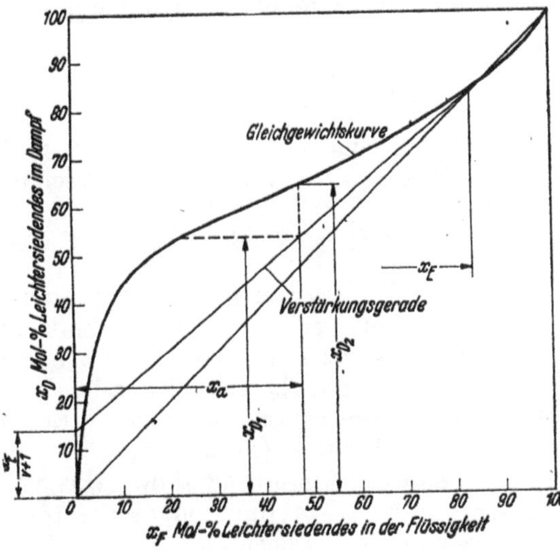

FIGURE 67

Concentrations for a rectifying plate handling ethanol-water mixtures

Mol % Leichtersiedendes in der Flüssigkeit—Mol per cent of low boiler in liquid
Mol % Leichtersiedendes im Dampf—Mol per cent of low boiler in vapor
Gleichgewichtskurve—Equilibrium curve
Verstärkungsgerade—Enrichment curve

densed. The reflux ratio for partial condensation, v in Figure 38, is equivalent to the plate reflux ratio v_B for the interchange process on rectifying plates.

These basic relations permit determination of the plate reflux ratio and the heat exchanged between liquid and vapor on a plate.[41] The latter is the product of the liquid condensed on a plate F_B and the molar heat of evaporation r. According to equation (69a), F_B is given by

$$F_B = D \times \frac{v_B}{1 + v_B} \tag{69b}$$

The amount of vapor flowing up the column is found from the relationship between the amount of distillate, E, and the reflux ratio, v. Because of the constancy of the molar heat of evaporation, the amount of reflux, F, in a column is constant. Thus

$$D = F + E = E\left(\frac{F}{E} + 1\right) = E\ (v + 1) \qquad (70)$$

From this, the heat exchanged on a rectifying plate for the case of a complete liquid mixing and perfect interchange is

$$Q = E\ (v + 1)\ r \frac{v_B}{v_B + 1} \qquad (71)$$

Although E, v, and r are constants in this equation, v_B varies from plate to plate. This variation may be found by plotting the sequence of steps between the equilibrium curve and enrichment line which gives the theoretical number of plates and values for the vapor concentrations from the vertical-horizontal intersections, from which the corresponding plate reflux ratios are determined as shown. Figure 68 shows this variation for an ethanol-water mixture as a function of the plate number. On the same figure a curve has been plotted to indicate the heat exchanged on the plates, Q, as computed from equation (71); this has been confirmed experimentally.

For concentration x_D, of a vapor entering a plate, and x_F, of a liquid leaving the same plate, determine the vapor temperature, t_D, and liquid temperature, t_F, in accordance with the dewpoint and bubble-point curves shown in Figure 17. The difference

$$\vartheta = t_D - t_F \qquad (72)$$

represents the temperature difference between liquid and vapor at the beginning of the heat and material interchange period. It is also shown on Figure 68 for the example selected. It is seen that the heat to be exchanged increases for high alcohol concentrations, while temperature difference, ϑ, decreases. To obtain perfect interchange, the exchanger surface must thus be greater for high alcohol concentrations than for low alcohol concentrations. It must be directly proportional to Q and inversely proportional to ϑ, so that the ratio Q/ϑ becomes the deciding factor for the interchange process.

As shown in Figure 68, this ratio increases rapidly for the upper interchanger plates of a column, i.e., for high alcohol concentrations.[42] However, as a rule, all plates in a rectifying column are built with equal dimensions. The paths of the Q and ϑ lines in Figure 68 explain the lowering of the enrichment ratio (in Figure 60) in this range of concentrations.

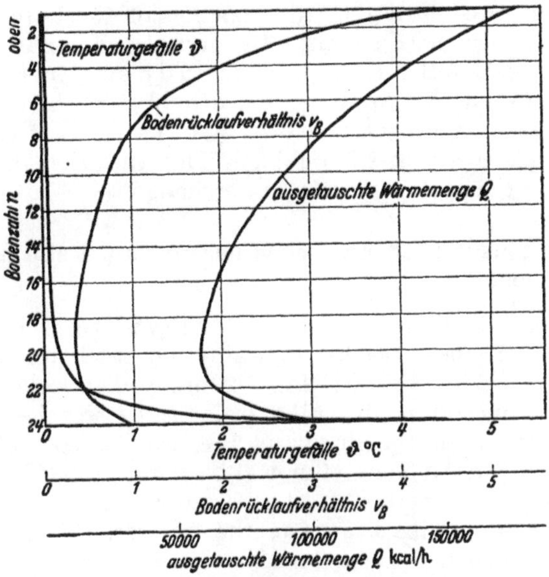

FIGURE 68

Variations of temperature drop, plate reflux ratio v_B and heat Q for a rectifying column, assuming perfect plates (Ethanol-water; reflux ratio $v = 4$, distillate concentration $x_E = 86$ mol percent)

Ausgetauschte Wärmemenge—Heat interchanged
Bodenrücklaufverhältnis—Plate reflux ratio
Temperaturgefälle—Temperature drop
Bodenzahl—Number of plates
Oben—Top of column

Certain circumstances indicate that the interchange of heat and material on a plate is a function of heat transmission, diffusion and turbulence. When a vapor bubble moves through the liquid on a plate, the heat and substance interchange at the surface of the bubble is limited to the boundary layers. At the front side of the bubble, viewed from the direction of flow, condensation takes place. Heat flows off through the boundary condensate layer. Therefore, temperature increases away from the liquid boundary layers parallel to the direction of flow. At the rear of the bubble, a vapor nucleus may form from vapor freshly generated. In the front portion of the bubble, vapor accumulates which is enriched by partial condensation, while in the rear portion there is vapor generated mainly by re-evaporation. If the flow is rapid enough, a bubble may thus contain both types of vapor, which means that a change of concentration take place within the bubble. This variation disappears, however, in the case of a perfect interchange.

VII. Heat consumption and reboil heat for a pot still and rectifying column

The heat requirements of a batch-type rectifier include, first of all, the heat needed to warm the batch up to boiling temperature prior to distilling. During distillation proper, heat is introduced into the still and the heat losses, i.e., the heat of condensation of the distillate, the heat to cool the liquid product and heat to form reflux will be compensated in the unit.

While the first heat quantities listed are easy to evaluate, determination of reflux heat is another matter. This represents that heat which has to be withdrawn from the vapors in the reflux condenser in order to liquefy the vapors. It controls not only fuel and coolant requirements, but also the number of plates, since it fixes the location of the enrichment line. The specific minimum reflux heat, Q_M, (kilocalories per mol) is the amount of heat which must be removed from the reflux condenser per mol of product, to produce that amount of reflux which corresponds to the minimum reflux ratio, v_M:

$$Q_M = v_M \times r \tag{73}$$

where r is specified by equations (45) and (47). The specific minimum reflux heat thus depends on v_M and on the still liquid composition. Therefore, if the ordinates of curves a, b, and c in Figure 65 are multiplied by r, Q_M is obtained as a function of the concentration of the low boiler in the still batch, for binary mixtures. The ordinates of these figures, which indicate the numerical value of the minimum reflux ratio, also represent the Q_M/r ratio, in accordance with equation (73).

The actual specific reflux heat, Q_S, (kilocalories per mol) which must be removed per unit quantity of product, in the reflux condenser, can be found for the actual reflux ratio, v, from equation (67)

$$Q_S = v \times r = v_M \times z \times r \tag{74}$$

The ordinates of the curves in Figure 65 give the ratio $Q_S/(z \times r)$. For a still concentration, x_{FB}, of 10 mol per cent alcohol and when $z = 1.6$ and $r = 9600$ kcal/mol it follows from curve b in Figure 65 that per mol of product

$$Q_S = 2.69 \times 1.6 \times 9600 = 41300 \text{ kcal}$$

must be removed to form the reflux. If the v_M values of Figure 65 are multiplied by $z \times r$, as well as the amount E of product, the actual total reboil heat is obtained

$$Q_R = E \times v \times r = E \times v_M \times z \times r \tag{75}$$

Curves a, b, and c in Figure 65 thus give the characteristic path of the total reboil heat as a function of the concentration of low boiler in the still batch.

1. Reboil heat with distillate of constant composition

During distillation the concentration of the batch varies, because the vapor liberated in the still has a composition different from that of the liquid. If there is more low boiler in the vapor than in the liquid, the concentration of low boiler in the batch decreases continuously. If, however, the composition of the distillate is constant, with x_E mol per cent low boiler, and if no changes are made in the setup of the equipment, the number of plates remaining the same, it is necessary to increase the reflux ratio during the run.

Assume that this increase is effected in such a manner that x_E remains constant. In this case, there are some important relationships between the quantity of the product E, the quantity of the batch in the still and the concentrations of the batch and of the product, which determine the amount of total reboil heat, which are discussed with the help of Figure 69, the equilibrium curve of a benzol-toluol mixture.

If the distillate composition is assumed to have an x_E value of 95 mol per cent benzol, point A on the diagonal is fixed. The rectifying effect of the reflux condenser is neglected. Let x_{F1} = concentration of low boiler in the still at the beginning of the run, and x_{F2} = concentration at the end of the run. The series of steps shown as a solid line between the equilibrium curve and the enrichment line that holds for the reflux ratio, v_1, at the beginning of the run indicates the number of theoretical plates equivalent to the effect of the column which enriches x_{F1} to the distillate composition x_E.

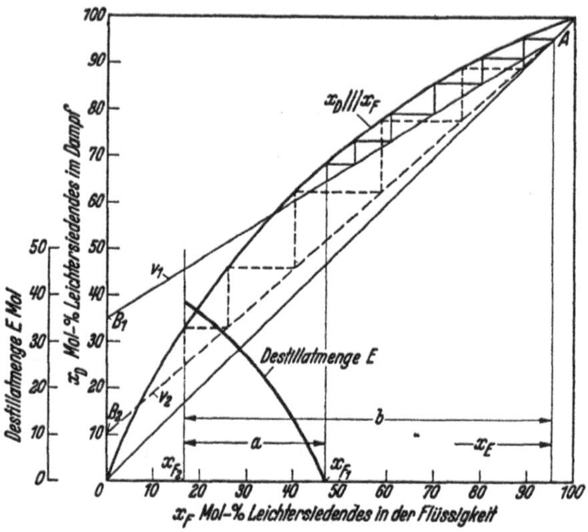

FIGURE 69

Determination of the quantity of distillate from a batch-type rectifying unit for constant product composition

Mol % Leichtersiedendes im Dampf—Mol per cent of low boiler in vapor
Mol % Leichtersiedendes in der Flüssigkeit—Mol per cent of low boiler in liquid
Destillatmenge—Quantity of distillate

At the end of the run, assume that the reflux ratio equals v_2 and that the corresponding still-liquid composition is x_{F2}. This is obtained by plotting for the same number of theoretical plates the series of steps between the equilibrium curve and the enrichment line for v_2, as shown by the dotted line in Figure 69. Then the following equations hold, if during the run the batch in the still decreases from its initial value, B_1 mols, to its final value, B_2 mols:

$$B_1 \times x_{F1} - B_2 \times x_{F2} = E \times x_E \quad (76)$$

$$B_1 - B_2 = E \quad (77)$$

which yields

$$E = B_1 \frac{x_{F1} - x_{F2}}{x_E - x_{F2}} \quad (78)$$

The ratio of the concentration differences in equation (78) equals the ratio of the distances a and b in Figure 69:

$$\frac{x_{F1} - x_{F2}}{x_E - x_{F2}} = \frac{a}{b} \tag{79}$$

Equation (78) may be solved for the amount of distillate over the run for any still concentration between the limits x_{F1} and x_{F2}, initial concentration x_{F1} remaining unchanged. Enrichment lines are drawn for the range between v_1 and v_2, and for the same number of theoretical plates the corresponding batch concentrations are determined (as shown in Figure 69), to give the final values x_{F1} and x_{F2}. For x_{F1}, E equals 0.

As an example, in Figure 69 the end of the run is defined at point B_2 given by reflux ratio v_2, for which the ordinate intersection o_2 = 10 mol per cent from this, equation (65) specifies a reflux ratio:

$$v_2 = \frac{x_E}{o_2} - 1 = \frac{95}{10} - 1 = 8.5$$

From equation (78), the amount of product obtained from an initial still batch of 100 mol = B_1 and for a final still concentration of x_{F2} is

$$E = 100 \times \frac{47 - 17}{95 - 17} = 38.5 \text{ mols},$$

where x_{F1} = 47 and x_{F2} = 17 mol per cent.

In the manner demonstrated, the amount of distillate, E, can be found for the same initial batch of 100 mol, for any final still concentration between x_{F1} and x_{F2}, the path of which is shown in Figure 69. It is seen that E diminishes in accordance with a definite curve. Therefore, in this case the total reboil heat is given by

$$Q_R = r \times \int_{v_1}^{v_2} v \times dE \tag{80}$$

The relation between E and v is known from Figure 69. As an example, for an ordinate intersection, o = 20, corresponding to a reflux ratio v, of 3.75, a still concentration of 28 mol per cent results. From this, equation 78 yields 28.6 mol of distillate.

In Figure 70, the quantities of distillate corresponding to certain reflux conditions has been plotted; among them the quantity of the product E = 28.6 mol corresponding to v = 3.75. The shaded strip having width dE represents the value $v \times dE$,

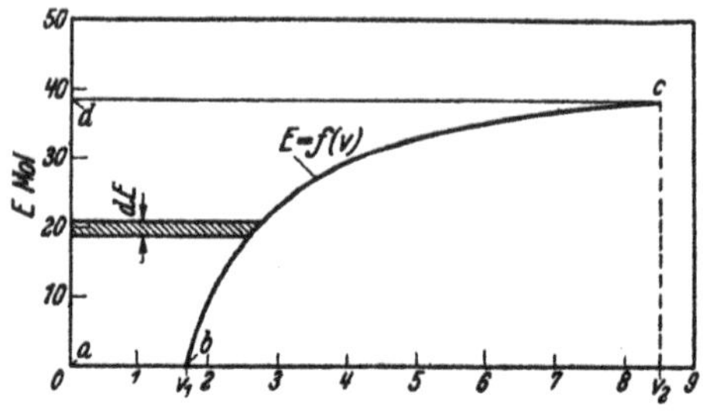

FIGURE 70

Quantity of product as a function of the reflux ratio

and the area yields the value of the integral of equation (80). For the previous example, according to Figure 70 it becomes

$$\int_{v_1=1.72}^{v_2=8.5} v \times dE = 122 \text{ mols}$$

The molar heat of evaporation (r) of the benzol-toluol mixture under consideration, on the basis of equations (45) and (47) and Table 3, is 7600 kcal/mol. Thus, to obtain 38.5 mols of distillate from an initial batch of 100 mols, a total reboil heat

$$Q_R = 7600 \times 122 = 927{,}000 \text{ kcal}$$

is required. According to equation (3), the average molecular weight of the product, μ, is 78.8, so that per kilogram of distillate the reboil heat is

$$\frac{927{,}000}{38.5 \times 78.8} = 305 \text{ kcal/kg}$$

2. Reboil heat with constant reflux ratio

If the reflux ratio of a batch-type distilling unit is held constant while continuous separation is effected, the composition of the distillate must necessarily vary. Any given concentration of low boiler in the still corresponds to a definite distillate composition at any point in the run.

The relation between these values is shown in Figure 71, in the equilibrium curve for the benzol-toluol mixture. The reflux condenser is assumed to operate in such a way that it

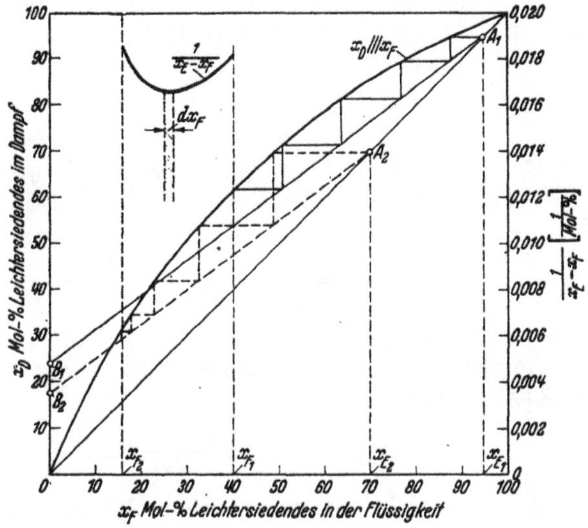

FIGURE 71

Interrelation between batch concentrations and product concentrations for a batch-type rectifier operating at constant reflux ratio

Mol % Leichtersiedendes im Dampf—Mol per cent of low boiler in vapor
Mol % Leichtersiedendes in der Flüssigkeit—Mol per cent of low boiler in liquid

has no separating effect; see Figure 52. The reflux ratio, v, is 3. The initial concentration of the distillate x_{E1} determines point A_1 on the diagonal, v fixes the slope $v/(v + 1)$ of the enrichment line and point B_1 the ordinate intersection $o_1 = 23.75$. The effect of the separating column is assumed to be that of four rectifying plates. Upon plotting the solid line in Figure 71 between equilibrium curve and enrichment line, the initial composition for the still, x_{F1} results. Distillation continues until the unit supplies a distillate of concentration x_{E2} corresponding to point A_2 on the diagonal, and, for the same reflux ratio, point B_2 on the ordinate axis. For the same four theoretical plates, the series of steps shown dotted in Figure 71 yields the final concentration x_{F2} in the still.

At any instant during the run the batch has a concentration x_F mol per cent, and the distillate x_E mol per cent of low boiler.

The amount of distillate, dE, obtained during an infinitesimal time under steady operating conditions, equals the decrease dB of the batch in the still

$$dB = dE \tag{81}$$

During the same period, the concentration of low boiler in the still decreases by dx_F, so that for the low boiler

$$B \times x_F = dB \times x_E + (B - dB)\,(x_F - dx_F) \tag{82}$$

From this

$$\frac{dB}{B} = \frac{dx_F}{x_E - x_F} \tag{83}$$

If B_1 = the number mols in the batch at the start of the run, and B_2 = the number of mols in the batch at the end of the run, equation (83) becomes

$$\ln \frac{B_1}{B_2} = \int_{x_{F2}}^{x_{F1}} \frac{dx_F}{x_E - x_F} \tag{84}$$

The integral on the right side of equation (84) may be solved graphically. For this purpose the value $1/(x_E - x_F)$ is plotted against the still concentration. To find this value, the batch concentration is determined for a distillate concentration between the limiting values x_{E1} and x_{E2} by the same method as used in Figure 71 for the final concentrations. In this way, the curve shown in the upper left-hand corner of Figure 71 results. The strip below it, having width dx_F, corresponds to the value $dx_F/(x_E - x_F)$ and the total area under this curve between values x_{F1} and x_{F2} corresponds to the value of the integral in the right-hand side of equation (84).

For the example under consideration, and for the concentrations $x_{E1} = 95$, $x_{E2} = 70$, $x_{F1} = 40$ and $x_{F2} = 15$ mol per cent

$$\int_{15}^{40} \frac{dx_F}{x_E - x_F} = 0.413$$

This means that, according to equation (84), the ratio of the initial to the final size of the batch in the still is

$$\frac{B_1}{B_2} = 1.51$$

When the initial batch, B_1, equals 100 mols, B_2 equals 66.2 mols and E equals 33.8 mols. The total reboil heat for this quantity of distillate becomes

$$Q_R = r \times v \times E = 7600 \times 3 \times 33.8 = 771{,}000 \text{ kcal}$$

and the reboil heat per 100 mols of distillate becomes

$$Q_R = 2{,}280{,}000 \text{ kcal}$$

The average concentration x_{EM} of the distillate arises from the equation

$$B_1 \times x_{F1} - B_2 \times x_{F2} = E \times x_{EM}$$

which gives

$$x_{EM} = \frac{B_1 \times x_{F1} - B_2 \times x_{F2}}{E} \tag{85}$$

For the example considered, x_{EM} becomes 87.1 mol per cent benzol. The mean molecular weight, μ, corresponding to this concentration, is 79.9 from equation (3). Thus, the reboil heat per kilogram of distillate is

$$\frac{2{,}280{,}000}{100 \times 79.9} = 286 \text{ kcal/kg}$$

In these computations, constant conditions were assumed. Thus, the liquid accumulating on the interchanging plates was not considered. As a matter of fact, the amount of this liquid depends largely on the design of the plates. They may be so constructed that they will not drain at the end of a run, the mass of liquid thus being stored for the next run.

Batch operation may be converted into continuous operation by arranging several stills in series along with their columns. The mixture to be separated then passes in a steady flow from one still to the next. At the top of each column a stream is withdrawn, while the highest boiler is concentrated in the last stage of the series of distilling units.

E. CONTINUOUS DISTILLING EQUIPMENT HAVING RECTIFYING AND STRIPPING SECTIONS

I. Determination of the number of plates

A more complete and continuous separation of liquid mixtures is possible if the kettle of a batch-type distilling unit (Figure 55) is replaced by a special rectifying column. Since the low boiler is removed in this added column, it is called a stripper. The lower part of the stripping column contains the heater.

The mixture to be separated is introduced continuously at an appropriate point between the stripping and the enriching sections, and is separated into the high-boiling *bottoms* and the low-boiling distillate or product, as shown in Figure 72.

An analysis of the conditions existing in a continuous rectify-

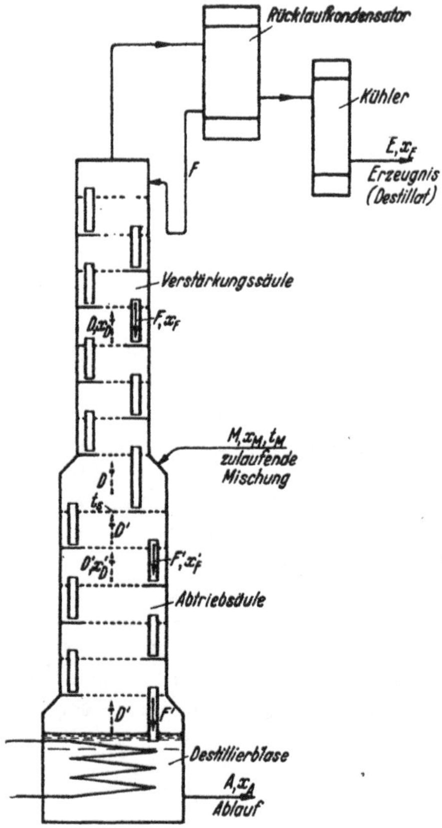

FIGURE 72

Continuous rectifier

Destillierblase—Still kettle
Abtriebsäule—Stripping column
Verstärkungssäule—Enriching column
Rücklaufkondensator—Reflux condenser
Kühler—Cooler
Ablauf—Drain
Zulaufende Mischung—Feed mixture
Erzeugnis (Destillat)—Product (Distillate)

ing unit is simplified considerably if equal molar heats of evaporation of the mixture to be separated are assumed, as illustrated in the separation of a binary mixture in the following.

All values referring to the stripper are designated as primes. Thus, D' and F' represent the vapor and liquid flowing between any two plates in the stripper in mol per hour; and x_D' mol per cent represents the concentration of low boiler in the vapor, while x_F' refers to the low boiler in the liquid. Since the molar heats of evaporation are considered constant, both D' and F' remain the same over the entire stripping column.

For the enriching column, equations (56) and (57) apply; this means that equation (60) for the enrichment line (line for the enriching column) also holds. Thus, according to Figure 73, the location of the enrichment line is fixed by points A and B, which result from the distillate concentration, x_E, and reflux ratio, v. In the case of the continuous unit, v is also the ratio of the liquid returning to the enriching column from the reflux condenser to the amount of distillate, as given by equation (59).

A quantity A, of bottom product, flows out of the bottom of the stripper. This material, having a concentration x_A, may actually be a valuable product.

$$F' = D' + A \tag{86}$$

and

$$F' \times x_F' = D' \times x_D' + A \times x_A \tag{87}$$

These two equations may be combined

$$x_D' = \frac{F'}{F' - A} \times x_F' - \frac{A}{F' - A} \times x_A \tag{88}$$

Since the only variables in this equation are x_D' and x_F', this equation represents a straight line, as shown in Figure 73, which indicates the relation between vapor composition and liquid composition between any two interchanger plates. This line is called the stripping line (line of the stripping column). In equation (88), where $x_D' = x_F'$, the ordinate value for the intersection of the 45° diagonal and the stripping line is

$$x_D' = x_F' = x_A' \tag{89}$$

In Figure 73, C is the intersection of the stripping line with the diagonal. Another point of interest is the intersection of the stripping and enrichment lines. To determine this, a relation

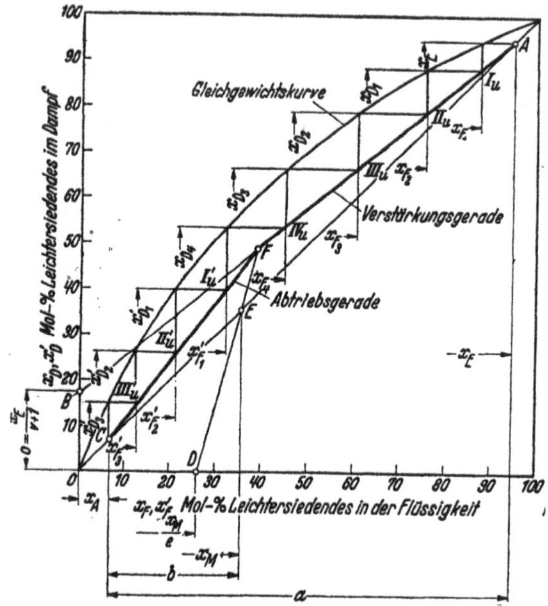

FIGURE 73

Determination of the enrichment and stripping lines, and of the theoretical number of plates in enriching and stripping columns

Mol % Leichtersiedendes in der Flüssigkeit—Mol per cent of low boiler in liquid
Mol % Leichtersiedendes im Dampf—Mol per cent of low boiler in vapor
Gleichgewichtskurve—Equilibrium curve
Verstärkungsgerade—Enrichment line
Abtriebsgerade—Stripping line

between the masses flowing in the enriching and in the stripping columns is required.

If M mol per hour of the mixture to be separated are fed into the top plate of the stripper (feed plate) at a temperature of t_M, corresponding to the boiling temperature of the liquid, the liquid overflowing in the stripper is the sum of the reflux of the enriching column and amount of mixture fed:

$$F' = F + M \tag{90}$$

If, however, the feed temperature is below the boiling temperature on the feed plate ($t_M < t_S$), a certain amount of vapor condenses on the feed plate and heats the feed to boiling temperature. The amount of condensate thus formed increases the reflux in the stripper

$$F' = F + e \times m \tag{91}$$

Let

q_M = enthalpy per mol of feed having temperature t_M

q_S = enthalpy per mol of liquid on the feed plate having boiling temperature t_S

Then

$$e = \frac{r + q_S - q_M}{r} = 1 + \frac{q_S - q_M}{r} \tag{92}$$

where r = heat of evaporation or condensation per mol.

From equations (60), (88), (91) and

$$x_D' = x_D \tag{93}$$

$$x_F' = x_F \tag{94}$$

the locus of the intersecting points of the enrichment and stripping lines follows:

$$x_D = \frac{e}{e-1} \times x_F - \frac{x_M}{e-1} \tag{95}$$

Since, in this equation, e and x_M are constant, it also represents a straight line. This intersects the horizontal axis (point D in Figure 73) at

$$x_F = \frac{x_M}{e} \tag{96}$$

which results when the value $x_D = 0$ is substituted into equation (95). Similarly, when $x_D = x_F$ is substituted in the same equation, intersection E with the diagonal results. This has the ordinate value

$$x_D = x_F = x_M \tag{97}$$

The line connecting D and E intersects the enrichment line at F, which is the second point required for the stripping line. The stripping line, then, is the line passing through C and F. Any point of this line gives the relation between the composition of the vapor and that of the liquid between any two plates in the stripper.

The theoretical number of enrichment plates required to separate a mixture continuously from an initial composition x_M into two mixtures having compositions x_E and x_A is found by using the methods outlined in Figure 57. Thus, the number of turning points, I_u, II_u, III_u, etc., in Figure 73 on the enrichment line obtained by drawing the series of steps between the equilibrium curve and enrichment line, beginning at A and ending at F yields the number of plates. The reflux condenser is considered to have no enriching action. However, thorough mixing of the liquid and complete interchange are assumed. This means that the vapor rising from a stripping plate is in phase equilibrium with the liquid leaving the same plate. The relation between a vapor rising from a plate and the liquid leaving the same plate is then given by a point on the equilibrium curve.

The theoretical number of stripping plates is determined analogously, frum the number of turning points I_u', II_u', III_u', etc., resulting if the last (lowest) turning point of the enrichment line is used as a start (Point IV_u in Figure 73) and the sequence of steps is continued between the equilibrium curve and stripping line up to point C. In this process it is assumed that the vapor liberated in the still is in phase equilibrium with the liquid in the still.

For the conditions shown in Figure 73, using the equilibrium curve for benzol-toluol mixtures, the theoretical number of enriching plates is 4, and the theoretical number of stripping plates is 3. In Figure 73,

$$x_{D3}' > x_{F3}' > x_A$$

which means that the composition of the reflux fed back to the still calandria lies between that of the charge in the calandria and that of the vapor rising from the charge in the still. To make this plain, the compositions corresponding to Figure 73 are located in the cross-section of the rectifier shown in Figure 74, and labeled to conform with Figure 73.

Figure 75 shows the constructions required to determine the number of theoretical plates required in a continuous rectifying unit to concentrate a 9 per cent by weight (or 3.78 mol per cent) ethanol-water mixture, at $t_M = 70°C$ feed temperature, to a distillate containing 94.2 per cent by weight (or 86.4 mol per cent) alcohol, and yield a practically spirit-free water residue.

According to equation (1a), the mean molecular weight of the mixture is 19.1. The temperature on the feed plate is 92°C. For this temperature and an alcohol concentration of 9 per cent, Figure 76 specifies an enthalpy of 89 kcal/kg.[43] At a tempera-

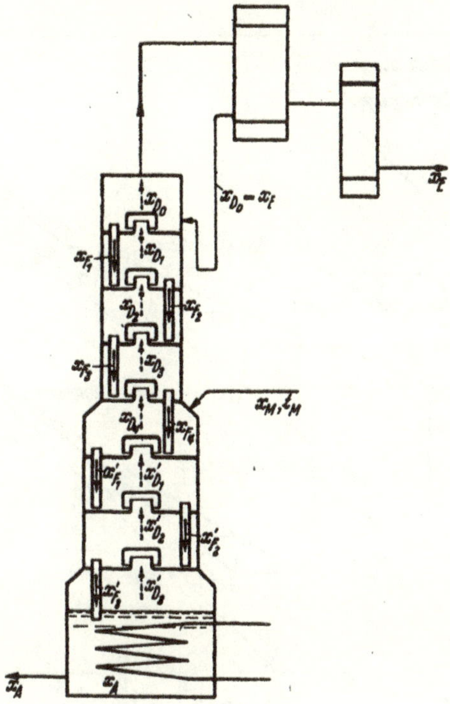

FIGURE 74

Location of the concentrations determined from Figure 73

ture of 70°C for the same alcohol concentration, the enthalpy is 66.5 kcal per kilogram. The molar heat of evaporation r, is 9600 kcal per mol.

Using equation (92), the numerical value of e becomes

$$e = 1 + \frac{19.1 \times 89 - 19.1 \times 66.5}{9600} = 1.048$$

In Figure 76, the ordinate corresponds to the enthalpy per kilogram of a liquid mixture, given as the sum of the heats required to raise the temperature of the individual components from 0°C to the temperature of the mixture. In establishing the numerical value of the enthalpy, allowance can be made for heats of mixing.

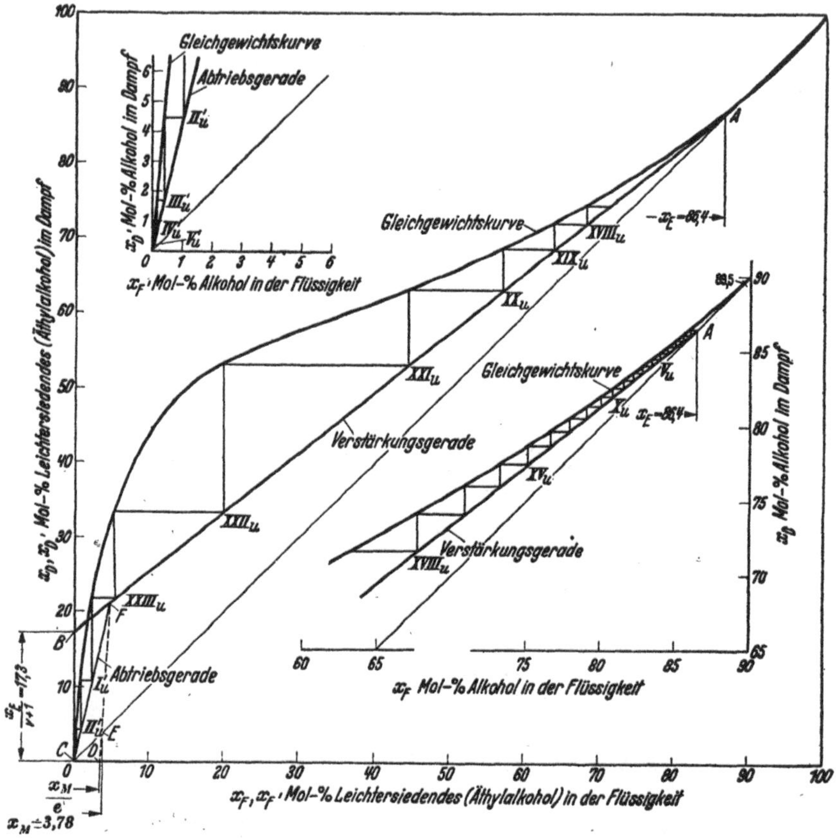

FIGURE 75

Separation of an ethanol-water mixture—Determination of the theoretical number of plates for the enriching and stripping columns

Determination of the theoretical number of plates for the enriching and stripping columns

Mol % Alkohol in der Flüssigkeit—Mol per cent alcohol in liquid
Mol % Alkohol im Dampf—Mol per cent alcohol in vapor
Gleichgewichtskurve—Equilibrium curve
Verstärkungsgerade—Enrichment line
Abtriebsgerade—Stripping line
Leichtersiedendes—Low boiler
Äthylalkohol—Ethanol

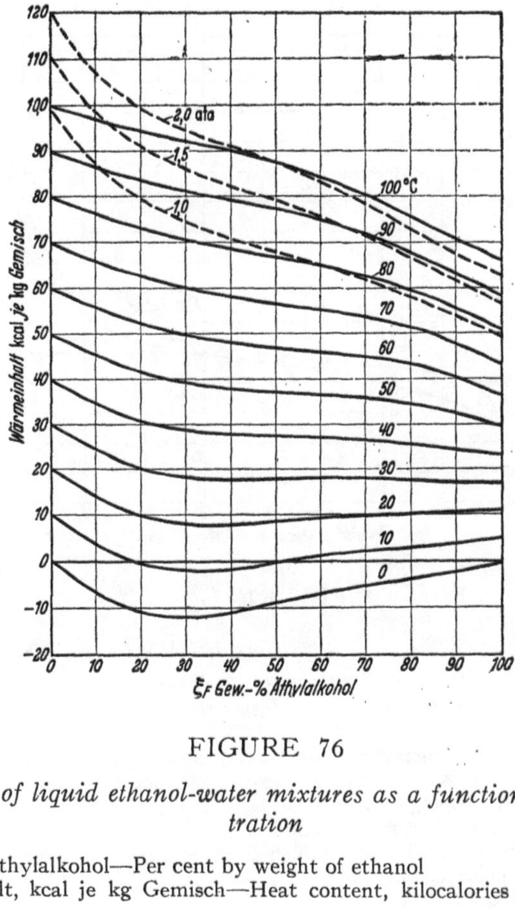

FIGURE 76

Enthalpy of liquid ethanol-water mixtures as a function of concentration

Gew. % Aethylalkohol—Per cent by weight of ethanol
Wärmeinhalt, kcal je kg Gemisch—Heat content, kilocalories per kilogram mixture

Figure 75 specifies 23 theoretical plates (n_t) for the enriching column, and 6 for the stripping column. The vapor and liquid compositions at different locations of the column can be found from the turning points of the sequence of steps drawn. These can be plotted against the number of plates in Figure 77. It is seen that the largest concentration differences between vapor and liquid are encountered on the plates near the bottom of the enriching column. The actual number of plates, n, of the enriching and stripping columns is obtained from the theoretical

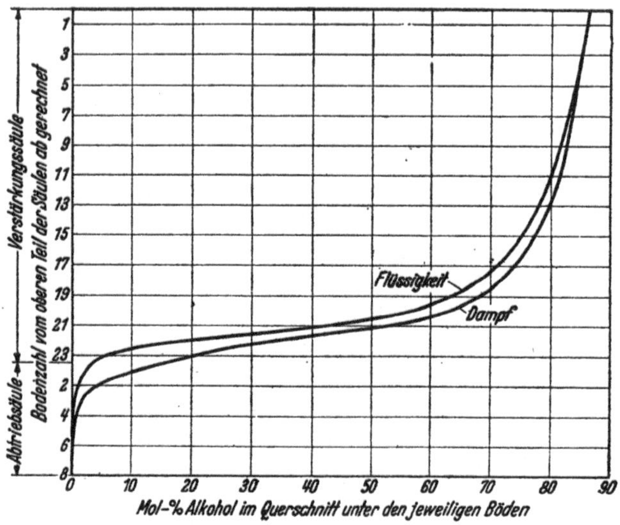

FIGURE 77

Relation between the concentrations, in liquid and in vapor, and number of plates in a continuous rectifier handling ethanol-water mixtures in accordance with Figure 75

Mol % Alkohol im Querschnitt unter den jeweiligen Böden—Mol per cent of alcohol in the cross-sections underneath the respective plates
Abtriebsäule—Stripping column
Verstärkungssäule—Enriching column
Bodenzahl vom oberen Teil der Säulen ab gerechnet—Plate number, counting from top of columns downwards
Flüssigkeit—Liquid
Dampf—Vapor

number, n_t, by dividing the latter by the average exchange ratio (enrichment ratio) s_M, as given by equation (64). (s_M = 0.6–0.8; see Section D, IV 3 a.)

II. Minimum reflux ratio of a continuous rectifying column for separating binary mixtures

In the case of a continuous unit the reflux ratio also determines both the number of interchanger plates required and the installation and operating costs. The reflux ratio is found from the minimum value v_M. With minimum reflux, the number of plates is infinite; the number of plates becomes finite as soon as the reflux ratio is increased a small amount above the minimum.

Figure 78 shows how the minimum reflux ratio is found for the case of a binary mixture having no azeotropic point, as for the ideal benzol-toluol mixture. The distillate composition x_E, and the compositions x_A of the bottoms, and of low boiler in the feed at temperature t_M are given.

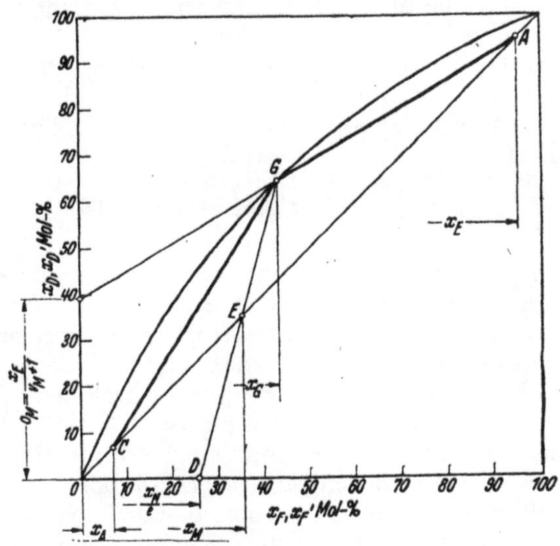

FIGURE 78

Graph for finding the minimum reflux ratio for separating an ideal binary mixture

From the temperature, *e* is obtained from equation (92), which fixes points D and E on Figure 78. The line connecting D and E, which is the locus of the points of intersection of the enrichment line and stripping lines intersects the equilibrium curve at point G, the abscissa of which is x_G. Lines AG and GC are the enrichment and stripping lines for the minimum reflux ratio. On drawing the sequence of steps between the equilibrium curve and the enrichment line mentioned, point G could be reached only by inserting an infinite number of stages.

This same situation holds for the sequence of steps traced for the stripping line and drawn between the latter and the equilibrium curve. At point G, the number of plates both for the lower part of the enriching column and for the top part of the stripping column becomes infinite. The extension of line AG intersects the ordinate axis at

$$o_M = \frac{x_E}{v_M + 1}$$

the numerical value of which to find the minimum reflux ratio can be determined from equation (65).

For instance, in the case shown in Figure 78, a distillate composition x_E, of 95 mol per cent has been taken, giving an ordinate intersection, at $o_M = 39$ mol per cent. The minimum reflux ratio is

$$v_M = \frac{95}{39} - 1 = 2.44 - 1 = 1.44$$

The second criterion applying to the minimum reflux ratio is also considered in Figure 78. If the minimum reflux ratio is increased by any small amount, the slope of the enrichment line through A becomes steeper; it intersects line DG between points E and G. This intersection between the enrichment and stripping lines yields a finite number of plates.

The second criterion for the minimum reflux ratio, however, for the case of high concentrations of low boiler in mixtures whose equilibrium curve has a point of inflection, can be met only by using a special procedure. This occurs in the case of ethanol-water mixtures.

This special procedure is explained in Figure 79. The composition of the distillate establishes point A on the enrichment line. Further, the locus of the intersections of enrichment and stripping lines is fixed by composition x_M and by the temperature of the feed. From these, e is computed by means of equation (92), and points D and E are determined. The line connecting these points intersects the equilibrium curve at G, the abscissa

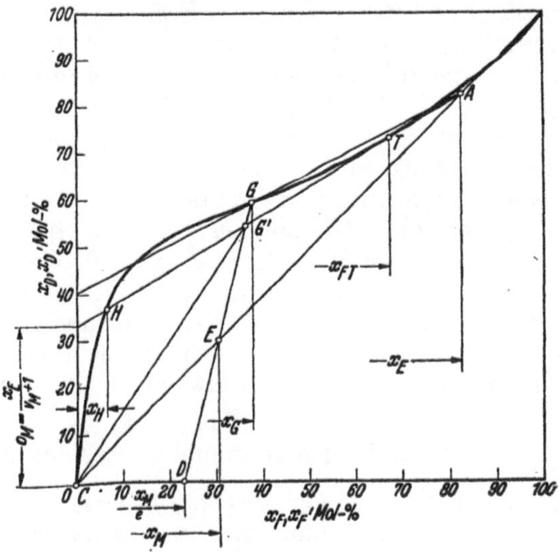

FIGURE 79

Graph for finding the minimum reflux ratio for the separation of a binary mixture whose equilibrium curve has a point of inflection

being x_G. Line AG intersects the equilibrium curve at three points.

This condition implies that the number of stages is still infinite when the reflux ratio is increased a small amount above the minimum. Therefore, the correct minimum reflux ratio must be computed from equation (65), using the ordinate intersection o_M. This is determined from the straight line drawn through A, which is tangent to the equilibrium curve at point T, the abscissa of which is x_{FT}.

If the reflux ratio, determined from this ordinate intersection o_M, is increased by a small amount, the slope of the enrichment line becomes steeper than the tangent AT, and intersects the equilibrium curve at one point only. The abscissa of this intersection is smaller than x_H.

This procedure must be used in all cases arising when the abscissa x_G of the intersection G lies between x_H and x_{FT}. For this range of concentrations, the reflux ratio thus has the same minimum. It increases for values of x_G below x_H, and becomes smaller than the constant value mentioned when x_G exceeds x_{FT}.

The numerical values determined from Figure 65 for the minimum reflux ratios for the normal benzol-toluol mixture and for an ethanol-water mixture hold for the batch unit as well as the continuous unit if the abscissas for x_G from Figures 78 and 79 are used, instead of the batch concentration x_{FB}.

The minimum reflux ratio remains constant over a certain x_G range, as seen from line *b* in Figure 65, only with distillate concentrations exceeding 70 mol per cent low boiler. For a distillate concentration of 70 mol per cent of alcohol, the minimum reflux ratio diminishes continuously as x_G increases, as shown by line *c*. This condition is general for normal mixtures, as shown in the figure by line *a* which holds for a distillate concentration, x_E of 98 mol per cent of low boiler.

III. The actual reflux ratio of a continuous rectifying unit

The operating cost, which includes that of heat and of cooling water, is lowest with the minimum reflux ratio, and increases as the reflux ratio increases, as shown in Figure 80. Size and first cost of heat exchangers in a rectifying unit vary in the same way.

However, the number of plates decreases as the reflux ratio increases, and since the first cost includes the heat exchangers, column, and accessories, a minimum value for the latter item exists, as shown in Figure 80.

The total cost is the sum of the operating and the investment costs; therefore, the total cost curve also exhibits a minimum value, as shown in Figure 80. This minimum cost corresponds to the most favorable actual reflux ratio. According to equation (67), the actual reflux is a multiple, $z \times v_M$, of the minimum reflux; the value of z is usually between 1.3 and 10.

In concentrating mash containing 9 per cent ethanol, when the minimum reflux ratio, as shown in Figure 65, is 2.7, a ratio between 4.0 and 4.5 would be used in designing actual equipment. In the atmospheric distillation and rectification of petroleum, an average reflux ratio of $v = 7$ is quite usual.

IV. Mass-concentration interrelations

In continuous operation, the total amounts of feed and low boiler introduced into the unit must be equal to the quantities of the corresponding substances leaving the unit.

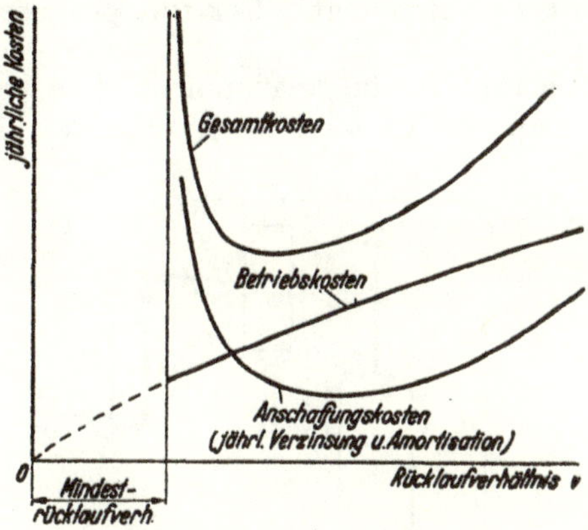

FIGURE 80

Investment and operating cost as a function of the reflux ratio, and determination of the optimum reflux ratio

Rücklaufverhältnis—Reflux ratio
Mindestrücklaufverh.—Minimum reflux ratio
Jährliche Kosten—Cost per annum
Gesamtkosten—Total cost
Betriebskosten—Operating cost
Anschaffungskosten—Investment
Jährl. Verzinsung u. Amortisation—Interest and amortization per annum

$$M = E + A$$
$$M \times x_M = E \times x_E + A \times x_A$$

The relation between mass and composition arises from these equations:

$$M = E \times \frac{x_E - x_A}{x_M - x_A} = E \times \frac{a}{b} \tag{98}$$

For a and b, see Figure 73.

V. Heat requirements

For the heat requirements in a continuous rectifier the reflux or reboil heat is the principal source. This is the heat that must be removed in the reflux condenser to produce the reflux. It is determined from equations (73), (74) and (75) which hold also for batch-type equipment. For this purpose, weight units may be used.

Figure 81 is helpful in the development of an equation to find the heat, Q, required to separate a mixture. It is assumed

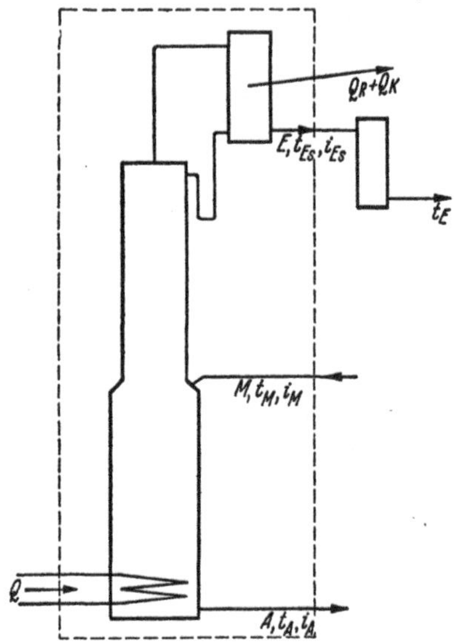

FIGURE 81

Heat balance of a continuous rectifier

that all of the vapor entering the reflux condenser is liquefied, yielding both the reflux and the product E in the liquid state and at a temperature t_{Es}°C.

The mixture to be separated, M, is fed to the unit at temperature t_M, while the bottoms A, leave at a temperature of t_A°C.

The heat quantities, i_{Es}, i_M and i_A refer, respectively, to the liquid product with boiling temperature t_{Es}, the feed and the bottoms. These quantities of heat must be added to heat the individual components from 0°C after mixing to the required temperature.

Considering this, the enthalpies include all of the heats of mixing which might influence the heat requirements. See equation (125).

For the hookup shown, the reflux condenser must handle both the reflux heat, Q_R, and the heat liberated during condensation of the product, Q_K. It is immaterial whether this heat is taken up by cooling water or by feed if the reflux condenser is operated as a feed heater.

Referring to the dotted section shown in Figure 81, the heat-balance equation neglecting heat losses, of a continuous rectifier may be written as follows:

$$Q = Q_R + Q_K + E \times i_{Es} + A \times i_A - M \times i_M \qquad (99)$$

This equation implies that the temperature, t_E, at which the product leaves the unit has no influence on the heat requirements; it merely determines the amount of cooling water needed.

Equation (99) applies whether the feed is preheated or not. Feed preheating merely affects the numerical values of t_M and i_M. If it is desired to take heat losses into account, appropriate corrections must be made on the right-hand side of equation (99).

Figure 76 gives enthalpies of liquid ethanol-water mixtures at different temperatures as a function of their compositions, including heats of mixing.[43] Figure 82 indicates the heat contents of liquid and gaseous petroleum fractions of different densities as a function of temperature.[44]

Values for oils which have undergone repeated cracking are given in the same figure. Although these have chemical compositions differing from petroleum fractions, their heat contents are in good agreement with those of non-cracked oils, especially at lower temperatures. Repeated tests indicate that the origin of the oil has little influence on the heat-content figures in Figure 82.[45]

In general, equation (99) also holds if the liquid mixture has more than two components. It is also valid if the product is liquefied in the condenser-cooler instead of in the reflux condenser. Again, heat content, i_{Es}, is based upon the condensation temperature.

Advantages of the continuous rectifier are its simple operation, production of a residue relatively free from low boiler, low heat consumption, and production of a distillate rich in low boiler. This is a consequence of the fact that the reflux ratio is kept constant, while for a batch rectifier it must be increased during operation, if the concentration of low boiler is to be the same as in the case of the continuous unit.

It is evident that the steam consumption of the continuous

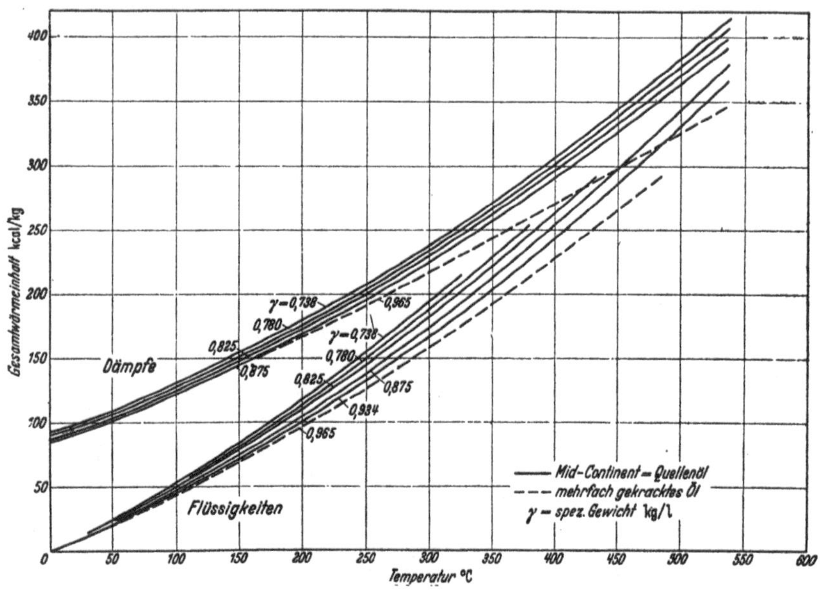

FIGURE 82

Enthalpies of petroleum products as a function of temperature

Temperatur—Temperature
Gesamtwärmeinhalt—Total heat content
Flüssigkeiten—Liquids
Dämpfe—Vapors
γ = Densities, kilogram per liter
——— Mid-continent petroleum types
- - - - Oils after undergoing repeated cracking

rectifier is considerably lower than that of the batch type for any given composition of the product and may be a fraction of that for the latter. Batch-type units are used in the case of small throughputs, or where fractionating need not be very close, and several fractions are desired. An advantage of the batch-type unit is that a large amount of solids may be allowed to settle in the kettle, without interfering with the operation.

Numerical example. Find the steam consumption per kilogram of distillate for the following conditions: Starting with an ethanol-water mixture containing 10 per cent by weight alcohol (ξ_M), a distillate with a ξ_E of 94 per cent by weight is desired.

The stillage is to have a negligible alcohol content ($\xi_A \approx 0$). The mixture enters at $t_M = 70°C$, while the boiling temperature $t_S = 92°C$. Reflux ratio $= 4$. The feed is heated from 25 to 70°C in the reflux condenser.

The mass of the mixture is found from equation (98):

$$M = 1 \times \frac{94 - 0}{10 - 0} = 9.4 \text{ kg per hour}$$

which yields a mass of bottoms,

$$A = 9.4 - 1 = 8.4 \text{ kg per hour}$$

From Table 3 and equation (47), the molar heat of evaporation (r) is 9600 kilocalories per mol. Equation (1a) gives the average molecular weight of the distillate

$$\mu_E = \frac{100}{\dfrac{94}{46.05} + \dfrac{6}{18.01}} = 42.1$$

from which the heat of evaporation per kilogram of distillate and reflux becomes

$$\rho = \frac{9600}{42.1} = 228 \text{ kcal per kilogram}$$

The same value is obtained on using equation (44), if the heat of evaporation of ethanol,[46,47] ρ_1, is taken as 208 kcal per kilogram and that of the water, ρ_2 as 540 kcal per kilogram. Figure 76 gives the heat contents corresponding to the temperatures of the product ($t_{Es} = 78.2°C$) and of the mixture ($t_M = 70°C$) as $i_{Es} = 53$ kcal per kilogram and $i_M = 66$ kcal per kilogram, for the concentrations involved.

Taking into account the slightly higher pressure in the still kettle, a heat content of the stillage, i_A of 104 kcal per kilogram is assumed. The feedplate temperature of 92°C, corresponding to the concentration specified by the abscissa of the intersection F of the enrichment and stripping lines (Figure 73), gives an e value of 1.048 from equation (92). Then the following numerical values correspond to the heats enumerated in equation (99)

$$\begin{aligned}
Q_R &= 1 \times 4 \times 228 = +912 \text{ kcal per hour}\\
Q_E &= 1 \times 228 = +228 \text{ kcal per hour}\\
E \times i_{Es} &= 1 \times 53 = +53 \text{ kcal per hour}\\
A \times i_A &= 8.4 \times 104 = +873 \text{ kcal per hour}\\
-M \times i_M &= 9.4 \times 66 = -621 \text{ kcal per hour}\\
\hline
Q &= 1445 \text{ kcal per hour}
\end{aligned}$$

If the steam used for heating has a heat of condensation of 540 kcal per kilogram, the steam requirement per kilogram of product per hour is equal to

$$\frac{1445}{540} = 2.68 \text{ kg per hour}$$

This table of heats indicates that to produce the reflux a comparatively large amount of heat must be removed, and also that much heat is contained in the bottoms leaving the unit. This is shown in Figure 83, drawn to scale. Similarly, the quantities involved are shown graphically in Figure 84.

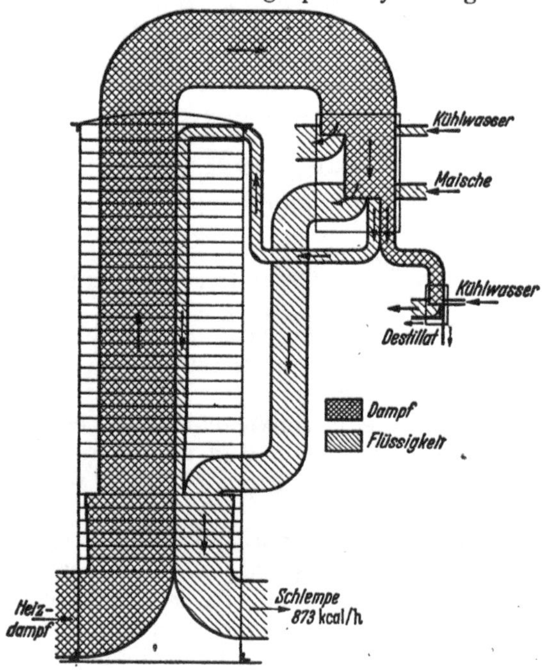

FIGURE 83

Heat balance diagram for a continuous rectifier

Dampf—Vapor
Flüssigkeit—Liquid
Heizdampf—Heating steam
Kühlwasser—Cooling water
Maische—Mash (*Beer*)
Destillat—Distillate
Schlempe—Slop (*Stillage*)

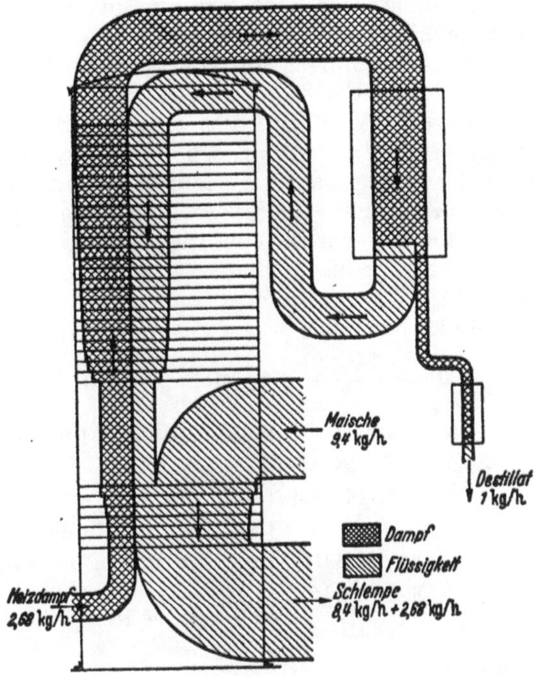

FIGURE 84

Material balance diagram of a continuous rectifier

Dampf—Vapor
Flüssigkeit—Liquid
Heizdampf—Heating steam
Maische—Mash (*Beer*)
Destillat—Distillate
Schlempe—Slop (*Stillage*)

The possibility of reducing the heat requirements of a rectifying unit by preheating the feed as far as feasible with the heat contained in the bottoms is evident. This is particularly promising when mashes of low concentration are worked up for their alcohol content. Then, the optimum feed temperature must be determined.

The more the feed is preheated, the larger becomes $M \times i_M$ as implied by equation (99). On the other hand, the recycled heat, Q_R, also increases, so that a minimum heat consumption results for a definite feed temperature. This optimum, t_M, is 80 to 85°C for mashes having low concentrations of 0.75 to 1.5 per cent by weight of ethanol.

Although masses have been given in kilograms and concentrations in per cent by weight, it is preferable to employ molar units when determining the heat requirements in order to calculate the number of stages and the column diameter.

Numerical example. For the conditions cited in the preceding example, the quantity of distillate is 180 l per hour. For a density of 0.81134 kg per cubic decimeter at 15°C, this corresponds to 146 kg per hour, or

$$E = \frac{146}{42.1} = 3.47 \text{ mols per hour}$$

Accordingly, the amount of vapor passing up the enriching column is constant and becomes

$$D = 3.47 \times (4 + 1) = 17.36 \text{ mols per hour}$$

The reflux, F, is $3.47 \times 4 = 13.89$ mols per hour. The molecular weight of water is 18.01, the mass rate of bottoms is

$$A = \frac{8.4 \times 146}{18.01} = 68.33 \text{ mols per hour}$$

and the feed rate is

$$M = A + D = 68.33 + 3.47 = 71.8 \text{ mols per hour}$$

This value can also be obtained from equation (98), where $x_E = 86.2$ mol per cent alcohol (corresponding to $\xi_E = 94$ per cent of weight) and $x_M = 4.17$ mol per cent alcohol (corresponding to $\xi_M = 10$ per cent of weight).

It may also be obtained from the ratio $(9.4 \times 146)/19.15$, where 19.15 = average molecular weight of the mixture containing 10 per cent by weight of alcohol. From equation (91), the amount of liquid flowing downward, F' is $13.89 + 1.048 \times 71.8 = 89.14$ mols per hour. The amount of vapor flowing through the stripping column thus is

$$D' = F' - A = 89.14 - 68.33 = 20.81 \text{ mols per hour}$$

VI. Reduction of heat requirements

1. Multistage arrangement

As in evaporator practice, the possibility of heating a unit with the heat which leaves another unit is present. Most of the waste heat accumulates in the reflux condenser. Therefore, the reflux condenser can be made to generate vapor to heat other units. To maintain the temperature drop required to produce the vapor in the reflux condenser, it is usually necessary to operate the primary rectifier at a higher pressure than is used in the unit operated with recovered heat.

This procedure is particularly adapted to rectifying processes where a column must be operated under pressure, so that the multistage principle may be applied, as in the case of the dehydration of alcohol by rectification of a mixture of ethanol, water, and dehydrating agent under pressure.[48]

The simplest type of multistage process is the rectifier used to distill liquid air. Here, two superimposed columns are used, the lower of which operates at 5 to 6 kilograms per square centimeter, and the upper at atmospheric pressure, as shown in Figure 85. This arrangement is also unique since the reflux condenser *d* (Figure 85) also acts as a vaporizer for the low pressure column.

Othmer has suggested operating only the stripping column of a continuous unit at high pressures, using high-pressure steam for heating. The vapor rising from the stripping column then passes to a heat exchanger where the vapor generated is fed to the overhead rectifying column, while the condensate is fed via a trap to the middle section of this rectifying column.

The condensate has a low-boiler concentration greater than that of the mixture fed to the stripping column. Consequently, the rectifying column above the heat exchanger may be operated with an unusually low reflux ratio, which in turn means a reduced column diameter and coolant consumption. However, this arrangement requires operating the stripping column at an elevated pressure. The quantity of heating steam is the same as for the usual setup, but the steam pressure is higher. Extra-low-pressure steam becomes available from the heat exchanger mentioned.

2. Vapor recompression; the heat pump

As in the case of evaporators operating with vapor recompression, the vapors rising from the top plate of a column may be compressed to boost the condensation temperature, and thus serve as a heating medium for the still. A portion of the condensate is fed back as reflux, while the remainder is the product.

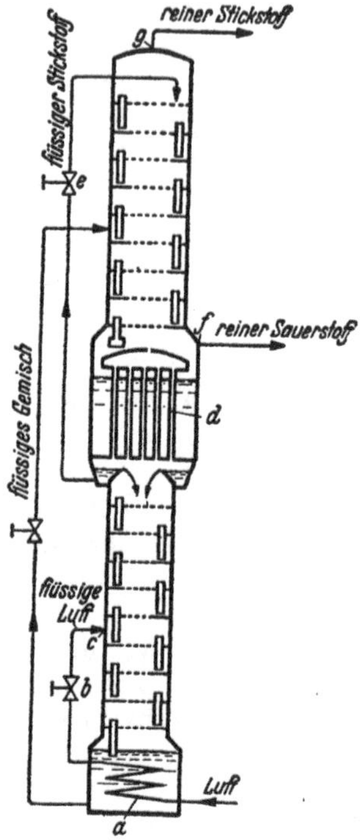

FIGURE 85

Two-column unit for separating liquid air

a, d Heat exchangers
b, e Throttling valves
c, f, g Material inlets and outlets
Luft—Air
Flüssige Luft—Liquid air
Flüssiges Gemisch—Liquid mixture
Flüssiger Stickstoff—Liquid nitrogen
Reiner Stickstoff—Pure nitrogen
Reiner Sauerstoff—Pure oxygen

The compression of the vapors must be carried to the point where the increase in the condensation temperature is sufficient to produce the temperature drop required for transmitting the heat to the still, since the difference in boiling temperatures between the liquid in the still and that on the top plate of the

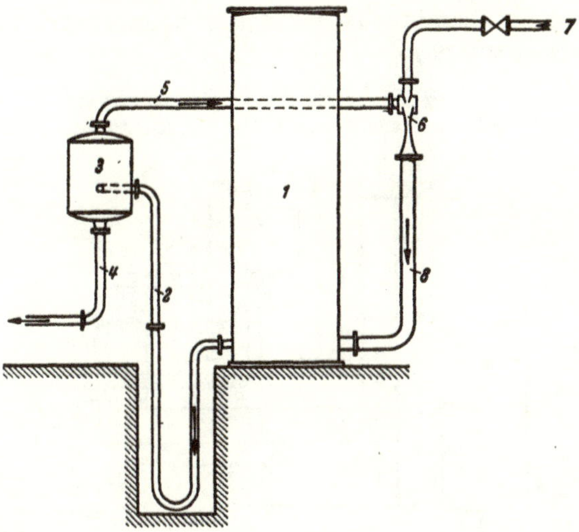

FIGURE 86

Utilization of the waste heat from a stripping column by means of vapor compression

The alcohol-free bottoms from stripping column *1* are passed via line *2* to flash chamber *3*, from which the spent water flows through line *4* to the sewer. The steam flashed off goes through line *5* to injector *6* which is fed with live steam from line *7*. The expanding steam and the compressed vapor drawn from *5* are passed through *8* and heat the lower section of the stripping column which separates a mixture of ethanol and water.

enriching column is important. Considering this latter temperature difference, heat recovery by vapor compression seems to offer little advantage in the case of liquid mixtures whose components have widely different boiling temperatures.

If the boiling temperatures of the individual substances are more nearly alike, the method outlined promises better success with vacuum operation, since for an equal temperature drop on both sides of the heating surface the consumption of live steam decreases as the boiling temperature in the still drops.[49]

The pressure drop resulting from the different saturation temperatures of the mixtures and the temperature drop for heat transmission can be eliminated by special methods, as outlined in Figure 86.[50]

VII. Layout of continuous rectifying equipment for handling binary mixtures

The arrangement and hookup of the different parts of a continuous rectifier affect not only the steadiness and convenience

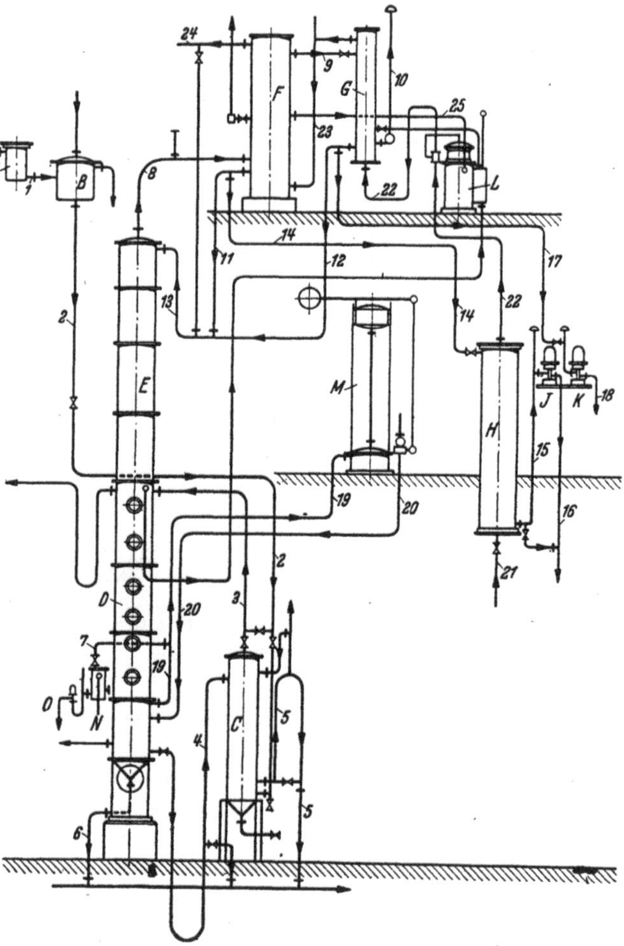

FIGURE 87

Continuous rectifier for the production of concentrated acetone from aqueous acetone solutions, with simultaneous separation of first fractions and solid residue

The acetone-water mixture from a crude storage tank passes via filter pot *A* and line *1* to flow controller *B* which maintains a uniform feed. From controller *B* the acetone-water mixture passes via line *2* to tubular preheater *C*. It is preheated by hot water leaving column *D* and goes through line *3* to the top section of column *D*. In the latter, the mixture flows downwards from plate to plate and is stripped of its acetone. The spent water leaves the lower part of column *D* by line *4* and transfers part of its heat to the acetone-water feed. The remaining liquid flows to the sewer through siphon *5*.

The lower part of column *D* is funnel-shaped and equipped with a gate valve to eliminate solid residues, which leave by line *6*. Tester cooler *N* and tester line *O* enable the operator to ascertain whether the spent water is completely acetone-free. For this purpose, a small amount of vapor is continuously withdrawn from the lower part of the column by line *7*, and liquefied in tester cooler *N*.

The condensate goes to the tester line *O*, where a hydrometer indicates the density of the condensate. Enriching column *E* is situated above stripping column *D*; this increases the acetone concentration of the vapors rising upwards to the desired degree, usually 99.5–100 per cent.

From enriching column, *E*, the acetone vapors pass through line *8* to condenser *F* where they are liquefied. Low-boiling products present, air and other non-condensable gases pass to vent condenser *G* via line *9*. In the latter, first fractions and residues of acetone vapors are liquefied, while non-condensed gases leave through line *10*.

The condensate formed in *F* and *G* is returned to the extent necessary for separation via lines *11*, *12* and *13* to the upper part of enriching column *E*. The balance of the hot condensate from *F* leaves line *14*; the distillate is cooled in acetone cooler *H* and proceeds through *15* to tailbox *J* and line *16* to a storage tank.

If first fractions should accumulate in vent condenser *G*, they may be withdrawn by line *17* to tailbox *K*, and passed through line *18* to a special storage tank.

The steam flow to stripping column *D* is controlled automatically. The operating pressure in the column is transferred via line *19* to steam controller *M*, the float of which actuates a throttle by a leverage to steam inlet line *20*.

The cooling water is at first passed through line *21* to cooler *H*, then passes upwards to line *22* into which the automatic water controller *L* is inserted; then it goes to vent condenser *G*, and, eventually through line *23* to condenser *F*. It leaves the latter through line *24*.

Water controller *L* receives its impulse from the pressure in condenser *F* via line *25*. As in the case of the steam controller, a float which moves up or down actuates a throttle by means of a leverage to water line *22*.

---

of operation, but the first cost of foundations and buildings as well. Usually the enriching column is superposed on the stripping column, as shown in Figure 87 for the case of a rectifier producing highly concentrated acetone.

The reflux condenser is separated from the column and is mounted at a higher level, while the bottoms are removed

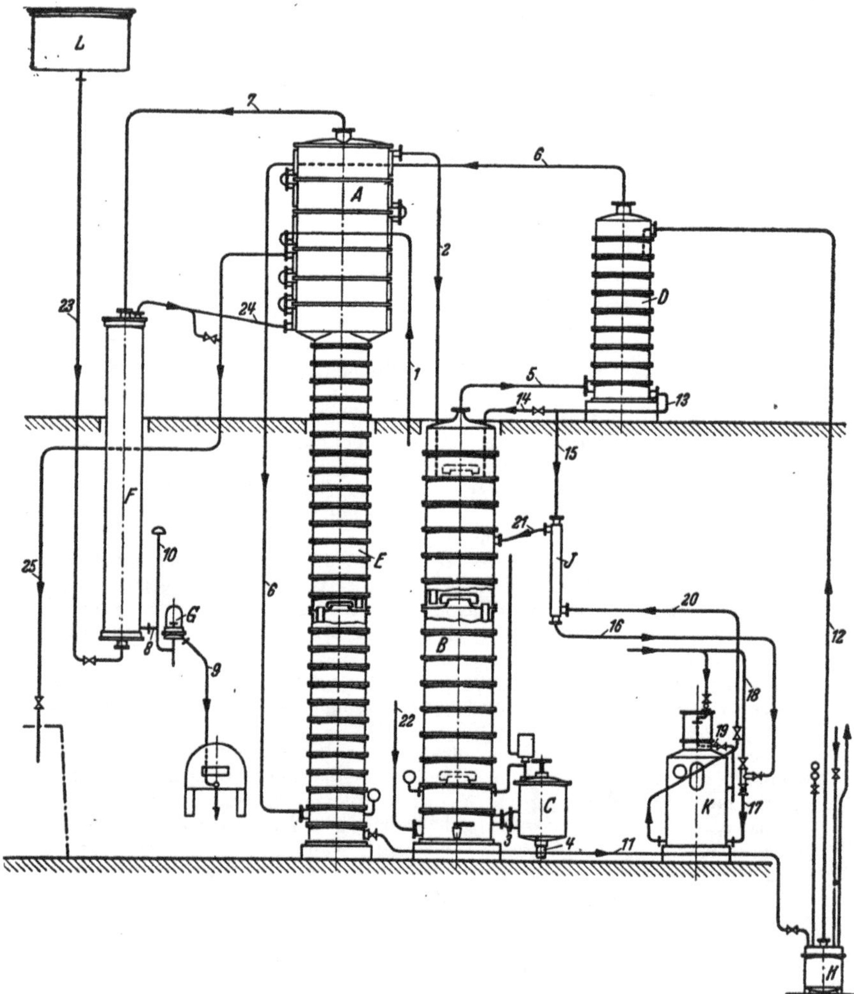

FIGURE 88

Rectifier for alcohol-water mixtures

The mash flows through line *1* to reflux condenser *A* from which it passes, via line *2* to stripping column *B*, the slop from which is drained by line *3*a the slop controller, and line *4*. The vapors leaving the stripping column pass through line *5* to section *D* of the enriching column and leave the latter through line *6* to pass to section *E* of the enriching column. The alcohol vapors from

E enter the superposed reflux condenser *A*, from which the reflux goes to the top enriching plate, while the remainder of the vapors go through line *7* to the condenser-cooler *F*. From the latter, the spirits pass through line *8*, tail-box *G*, and piping *9*, while gases are vented by *10*.

The reflux from column section *E* goes via line *11* to a liquid siphon, which passes the mixture through line *12* to the top plate of column *D*. The liquid is withdrawn from the bottom section and split up into two streams. The bulk of the liquid reaches the top plate of the stripping column through line *14*. The balance is passed through line *15* to a heat exchanger *J*, from which it goes through *16* and *17* to fusel oil separator *K*, after cold water has been added through line *18*.

The fusel oil is collected through line *19*, and the fusel-free alcohol-water mixture is passed through line *20* to heat exchanger *J* and further through *21* to the top section of the stripping column. Through line *22*, heating steam is blown into the lower portion of the stripping column.

Cooling water is taken from tank *L*; it flows through *23* to the condenser-cooler, through *24* to the reflux condenser which it leaves by line *25*.

---

continuously by a siphon, so that a constant liquid level is maintained in the distilling unit. This arrangement is feasible since a pressure drop exists across the rectifier plates so that the absolute pressure is higher in the lower part of the unit.

When the available headroom is limited, the column, consisting of the enriching and stripping sections, is divided into two parts operating in series. Best utilization of space results when the two parts are of equal height.

Figure 88 shows the general arrangement of such a unit drawn to scale. In contrast to the arrangement shown in Figure 87, the reflux condenser is superposed on the enriching column and acts as a dephlegmator, while the effluent level is controlled by a float. Since the effluent consists of water, the heating steam is introduced by a sparger into the bottom section of the unit.

VIII. Special cases

1. The mixture to be separated is fed at its boiling temperature

The determination of the enrichment and stripping lines is especially simple if the mixture is fed at a temperature, t_M, equal to the boiling temperature, t_S, of the liquid on the feed-plate. Evidently, the rectifying plates are most efficient when the concentration of low boiler in the feed lies between that of the bottoms from the lowest plate of the enriching column and the overflow from the top plate of the stripping column.

It is assumed that the feedpoint is located correctly, so that the concentration of the mixture to be separated practically equals that of the liquid on the feedplate ($t_M = t_S$); it follows that the heat contents (q_M and q_S) per mol of mixture to be separated and liquid on the feedplate are identical

$$q_M = q_S \tag{100}$$

When this equation is substituted into equation (92)

$$e = 1 \tag{101}$$

In that case, abscissa x_M of the feed concentration corresponds both to the point of intersection of the enrichment and stripping lines for both the minimum and the actual reflux ratio, as shown in Figure 89. Points D, E, F and G (Figures 78 and 79) have the same abscissa values, x_M, and thus are located on a perpendicular through the concentration x_M. This vertical then takes the place of the line specified by equation (95).

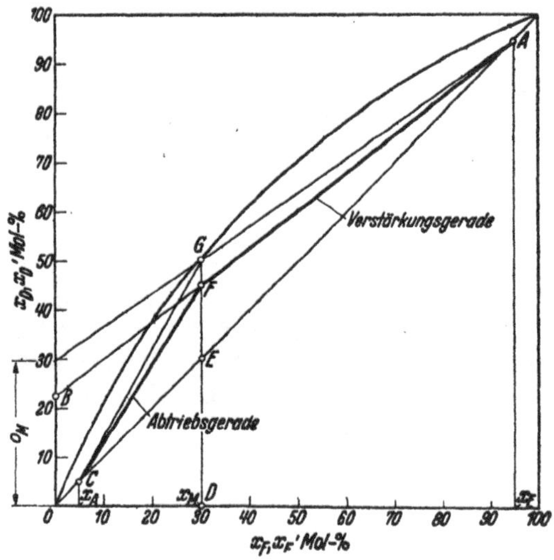

FIGURE 89

Determination of the minimum reflux ratio and the intersection of the enrichment and stripping lines in the case where the mixture is fed at its boiling temperature

Verstärkungsgerade—Enrichment line
Abtriebsgerade—Stripping line

The stripping line then has a steeper slope, and the enriching effect of the stripping plates is less than in the case of cold feed.

For the case where $t_M = t_S$ the minimum reflux ratio, v_M, is found by the method explained in Figures 78 and 79. Figure 89 shows the ordinate section (o_M) used to determine the minimum reflux ratio for an ideal mixture.

2. The simple stripping column

If the problem is merely to strip the low boiler from a binary mixture, this liquid mixture is passed to the top plate of a stripper; the resulting vapors are not separated any further. A liquid is recovered from the still, which in some cases may be the product itself.

Figure 90 shows a sulfuric acid concentration plant as an example. The direct-fired still is constructed of cast-iron alloy, and the product is recovered as a highly concentrated acid.

Computations for the stripper are made by means of equations (86), (87), (88) and (89). The last equation states that point C on the stripping line is given by concentration x_A mol per cent of low boiler in the bottoms, as shown in Figure 91.

Again, the quantities of vapor and liquid passing through the stripping column remain constant only if the molar heat of evaporation is independent of the composition of the mixture; this is assumed to be true in this case.

Another convenient point H on the stripping line is obtained most readily by finding the intersection with the line drawn parallel to the abscissa for $x_D' = 100$ mol per cent.

If the quantities of feed, M, and bottoms, A, are known, this intersection arises on applying equation (88):

$$x_F' = 100 - \frac{A}{M \times e}(100 - x_A) \tag{102}$$

where e, found from equation (92), takes into account the feed temperature. The amount of liquid leaving the top plate is

$$F' = M \times e \tag{103}$$

If a vapor quantity D_M containing x_{DM} mol per cent of low boiler is released from the top plate

$$M = D_M + A \tag{104}$$

and

$$M \times x_M = D_M \times x_{DM} + A \times x_A \tag{105}$$

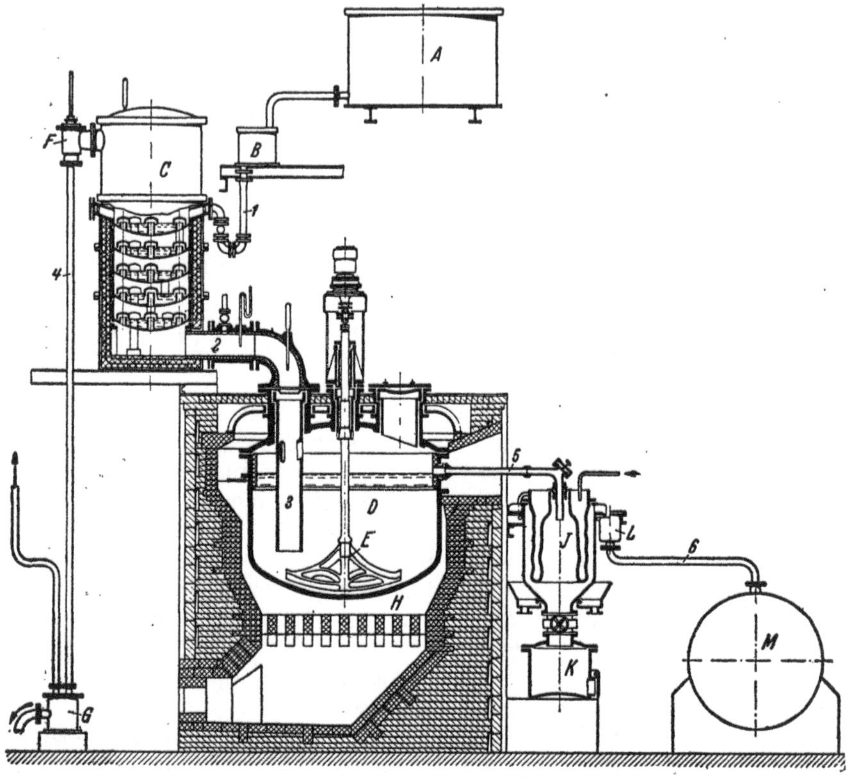

FIGURE 90

Sulfuric acid concentrating plant

The mixture of sulfuric acid and water flows from an overhead tank *A* to a proportioning vessel *B*, from which it flows at a constant rate via line *1* and siphon to column *C*. The latter is made of silicon cast iron and equipped with bubble caps. The pre-concentrated acid from the column passes through connection *2* and feedpipe *3* to the lower part of kettle *D* fitted with agitator *E*.

The vapors liberated in *D* flow counter-currently to the liquid through connection *2* and column *C*, thus being freed of the sulfuric acid. The water vapor is drawn off by injector *F* and condenses; the liquid and gases pass through line *4* into a collector *G*, from which the spent water flows off and gases are vented.

Kettle *D* is set in a furnace *H* usually fitted with gas or oil heating. The concentrated acid from the kettle (up to 98 per cent weight) passes through line *5* into a pot cooler *J* connected to a sludge separator *K*. From the cooler, the acid goes via a sludge settler *L* and line *6* to storage tank *M*.

Therefore

$$\frac{M}{D_M} = \frac{x_{DM} - x_A}{x_M - x_A} \tag{106}$$

and

$$\frac{M}{A} = \frac{x_{DM} - x_A}{x_{DM} - x_M} \tag{107}$$

These equations are essential in determining mass rates from concentrations. Since the relations between vapor and liquid concentrations are given by a point of the stripping line *CH* (Figure 91), equations (104) and (105) specify the following relations between x_{DM} and x_M:

$$x_{DM} = \frac{M}{D_M} \times x_M - \frac{A}{D_M} \times x_A = \frac{M}{M - A} \times x_M - \frac{A}{M - A} \times x_A \tag{108}$$

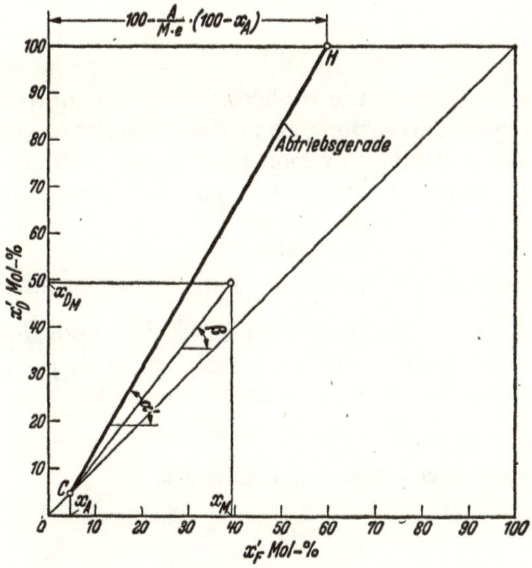

FIGURE 91

Operating line of a pure stripping column

Mol %—Mol per cent
Abtriebsgerade—Stripping line

Vapor of concentration x_{DM} is considered to be in phase equilibrium with the liquid leaving the top plate. Equation (108) states that x_{DM} is cut off by x_M on a straight line which passes through point C of the diagonal (Figure 91), has the abscissa value x_A, and a slope given by

$$\tan \beta = \frac{M}{D_M} \tag{109}$$

while the stripping line makes an angle α' as shown in Figure 91, which is specified by

$$\tan \alpha' = \frac{M \times e}{D_M + M(e-1)} \tag{110}$$

while the following relationship exists for α' and β

$$\tan \alpha' = \frac{e \times \tan \beta}{1 + (e - 1) \tan \beta} \tag{111}$$

If the feed temperature is that of the feedplate ($t_M = t_S$), e becomes unity and

$$\alpha' = \beta \tag{112}$$

Although constancy of the molar heat of evaporation has been assumed in these derivations, equations (103) through (108) also hold if the molar heat of evaporation varies with concentration. The straight stripping line must then be replaced by a slightly curved line (see Section E XI).

3. Vapor feed in a continuous rectifying column

If the mixture to be separated is fed in vapor form, with a low boiler concentration of x_{DM} mol per cent, q_M in equation (92) is replaced by the total heat of the vapor

$$q_M = q_S + r \tag{113}$$

Thus the thermal condition of the feed is

$$e = 0$$

and

$$\frac{x_{DM}}{e} = \infty \tag{114}$$

x_M in equation (95) is replaced by x_{DM}. This concentration locates point E of the diagonal (Figure 92), while the intersection F of the stripping and enrichment lines is specified by

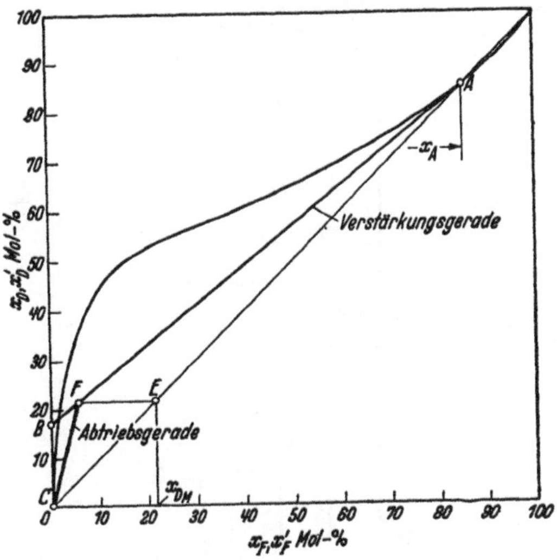

FIGURE 92

Separation of a binary vapor
Enrichment and stripping lines

Mol %—Mol per cent
Verstärkungsgerade—Enrichment line
Abtriebsgerade—Stripping line

the horizontal parallel drawn through point E to intersect the enrichment line.

The special cases of a stripping column and that of the charging of a continuous apparatus with a vapor feed occur simultaneously in the case of the two-stage rectifier and low-wine column encountered in the alcohol industry and shown diagrammatically in Figure 93.

The feed flows into a stripper "beer still," from which the intermediate product A' is withdrawn, while the vapor rising from the top plate of the stripping column, D_M, is fed to a fractionating column.

In the fractionating column, the vapors are freed of the low boiler in the enriching column, which rests on the stripping column (low-wine column), and from which the effluent liquid $A^{\times}$ flows. The upshot of this operation is that the output is divided into two partial streams, A' and $A^{\times}$, and both the

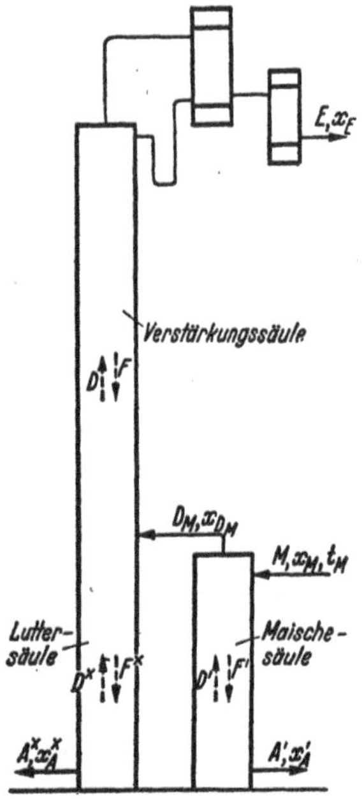

FIGURE 93

Two-stage rectifier featuring a low-wine column and a mash column ('Beer still')

Maischesäule—Mash column
Luttersäule—Low-wine column
Verstärkungssäule—Enriching column

stripper and the low-wine column must be heated. In the alcohol industry, steam is injected directly below the bottom plates of both the stripper and the low-wine column.

Numerical example. The mixture described in the last numerical example (Section E V) is to be separated in a two-stage low-wine rectifying unit. (Product $E = 3.47$ mols per hour, feed $M = 71.8$ mols per hour, bottoms $A = 68.33$ mols per hour, reflux in enriching column $F = 13.89$ mols per hour, in stripping column $F' = 89.14$ mols per hour, vapor in enriching column $D = 17.36$ mols per hour, in stripping column $D' =$ 20.81 mols per hour.)

In the following discussion, values referring to the low-wine column below the enriching column are indicated by a $\times$ mark, while values referring to the beer still are indicated by a $'$ mark.

The sum of the two partial streams of bottom product is

$$A' + A^{\times} = 68.33 \text{ mols per hour}$$

The value of A' obtained from this relation is substituted into

$$F' = M \times e = 71.8 \times 1.048 = 75.25 = D' + A'$$

whence:

$$A^{\times} = D' + 68.33 - 75.25 = D' - 6.92$$

Combining this with the following relation, which holds for the low-wine column

$$F^{\times} = F = 13.89 = A^{\times} + D^{\times}$$

leads to

$$D' + D^{\times} = 20.81$$

If, furthermore, it is assumed that in the x_F vs. x_D diagram the operating lines for the low-wine column and beer still have the same slope

$$\frac{F^{\times}}{D^{\times}} = \frac{13.89}{D^{\times}} = \frac{F'}{D'} = \frac{75.25}{D'} \quad \text{and} \quad \frac{D'}{D^{\times}} = 5.42$$

The vapor rising in the beer still is

$$D' = 17.56 \text{ mols per hour}$$

and that passing through the low-wine column is

$$D^{\times} = 3.25 \text{ mols per hour}$$

These rates of vapor flow determine the heating-steam requirements in mols per hour for the beer still and low-wine columns. The amounts of reflux in the beer still and low-wine columns being equal, the amount of vapor rising from the top plate of the mash column is:

$$D_M = D - D^{\times} = 17.36 - 3.25 = 14.11 \text{ mols per hour}$$

This may also be obtained from the relation

$$D_M = D' - (e - 1) \times M = 17.56 - 0.048 \times 71.8 = 14.11 \text{ mols per hour}$$

In the determination of the column cross-section, the ratio of the amounts of vapor passing through the enriching and low-wine columns is also important

$$\frac{D}{D^{\times}} = \frac{17.36}{3.25} = 5.34$$

The effluent from the low-wine column is

$$A^{\times} = D' - 6.92 = 17.56 - 6.92 = 10.64 \text{ mols per hour}$$

and that from the beer still is

$$A' = 68.33 - 10.64 = 57.69 \text{ mols per hour}$$

The slope of the operating lines for the beer still and low-wine columns is given by

$$\frac{F'}{D'} = \frac{75.25}{17.56} = \frac{F^{\times}}{D} = \frac{13.89}{3.25} = 4.29$$

The composition x_{DM} of the vapor leaving the beer still and entering the enriching column may be found graphically from equation (108) and (109), or analytically from equation (106). In the latter, for $x_A' = 0$:

$$\frac{71.8}{14.11} = \frac{x_{DM}{}^{-0}}{4.17^{-0}}$$

$$x_{DM} = 21.3 \text{ mol per cent alcohol.}$$

Equation (107) gives the same result.

4. Mixed liquid and vapor feed

In the case where the feed is introduced as a mixture of liquid and vapor, q_M of equation (92) also refers to the heat content per mol of mixture. Since the feed contains vapor, q_M exceeds q_S, and e is less than unity. The abscissa section x_M/e of point D (see Figure 73) is then larger than x_M. The connecting line DE gives intersection F with the enrichment line. The abscissa of F has a lower value than x_M; this point determines the location of the stripping line.

IX. Location of the feedpoint

In the method for determining the number of stages which has been explained in connection with Figure 73, it had been assumed that the abscissa of the intersection F of the enrichment and stripping lines is located between x_n (the concentration of low boiler in the overflow to the feedplate) and x_m (composition of the liquid leaving the feedplate).

If the feed is at its boiling temperature and has the composition

$$x_m < x_M < x_n \tag{115}$$

This relation leads to the best operation, resulting in maximum separation with a given number of plates in the enriching and stripping columns.

If the feedpoint is located too low, as shown in Figure 94, the lower plates do not separate at their highest efficiency. The

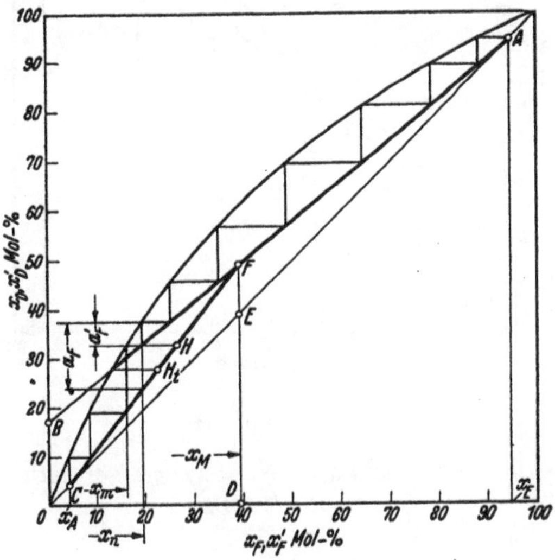

FIGURE 94

Changes in concentration in a continuous rectifier with the feedpoint placed too low

bottom plate produces a theoretical enrichment, a_F', compared to an enrichment, a_F, with the feedpoint at its proper location.

The liquid resulting from mixing the overflow leaving the bottom enriching plate with the feed has a composition specified by point H in Figure 94. Under extreme conditions, when the number of stages in the enriching column is infinite, this point is replaced by point H_t. This is the intersection of the stripping line and the horizontal drawn through the intersection of the equilibrium curve and the enrichment line. When the feedpoint is too low

$$x_m < x_n < x_M \tag{115a}$$

When the feedpoint is too high, the conditions illustrated in Figure 95 appear. The top stripping plate then produces the

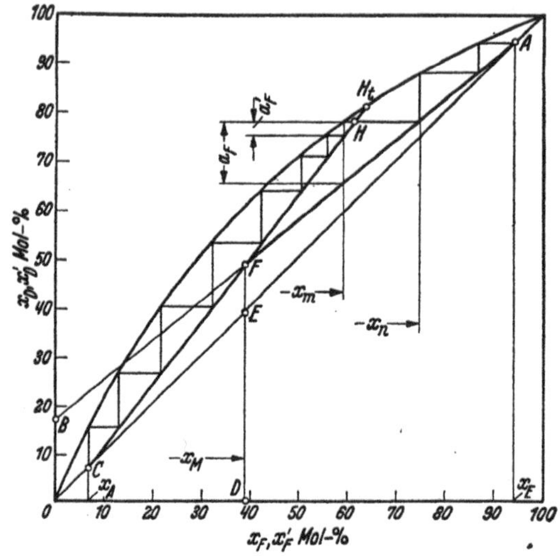

FIGURE 95

Changes in concentration in a continuous rectifier with the feedpoint placed too high

theoretical enrichment a_F', which is smaller than the enrichment a_F, which could be obtained with the same plate located in the enriching column. Before any interchange occurs on the feedplate, the mixture has a concentration which is determined by the abscissa of point H. This is displaced toward H_t, if the number of plates of the stripper becomes infinite. If

$$x_M < x_m < x_n \tag{115b}$$

the best operation does not result either.

In addition, the best results are not obtained if the masses handled in actual operation are different from those used in the determination of the number of plates.

X. Heat losses

The question naturally arises as to whether it is of value to lag a rectifier. Heat losses to the surroundings require adding more heat in the calandria. On the other hand, such heat losses reduce the consumption of cooling water.

As a consequence of the heat losses through the rectifier shell, a portion of the vapors passing up the column condense at the shell; however, this portion is small. The condensate mixes with the overflow, increasing the latter. This increase of overflow occurs in every zone between contiguous plates. Thus the bottom plate of an enriching column receives the overflow from the condenser to the top plate, plus the amount of condensate corresponding to the heat losses.

The increase of reflux leads to more complete enriching on the plates, since, as is evident in the x_F vs. x_D plot, an increase in the reflux rate lowers the enrichment line. Further, since condensate that forms above a plate overflows to the plate below, a different and definite enrichment line applies to each plate. The increase in the enriching effect of a plate due to condensation is more pronounced for the lower plates than for the upper ones. The overflow passing from the bottom plate, however, determines the heat requirements. Thus, if optimum column efficiency is desired for a given heat consumption, the column must be lagged. The heat which would otherwise be lost in an unlagged column is abstracted from the vapors in the reflux condenser.

In a lagged column it is possible to deliver the maximum amount of overflow to all of the plates in an enriching column, while in an unlagged column only the bottom plate operates with maximum reflux.

For large rectifying columns, particularly those not protected by buildings, heat savings obtained by lagging amply counterbalance the added cooling required in the reflux condenser. Of course, the heat liberated in the reflux condenser can be used for preheating in many cases. As a general rule, industrial-scale rectifiers should always be lagged.

XI. Variation of the molar heat of evaporation in the interchanger column

The relationship between liquid and vapor composition in the horizontal section between two plates of a column is a straight line in the x_F vs. x_D plot only when the molar heat of evaporation is independent of the concentration, and when heats of solution are negligible.

If heats of solution are present, which is true in the case of non-ideal mixtures, they are usually negligible in comparison with the heat exchanged on the plates between the two phases.

However, there are cases where the molar heats of evaporation of the individual substances to be separated are so different that this variation must be considered. In these cases, the straight lines representing the enrichment and stripping lines are replaced by curves.

In Figure 96, F and D represent the rate of flow of liquid and vapor (in mols per hour) through any section considered. Let

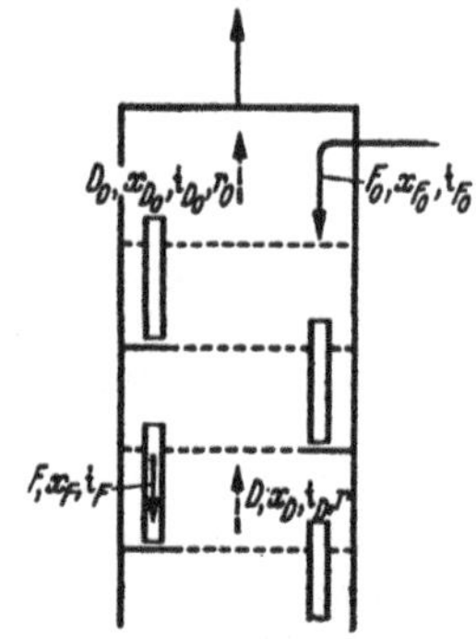

FIGURE 96

Symbols used in equation (116) and following

these streams have the concentrations x_F and x_D of low boiler and heat contents i_F and i_D kilocalories per mol of mixture. In some other section, such as that above the top plate, F_0 and D_0 refer to the flow rates (in mols per hour) associated with the concentrations x_{F0} and x_{D0} and heat contents i_{F0} and i_{D0}. Then

$$F_0 + D = F + D_0 \tag{116}$$

$$F_0 \times x_{F0} + D \times x_D = F \times x_F + D_0 \times x_{D0} \tag{117}$$

$$F_0 \times i_{F0} + D \times i_D = F \times i_F + D_0 \times i_{D0} \tag{118}$$

It is assumed that the heat content of the mixture results from adding the heat contents of the components, i_{F1}, i_{F2}, i_{D1} and i_{D2}:[51]

$$i_F = \frac{x_F}{100} \times i_{F1} + \frac{100 - x_F}{100} \times i_{F2} \tag{119}$$

$$i_D = \frac{x_D}{100} \times i_{D1} + \frac{100 - x_D}{100} \times i_{D2} \tag{120}$$

If the superheat of the vapor, which is negligible as compared to the total heat of the vapor, is omitted

$$i_{D1} - i_{F1} = r_1 \quad (121)$$

and

$$i_{D2} - i_{F2} = r_2 \quad (122)$$

where r_1 and r_2 are the heats of evaporation of the pure constituents. Upon substituting equations (121) and (122) into (120)

$$i_D = \frac{x_D}{100}(i_{F1} + r_1) + \frac{100 - x_D}{100}(i_{F2} + r_2) \quad (123)$$

Further, applying equation (45)

$$\frac{x_D}{100} \times r_1 + \frac{100 - x_D}{100} \times r_2 = r \quad (45b)$$

$$\frac{x_{D0}}{100} \times r_1 + \frac{100 - x_{D0}}{100} \times r_2 = r_0 \quad (45c)$$

Here, r and r_0 represent the heats of evaporation per mol of the mixture having concentrations x_D and x_{D0}. Further, on substituting equations (119) and (123) into equation (118):

$$D \times r = D_0 \times r_0 \quad (124)$$

Therefore, the product of the vapor mass rate and the molar heat of evaporation is constant for any section of the column. If, for instance, the heat of evaporation increases in the lower section of the column, the mass flow of vapor decreases in that section.

In determining the curved enrichment and stripping lines to replace the straight ones, computation is first made for the correlation between the molar heat of evaporation and x_D or x_D' for the enriching and stripping columns from equation (45). If the vapor is completely liquefied in the reflux condenser, $x_{D0} = x_{F0}$. This relation immediately establishes a point of the enrichment line on the diagonal of the x_F vs. x_D plot.

The rate of production of product E and the reflux ratio $v_0 = F_0/E$ specify the mass of vapor rising from the top plate as $D_0 = E + F_0$. Equation (124) gives D as a function of x_D. The path of the enrichment curve is found by applying equation (117)

$$x_F = \frac{D \times x_D - D_0 \times x_{D0} + F_0 \times x_{F0}}{F} \quad (117a)$$

From this the value of x_F corresponding to an assumed value of x_D may be computed.

These relations also hold for the stripping column. The symbols referring to the latter are indicated by the ′ mark. Values F_0' and D_0', in this case, refer to the section below the top stripping plate.

Example. Find the enrichment and stripping lines for an ideal mixture having the equilibrium curve shown in Figure 97.

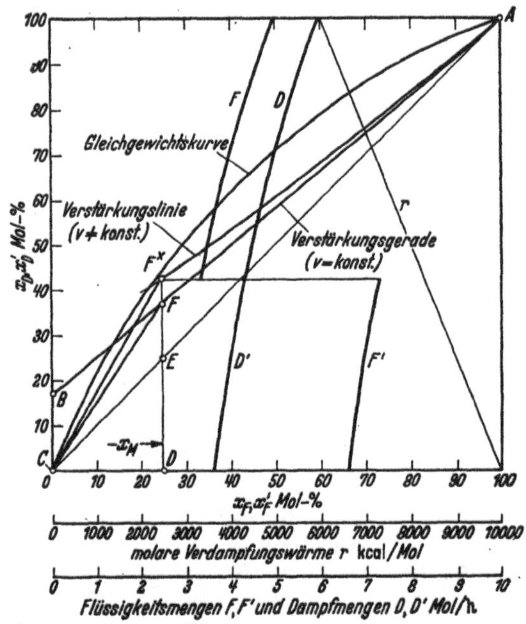

FIGURE 97

Determination of the enrichment and stripping lines for a continuous rectifier

Gleichgewichtskurve—Equilibrium curve
Verstärkungslinie ($v \neq$ konst.)—Enrichment line for v = non-constant
Verstärkungslinie (v = konst.)—Enrichment line for v = constant
Molare Verdampfungswärme—Heat of vaporization per mol
Flüssigkeitsmengen und Dampfmengen—Quantities of liquid and vapor

The heat of evaporation of the high and low boiler are $r_1 = 6{,}000$ and $r_2 = 10{,}000$ kcal per mol respectively. Base computations on a production rate of 1 mol per hour. Assume that the condenser operates with a reflux ratio, v_0 of 5. Then, the mass of vapor rising from the top plate is $D_0 = F_0 + 1 = 5 + 1 = 6$ mols per hour. The feed is at its boiling temperature and has a low-boiler concentration, x_M, of 25 mol per cent. The product consists of low boiler only, while the residue consists of high boiler only, i.e., $x_E = 100$, $x_A = 0$.

If the heat of evaporation were constant, the enrichment and stripping lines would be represented by AF and CF in Figure 97. On the basis of the data given, however, the heat of evaporation, as specified by equation (45), corresponds to line r in Figure 97.

For $x_D = 100$ mol per cent, $r_1 = r = r_0 = 6{,}000$ kcal per mol. Thus, for $x_D = 60$ mol per cent, r is 7,600 kcal per mol. Applying equation (124) leads to a vapor concentration of

$$D = 6 \times \frac{6000}{7600} = 4.74 \text{ mols per hour}$$

Analogously, other D values may be determined over the entire range of x_D, as shown in Figure 97. Since the feed is at boiling temperature, the curves obtained in the low-concentration range in the x_F vs. x_D plot may be considered as applying immediately to the stripper.

In the enriching column, $F = D - 1$, which gives the curve for F as a function of x_D, as shown in Figure 97. For $x_D = 60$ mol per cent, $D = 4.74$ mols per hour, and $F = 4.74 - 1 = 3.74$. The point on the enrichment line corresponding to this value of x_D is

$$x_F = \frac{4.74 \times 60 - 6 \times 100 + 5 \times 100}{3.74} = 49.4 \text{ mol per cent low boiler}$$

Connecting the points computed in this fashion yields the enrichment line which intersects the vertical axis at $F^{\times}$ at $x_M =$ 25 mol per cent. Compared to the enriching column, the reflux in the stripper is increased by

$$M = 1 \times \frac{100 - 0}{25 - 0} = 4 \text{ mols per hour} \quad \text{[See equation (98).]}$$

so that the liquid from the top plate of the stripper is

$$F_0' = 3.3 + 4 = 7.3 \text{ mols per hour}$$

F is 3.3 mols per hour for a concentration, x_D, of 42 mol per cent, indicated by intersection $F^{\times}$. The amount of vapor from the stripping column corresponding to the same point is

$$D_0' = 7.3 + 1 - 4 = 4.3 \text{ mols per hour}$$

This result is also obtained from the curve previously computed to show the relation between the rate of vapor flow and the concentration of low boiler in the vapor. The point on the stripping line corresponding to $x_D' = 30$ mol per cent, when $D' = 40.8$ and $F' = 70.5$ mols per hour is obtained from equation (117a)

$$x_F' = \frac{40.8 \times 30 - 4.3 \times 42 + 7.3 \times 25}{70.5} = 17.4 \text{ mol per cent}$$

In this way, any number of points on the stripping line may be computed. Instead of lines AF and CF, $AF^{\times}$ and $CF^{\times}$ are used.

The number of plates is then found from the number of steps in the sequence drawn between these lines and the equilibrium curve. It is seen that the increase in the heat of evaporation in the lower part of the unit increases the number of plates.

Conditions are the reverse if the heat of evaporation of the low boiler is greater. In contrast to the conditions illustrated in Figure 97, the enrichment lines are located below AF and CF. With cold feed, the increase in overflow in the stripper is $M \times e$, where e is determined by means of equation (92). The minimum reflux has been reached if $F^{\times}$ is situated on the equilibrium curve.

Even when the molar heat of evaporation is variable, Equation (99) can be applied for determining the heat requirements.

In deriving equation (124), it has been assumed that the heat content is independent of the temperature. This is usually true. However, if greater accuracy is desired, r_1 and r_2 must be increased by the product of the specific heat and mass of the liquid, multiplied by the temperature difference between liquid and vapor in the section of the column under consideration.*

F. TREATMENT OF RECTIFICATION USING ENTHALPY–CONCENTRATION DIAGRAMS

The methods previously discussed, which make use of the equilibrium curve and operating line (or, with variable heats of evaporation, operating lines), are in most cases satisfactory for determining the reflux ratio, number of plates, and other pertinent operating data.

The heats of mixing have been neglected until now. Since these heats are insignificant in comparison with the heats of evaporation, this assumption is usually valid. For example, when an ethanol-water mixture of 70 per cent by weight alcohol

*Personal communication from H. Hausen.

content is prepared from the individual components at 80°C, the heat liberated (isothermal heat of mixing) is only 3 kcal per kilogram of mixture.

However, variable heats of mixing can be included with the heats of evaporation of the individual materials when the thermal properties of the liquid and vapor are known. In this case, a diagram is used which plots the per cent by weight of low boiler, ξ, of a binary mixture on the abscissa, and the heat content, i, of the mixture along the ordinate.

Since the molar heats of evaporation of the two substances to be separated may be different, there is no point in using mol units. Thus, kilograms and per cent by weight are employed.

If the isothermal heat of mixing is q_t kilocalories per kilogram of mixture, and i_1 and i_2 the heat contents of the two pure components at the temperature of the mixture, ξ, with i taken as the heat content of the mixture

$$i = \frac{\xi}{100} \times i_1 + \left(1 - \frac{\xi}{100}\right) \times i_2 + q_t \qquad (125)$$

Figure 98 shows the i vs. ξ plot for ethanol-water mixtures.[52] Points on the boiling-point curve indicate the heat content of boiling liquids, and points on the dewpoint curve, the heat contents of condensing vapors. The points indicating superheated vapors are located above the dewpoint curve; the region of pure liquid is below the boiling-point curve and the region of saturated vapor is between the boiling-point and dewpoint curves.

Points which are at equilibrium on both curves are connected by lines which, at the same time, represent isothermals. The auxiliary line shown permits the determination of the isothermals, as shown in the figure for the 80°C-isothermal by the dotted line *ABC*.

In Figures 99 and 100, the enthalpy-concentration diagram for ammonia-water and oxygen-nitrogen mixtures has been shown for different pressures.[52,52a] In the second diagram, the enthalpy refers to 1 mol (not to 1 kg), and the concentration has been given in mol per cent (rather than per cent by weight) in the interest of simplicity. The numerical values of the mol percentages are also true for the per cent by volume of the material considered, in vapor or liquid form.

This mode of presentation offers the advantage of yielding exact values of several physical variables. Its disadvantage is that it is less convenient to use, especially where only the

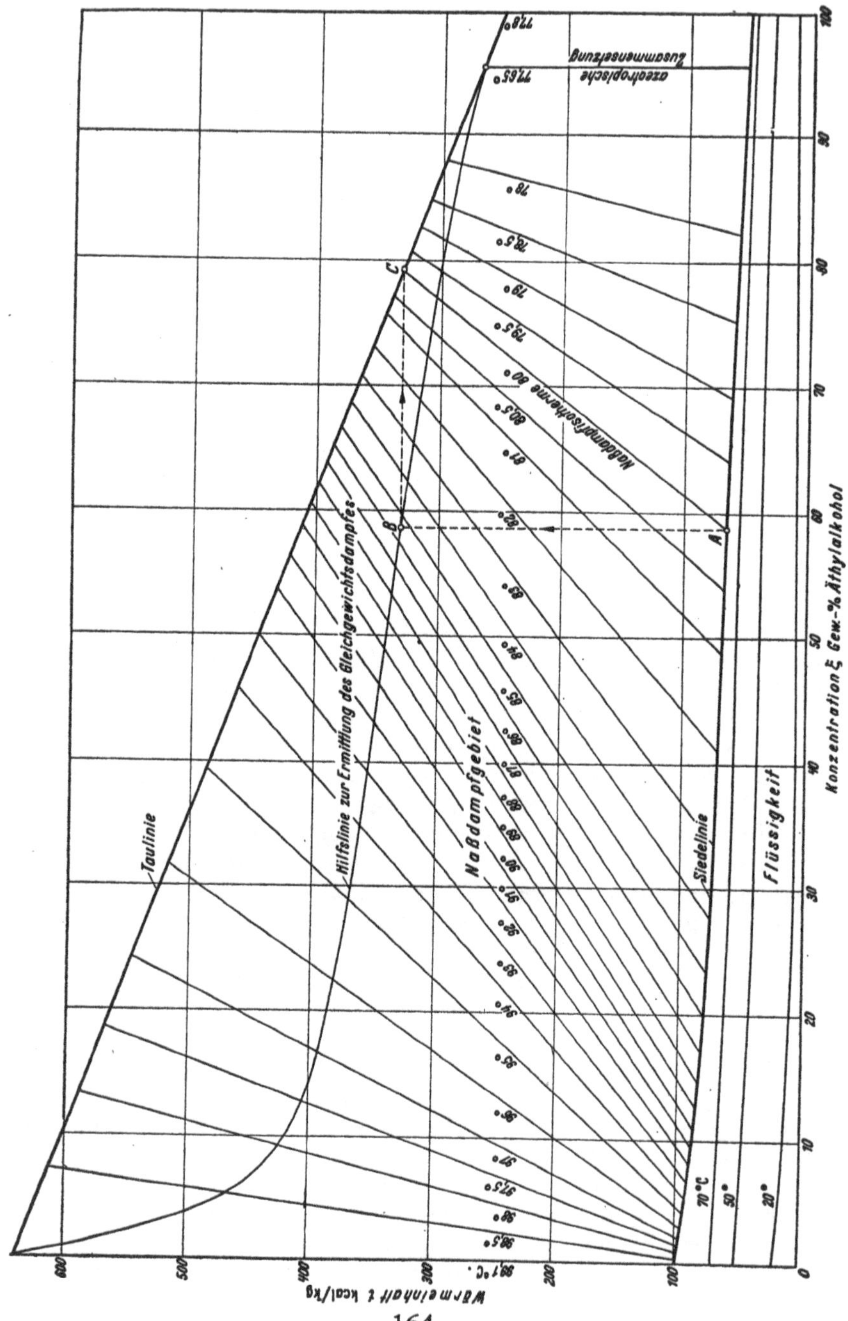
Taulinie
Hilfslinie zur Ermittlung des Gleichgewichtsdampfes
Naßdampfgebiet
Naßdampfisotherme 80°
Siedelinie
Flüssigkeit
azeotropische Zusammensetzung
A
B
C
99,1 °C
98,5°
98°
97,5°
97°
96°
95°
94°
93°
92°
91°
90°
89°
88°
87°
86°
85°
84°
83°
82°
81°
80,5°
79,5°
79°
78,5°
78°
77,65°
77,8°
70 °C
50°
20°
Wärmeinhalt i kcal/kg
100
200
300
400
500
600
Konzentration ξ Gew.-% Äthylalkohol
0
10
20
30
40
50
60
70
80
90
100

FIGURE 98

Graph of i vs. ξ (Bosnjakowic) for a mixture of ethanol and water, at a pressure of 735.5 mm Hg

Wärmeinhalt—Enthalpy
Gew. % Äthylalkohol—Per cent by weight ethanol
Flüssigkeit—Liquid
Siedeline—Boiling-point curve
Azeotropische Zusammensetzung—Azeotropic composition
Nassdampfgebiet—Zone of moist vapors
Nassdampfisotherme—Moist vapor isothermal
Hilfslinie zur Ermittlung des Gleichgewichtsdampfes—Auxiliary curve for determination of the equilibrium vapor
Taulinie—Dewpoint curve
Konzentration—Concentration

---

number of plates required is of interest, rather than the variation of certain thermodynamic properties.

I. The rectifying column

On the basis of heat and material balances certain relations result which are particularly easy to represent in i vs. ξ diagram. The quantities which enter sections a and b at constant conditions (Figure 101), will also leave them

$$F_a + D_b = F_b + D_a$$

or

$$D_a - F_a = D_b - F_b \qquad (126)$$

Equation (126) states that the difference $D - F$ between vapor and liquid quantities is constant in any given section. It is equal to the quantity E of product

$$D - F = E = \text{a constant} \qquad (127)$$

Analogously, for the quantities of low-boiler

$$D \times \xi_D - F \times \xi_F = E \times \xi_E = \text{a constant} \qquad (128)$$

ξ_D and ξ_F represent the concentrations, and i_D and i_F the heat contents, of the vapor and liquid forms, respectively, of the low boiler in any horizontal section between two plates.

Assume that the recycled heat Q_R is abstracted in the reflux condenser, the product in vapor form leaving with ξ_E per cent

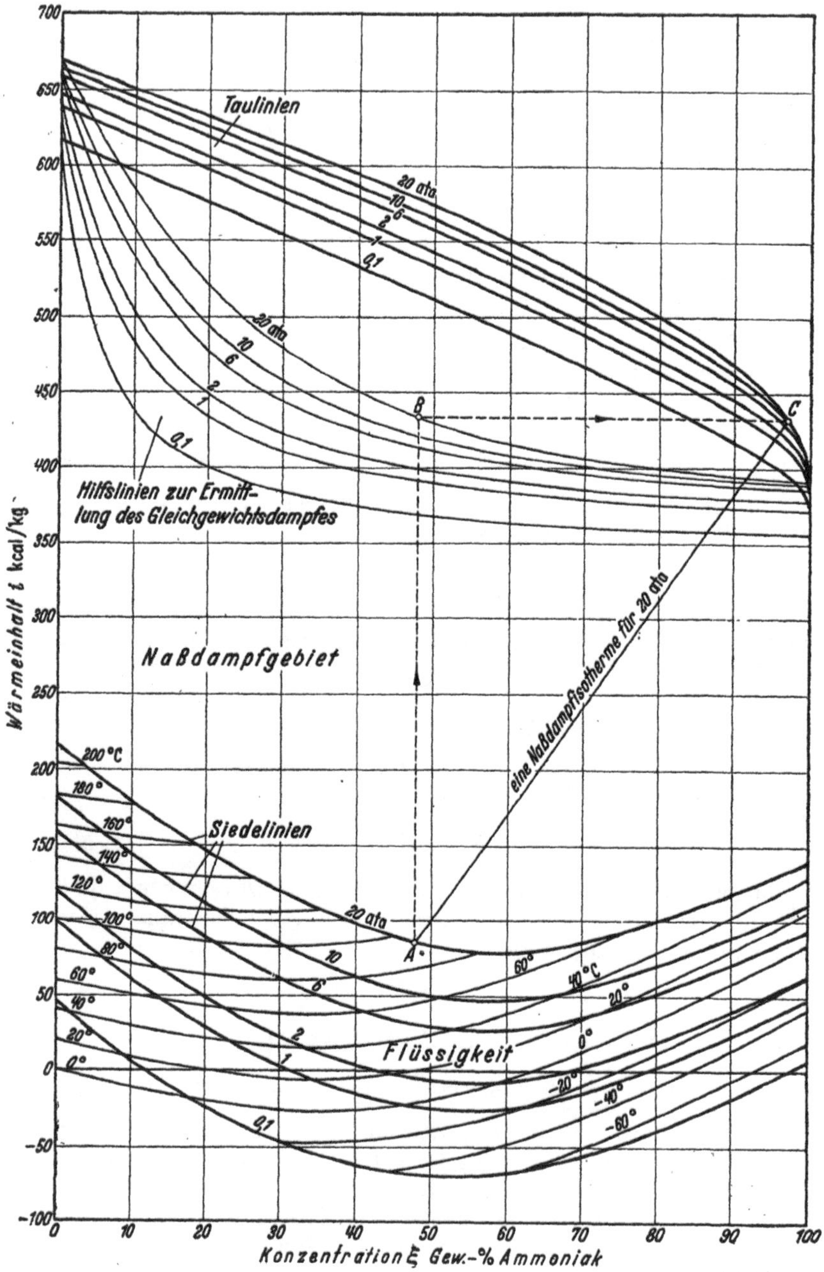
Taulinien
20 ata
10
6
2
1
0,1
20 ata
10
6
2
1
0,1
B
C
Hilfslinien zur Ermittlung des Gleichgewichtsdampfes
Naßdampfgebiet
eine Naßdampfisotherme für 20 ata
Siedelinien
200 °C
180°
160°
140°
120°
100°
80°
60°
40°
20°
0°
20 ata
10
6
2
1
0,1
A
60°
40 °C
20°
0°
-20°
-40°
-60°
Flüssigkeit
Wärmeinhalt i kcal/kg
Konzentration ξ Gew.-% Ammoniak

FIGURE 99

i vs. ξ (Bosnjakowic) for a mixture of ammonia and water at different pressures

Konzentration, Gew. % Ammoniak—Concentration, per cent by weight of ammonia
Wärmeinhalt—Enthalpy
Flüssigkeit—Liquid
Siedelinien—Boiling-point curves
Eine Nassdampfisotherme für 20 ata—One moist-vapor isothermal for 20 kg/cm² absolute
Nassdampfgebiet—Zone of moist vapors
Hilfslinien zur Ermittlung des Gleichgewichtsdampfes—Auxiliary curves for determining the equilibrium vapor
Taulinien—Dewpoint curves

---

by weight of low boiler and a heat content of i_E kilocalories per kilogram. Then

$$D \times i_D - F \times i_F = E \times i_E + Q_R = \text{a constant} \qquad (129)$$

The specific recycle heat follows from

$$\frac{Q_R}{E} = q_R \text{ (in kilocalories per kilogram)} \qquad (130)$$

Combining equations (125) to (129)

$$i_F + \frac{\xi_E - \xi_F}{\xi_D - \xi_F}(i_D - i_F) = i_E + q_R \qquad (131)$$

which can readily be represented by plotting i against ξ. Any variation in the conditions of vapor and liquid in the interchanger column must be in accordance with equation (131). Geometrically speaking, this equation says that any straight line connecting points D and F of vapor and liquid, which correspond to each other in the i vs. ξ plot will pass through a fixed point P_v, the pole, having the ordinates ξ_E and $i = i_E + q_R$ (Figure 102).

Point E, with ordinates i_E and ξ_E, corresponding to the vapor remaining from the reflux condenser is located on the dewpoint line. If, contrary to the situation shown in Figure 101, all of the vapor passing to the condenser were condensed, point E (Figure 102) with the abscissa value ξ_E would not be located on the dewpoint line, but on the boiling-point line, if the liquid leaves the condenser at condensation temperature.

If the liquid has been cooled below this temperature, point E

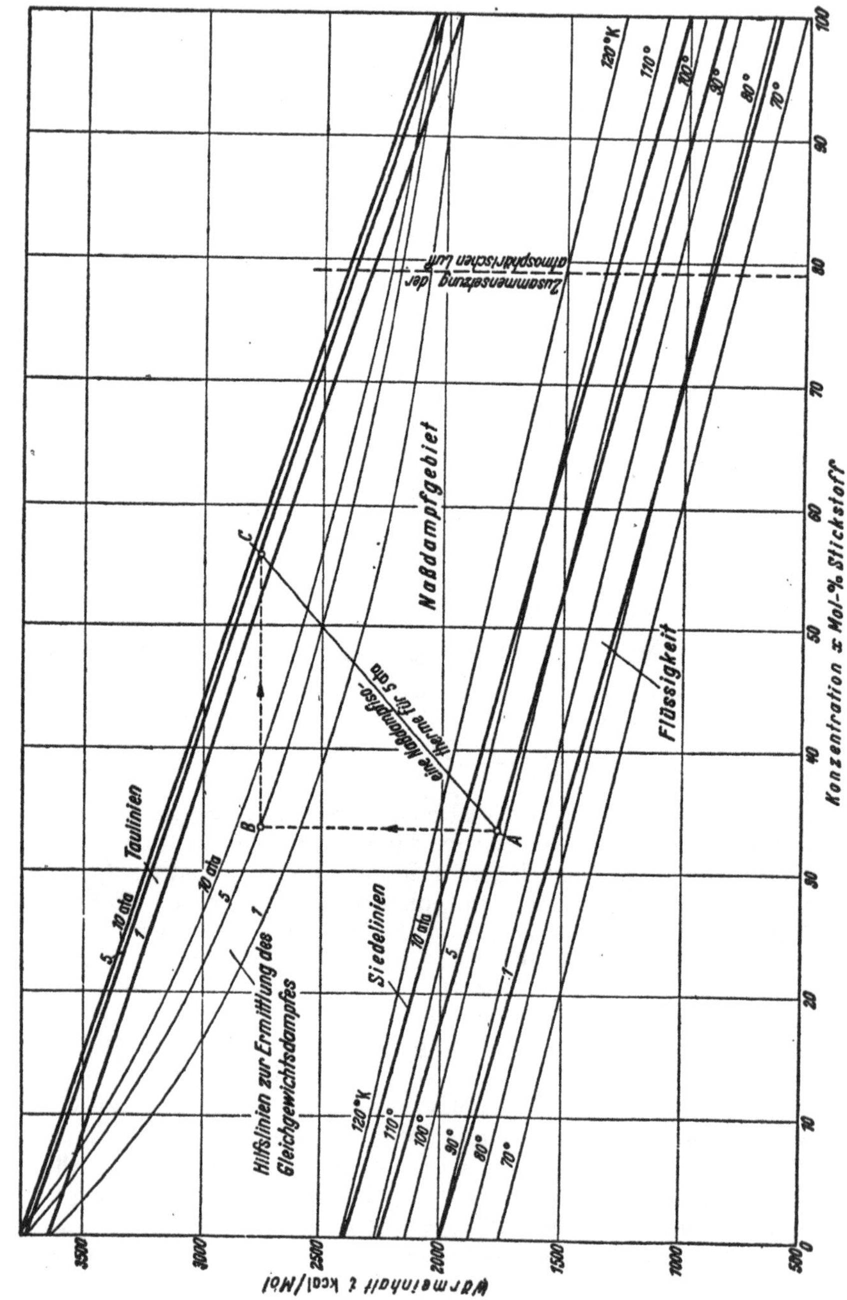
Taulinien
5
10 ata
1
10 ata
5
1
Hilfslinien zur Ermittlung des Gleichgewichtsdampfes
Siedelinien
10 ata
5
1
120°K
110°
100°
90°
80°
70°
A
B
C
eine Naßdampfisotherme für 5 ata
Naßdampfgebiet
Flüssigkeit
Zusammensetzung der atmosphärischen Luft
120°K
110°
100°
90°
80°
70°
Konzentration x Mol-% Stickstoff
Wärmeinhalt i kcal/Mol
0
10
20
30
40
50
60
70
80
90
100
500
1000
1500
2000
2500
3000
3500

FIGURE 100

i vs. x values (Bosnjakowic) for nitrogen-oxygen mixtures

(Heats of mixing are neglected)

Wärmeinhalt—Enthalpy
Mol % Stickstoff—Mol per cent nitrogen
Flüssigkeit—Liquid
Zusammensetzung der atmosphärischen Luft—Composition of atmospheric air
Nassdampfgebiet—Zone of moist vapors
Siedelinien—Boiling-point curves
Eine Nassdampfisotherme für 5 ata—One moist vapor isothermal for 5 kg/cm² absolute
Hilfslinien zur Ermittlung des Gleichgewichtsdampfes—Auxiliary curves for the determination of the equilibrium vapor
Taulinien—Dewpoint curves
Konzentration—Concentration

---

with ξ_E will be located below the boiling line. In the two latter cases, Q_R will stand not only for the recycled heat abstracted to provide reflux, but also for the sum of the latter and the heat of condensation of the product. If the condensate should be undercooled, the corresponding heat must also be added.

By the way, the location of pole P_v could have been indicated directly from the fact that the phase point of a mixture is located on a straight line connecting the points of the two original mixtures, without resorting to equation (131). In Figure 101, a closed zone can be traced, into which vapor with i_D and ξ_D passes the reflux leaving with i_F and ξ_F, and the distillate leaving with ξ_E. Since the heat Q_R is also abstracted from the reflux condenser, the ordinate value, $i_E + q_R$, corresponds to the abscissa value, E. The lines FDP_v are called cross-section lines.

Equations (127), (128) and (129) contain another significant relationship between the quantities and the distances between points F, D and P_v of Figure 102. If the quantity of vapor flowing between two plates, D, is represented by the distance between points F and P_v (Figure 102), distance $\overline{FD}$ indicates the quantity of distillate, E, and distance, $\overline{DV_p}$ indicates the quantity of reflux F passing through the same horizontal section as D.

1. Finding the Number of Plates in a Rectifying Column

The cross-section line gives the relationship between the conditions of the vapor and liquid phases between two plates.

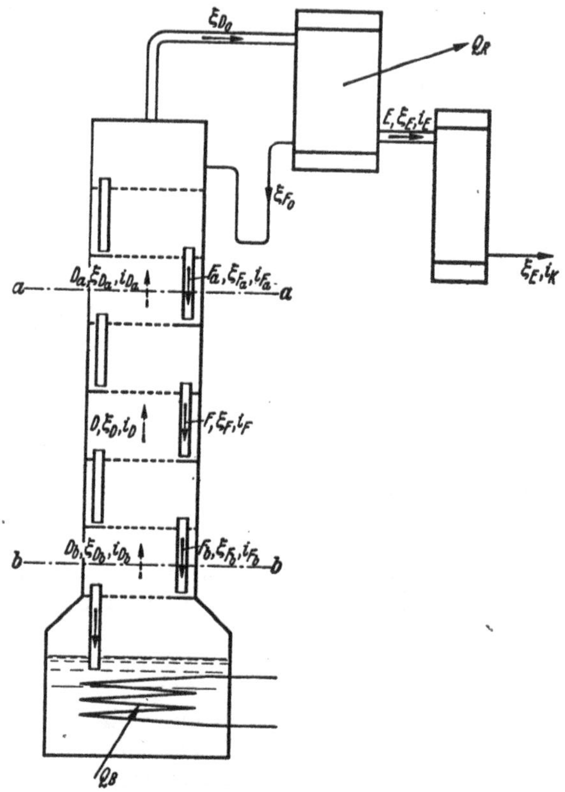

FIGURE 101

Rectifying column

Investigating conditions along the rectifying column and determining the number of plates becomes possible if another relation between the condition of the vapor rising from a plate on the one hand, and that of the liquid of the same plate on the other is given.

It is assumed that mixing of the liquid is satisfactory, that interchange of materials is perfect and that no liquid is entrained in the vapor. Then, the vapor rising from a plate will be in equilibrium not only with the liquid on the plate, but also with the effluent from the same plate. (ξ_{D2} / / / ξ_{F1}; ξ_{D3} / / / ξ_{F2}, etc., see Figure 103.)

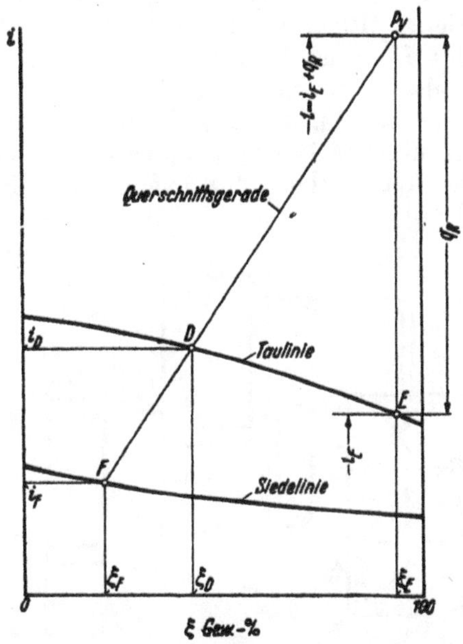

FIGURE 102

The cross-section line of an enriching column

Taulinie—Dewpoint curve
Siedelinie—Bubble-point curve
Querschnittsgerade—Cross-section line
Gew. %—Per cent by weight

The method of finding the conditions along an interchanger column and for determining the theoretical number of plates is explained by Figures 103 and 104. The condition of the liquid with ξ_B per cent by weight of low boiler in the still (See Figure 103.) is represented by point B on the boiling line (Figure 104). Point D_1 (Figures 104 and 103) of the vapor rising from the still must be located on isothermal t_1 passing through B, due to the phase equilibrium and also on the dewpoint line, since it it is supposed to be saturated.

If D_1 is connected to pole P_v by a straight line, this will yield point F_1 on the boiling line, which indicates the condition of the liquid leaving the bottom plate and passing into the still.

Vapor D_2 rising from the bottom plate is in equilibrium with effluent F_1 (Figure 103) of the same plate. Therefore, in Figure 104 point D_2 must be located on isothermal t_2 passing through F_1 and also on the dewpoint line.

Analogously, points F_2, D_3, F_3, D_4 and F_4 are determined from D_2. The difference $\xi_E - \xi_{D_4}$ represents the enrichment by partial condensation in the reflux condenser. (See section B, IV and D, IV.)

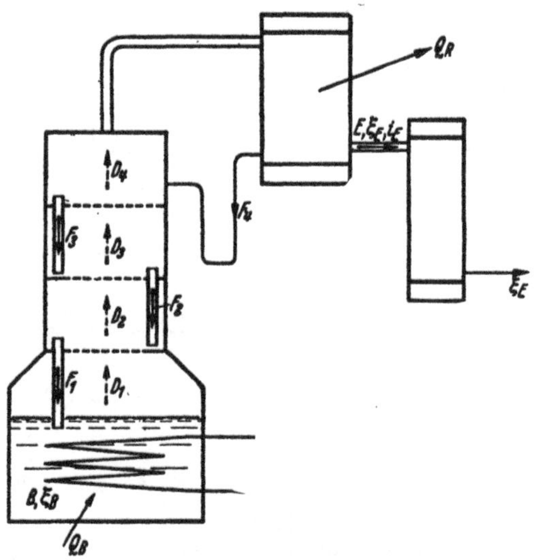

FIGURE 103

Location of the conditions appearing in Figure 104 in the enriching column

If the latter is neglected, or if all of the vapor entering the condenser is liquefied, and the mass of condensate is afterwards split up into a stream of reflux and a stream of distillate, the sequence of lines appearing in Figure 104 must be plotted until the point representing the vapor rising from the top plate coincides with point E in Figure 104. The vapor flowing to the condenser, the reflux to the top plate, and the product then have the same concentration, ξ_E.

The theoretical number of plates n_t follows from the number of corner points F_1, F_2, F_3, etc., in Figure 104. The actually required number of plates n then is

$$n = \frac{n_t}{s_M}$$

where s_M = average interchange ratio of the plates; this is usually of the order of magnitude 0.7, but may vary within limits. (See section D, IV, 2.)

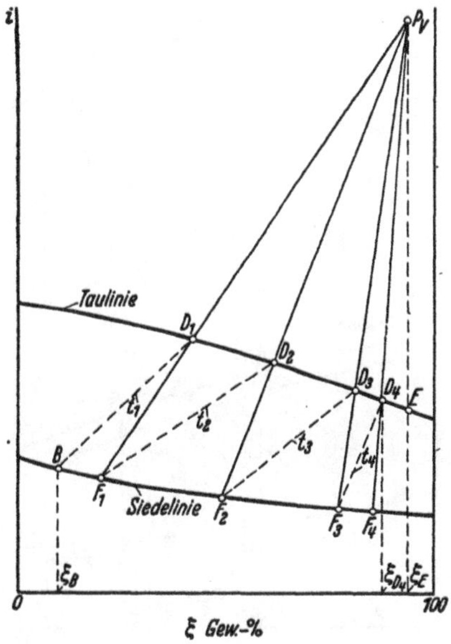

FIGURE 104

Determination of the theoretical number of plates of an enriching column

Gew. %—Per cent by weight
Taulinie—Dewpoint curve
Siedelinie—Boiling-point curve

The advantages of the i vs. ξ plot become evident if knowledge of the actual conditions in a column is desired. It permits a more thorough analysis of the physical properties of the two phases both separately and in their mixed condition. As a matter of fact, there actually is never a perfect interchange of material on a rectifying plate, which means that equilibrium between vapor and liquid on a plate is never reached. Furthermore, some liquid is always entrained by the vapor from one plate and carried to the next above. Even under these conditions, the cross-section line holds absolutely.

However, in this case the mass D, which in equations (127), (128) and (129) referred to pure vapor free of liquid, will refer to a mixture of vapor and entrained liquid, i_D will be enthalpy

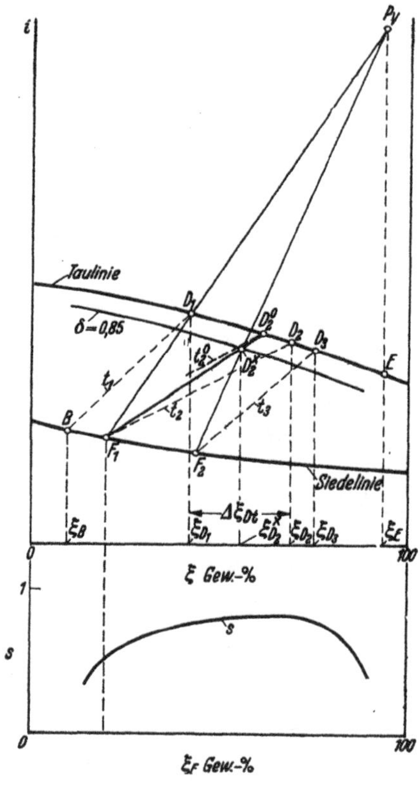

FIGURE 105

Determination of the actual conditions in an interchanger column

Gew. %—Per cent by weight
Taulinie—Dewpoint curve
Siedelinie—Boiling-point curve

per kilogram of vapor-liquid mixture, and ξ_D per cent by weight of low boiler in the vapor-liquid mixture.

In Figure 105, point F_1 corresponds to the condition of the effluent of a given plate, such as the bottom plate (See Figure 103) of a column, and D_1 refers to the vapor flowing to the same plate from the still. The dry saturated vapor rising from the plate under consideration may, under ideal conditions, reach a state indicated by point D_2 located on isothermal t_2 and the dewpoint line. (See Figure 105.) The theoretical enrichment of the vapor is thus given by

$$\Delta\xi_{Dt} = \xi_{D2} - \xi_{D1}$$

Actually, the vapor-liquid mixture rising from the bottom plate contains only $\xi_{D2}^{\times}$ per cent by weight of low boiler, so that the actual enrichment is only

$$\Delta\xi_D = \xi_{D2}^{\times} - \xi_{D1}$$

The interchange ratio s in this case is defined by

$$s = \frac{\Delta\xi_D}{\Delta\xi_{Dt}} \tag{132}$$

In Figure 105, s is plotted against the final plate concentration (effluent concentration). Thus, an s of 0.5 corresponds to ξ_{F1}, so that $\xi_{D2}^{\times}$ is located between ξ_{D1} and ξ_{D2}. As the vapor rising from a plate contains liquid, its condition is that of a wet vapor located between the boiling-point and the dewpoint line. Assume that it contains 85 per cent pure vapor and 15 per cent of liquid; its condition is then described by some point on the line of $\delta = 0.85$ (Figure 105). Its intersection with the vertical passing through $\xi_{D2}^{\times}$ yields point $D_2^{\times}$ of the true condition of the mixture. This consists of a liquid of condition F_1 and of a vapor of condition D_2^0, where D_2^0 is located on the dewpoint line and on the line going through points F_1 and $D_2^{\times}$. The distance between points D_2^0 and D_2 affords a yardstick for the degree of perfection of the heat and material interchange.

The isothermal starting at D_2^0 corresponds to a temperature t_2^0. The remaining temperature drop between the vapor and entrained drops of liquid then equals $t_2^0 - t_2$.

The cross-section line passing through P_v and $D_2^{\times}$ yields point F_2 on the boiling-point line for the liquid on its way to the bottom plate.

In this manner, the actual conditions can be determined from plate to plate, and the actual number of plates is found from the number of the points F_1, F_2, etc. The penultimate plate receives a mixture with concentration $\xi_{D2}^{\times}$, and the theoretical enrichment is given by $\xi_{D3} - \xi_{D2}^{\times}$; see also Figure 105.

2. Determination of the minimum reflux

The minimum reflux and the minimum reflux ratio are characterized by the fact that the number of plates is infinite; increasing the minimum reflux by any small amount will yield a finite number of plates.

The reboil heat, and therefore the reflux, become less as the sum $i_E + q_R$ becomes smaller (Figure 102), that is, as the level of pole P_v in the i vs. ξ plot becomes lower.

By shifting pole P_v in Figure 104 downwards, the theoretical number of plates increases, and the cross-section lines approach

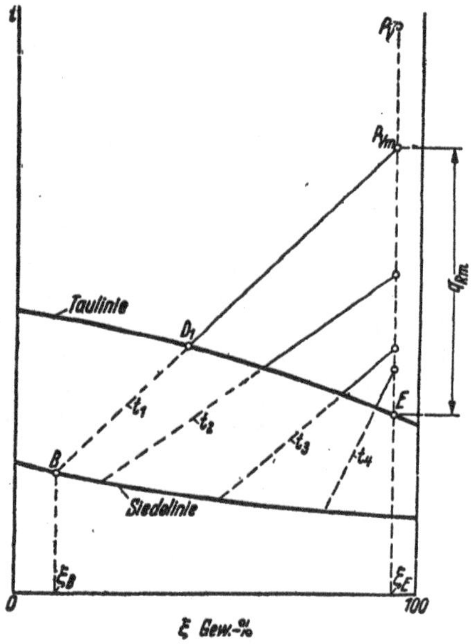

FIGURE 106

Determination of the minimum reboil heat

Taulinie—Dewpoint curve
Siedelinie—Bubble-point curve
Gew. %—Per cent by weight

the isothermals more and more. In the extreme case, they merge. If the still-concentration equals ξ_B in accordance with point B in Figure 106, the cross-section line indicating the minimum reflux heat will merge with isothermal t_1 passing through point B. On the vertical with the abscissa value ξ_E the cross-section line will cut off the ordinate value $i_E + q_{Rm}$ at point P_{vm}, where q_{Rm} = minimum reflux heat per kilogram of distillate (specific minimum reflux heat). With the latter, the number of plates becomes infinite. It becomes finite if q_{Rm} increases to any extent, because all of the isothermals for temperatures below t_1 will intersect the vertical through ξ_E at points located below P_{vm}.

The latter condition is not always complied with. If, for instance, the state of an ethanol-water mixture to be separated is given by point B (Figure 107), isothermal t_1 through B will

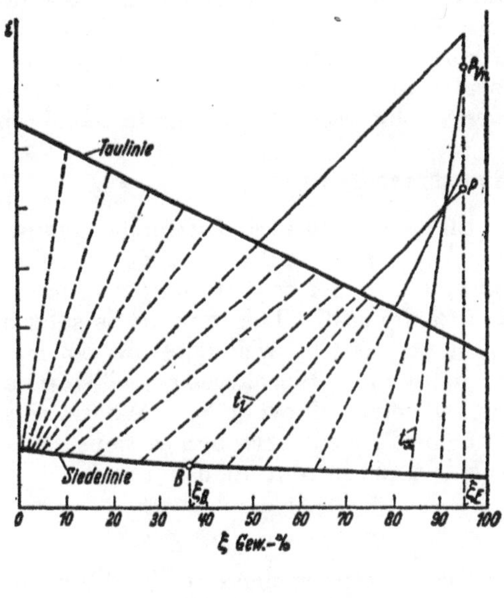

FIGURE 107

Determination of the minimum reboil heat for a binary mixture with a point of inflection in its equilibrium curve

Taulinie—Dewpoint curve
Siedelinie—Bubble-point curve
Gew. %—Per cent by weight

supply intersection P. The latter, however, will not yield q_{Rm} because there are isothermals for temperatures below t_1 which intersect the vertical through ξ_E in a point located above P. Thus, isothermal t_x supplies intersection P_{vm}. Then, that isothermal on the right-hand side of t_1 in the i *vs.* ξ plot must be found which yields the highest located intersection on the vertical through ξ_E. This point establishes the q_{Rm} value, since in this way only any small increase of q_{Rm} will yield a finite number of plates which permits a separation of ξ_B to ξ_E.

The actual reflux must be selected in such a fashion that the sum of annual operating cost and investment is a minimum. As a rule

$$q_R = (1.3 \text{ to } 5 \text{ to } 10) \times q_{Rm}$$

In alcohol-distilling units for rectification, q_R usually equals approximately 1.5 times q_{Rm}.

If all of the vapor in the reflux condensate is liquefied, and if the condensate is passed to the column, which means that no product is withdrawn from the unit, ($E = 0$), q_R becomes infinite. Pole P_v is located at an infinite distance, and the cross-section lines in the i vs. ξ plot become verticals to the abscissa.

II. The continuous rectifying unit

The kettle still is replaced with a stripping column; see Figure 108. Mixture M, with an enthalpy of i_M kilocalories per kilogram and containing ξ_m per cent by weight of low boiler, is fed continuously into the still. This mixture is split into product E, containing ξ_E per cent by weight of low boiler, and effluent A, with an enthalpy of i_A kilocalories per kilogram and containing ξ_A per cent by weight of low boiler. Whatever comes within the region of the balance zone through sections a and b in Figure 108, must leave it again in continuous operation

$$F_a + D_b = F_b + D_a$$

$$D_a - F_a = D_b - F_b \tag{133}$$

Thus, the difference between mass of liquid and mass of vapor is constant in any section of the stripping column, and equal to the amount of effluent

$$F - D = A \tag{134}$$

An analogous statement holds for the mass of low boiler flowing through a cross-section of the stripping column (Figure 108)

$$F \times \xi_F - D \times \xi_D = A \times \xi_A \tag{135}$$

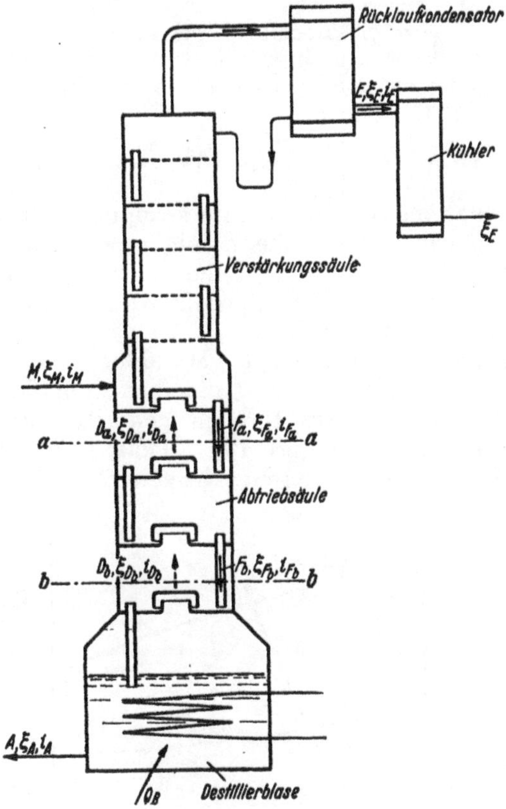

FIGURE 108

Continuous rectifier

Destillierblase—Still kettle
Abtriebsäule—Stripper column
Verstärkungssäule—Enriching column
Rücklaufkondensator—Reflux condenser
Kühler—Cooler

In writing the heat balance of the stripping column, the heat for operating the still, Q_B, must be taken into account

$$F \times i_F - D \times i_D = A \times i_A - Q_B \qquad (136)$$

Introducing the specific operating heat requirement into the equation

$$\frac{Q_B}{A} = q_B \text{ kilocalories per kilogram} \tag{137}$$

which is the heat required per kilogram of effluent, from equations (134) to (137) follows

$$i_D - \frac{\xi_D - \xi_A}{\xi_D - \xi_F} \times (i_D - i_F) = i_A - q_B \tag{138}$$

Even with varying conditions in the stripping column, equation (138) will remain valid. The points representing liquid F, with ordinates ξ_F and i_F and of the vapor D, with ordinates ξ_D and i_D as well as point P_A with ordinates ξ_A and $(i_A - q_B)$ are located on the cross-section line.

Since the conditions of F and D vary along the column, equation (138) represents in the i vs. ξ diagram a bundle of straight lines each of which passes through pole P_A with ordinates ξ_A and $i = i_A - q_B$; see Figure 109. Thus, pole P_A is located q_B below point A which represents the effluent.

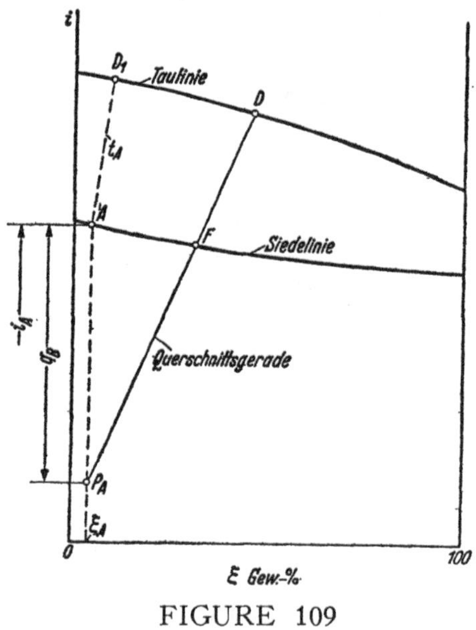

FIGURE 109

The characteristic of the stripper column

Taulinie—Dewpoint curve
Siedelinie—Boiling-point curve
Querschnittsgerade—Characteristic (Cross-section line)
Gew. %—Per cent by weight

Furthermore, equations (134) to (136) supply a relation between masses of liquid and vapor which can be easily illustrated in an i vs. ξ diagram. In Figure 109, distances $\overline{DP_A}$ and $\overline{FP_A}$ or $\overline{FD}$ are in the same ratio as the masses of liquid F and vapor D, or effluent A.

If a state of thorough agitation of the contents of the still prevails, as it does when they are boiling, it may be assumed that the vapor liberated in the still, which is represented by point D_1 in the i vs. ξ plot, is in equilibrium with the contents of the still. Therefore, D_1 must be located on isothermal t_A through A and on the dewpoint line; see Figure 109.

The material balance of the entire unit (Figure 108) supplies

$$M = E + A \tag{139}$$

$$M \times \xi_M = E \times \xi_E + A \times \xi_A \tag{140}$$

from which follows

$$M = E \times \frac{\xi_E - \xi_A}{\xi_M - \xi_A} \tag{141}$$

and

$$E = A \times \frac{\xi_M - \xi_A}{\xi_E - \xi_M} \tag{142}$$

The heat balance of the unit (Figure 108) yields

$$M \times i_M + Q_B = E \times i_E + Q_R + A \times i_A \tag{143}$$

From equations (141), (142), (143) follows, if the relations in equations (130) and (137) are considered:

$$\frac{(i_E + q_R) - i_M}{\xi_E - \xi_M} = \frac{i_M - (i_A - q_B)}{\xi_M - \xi_A} \tag{144}$$

Equation (144) states that, in an i vs. ξ plot, point M of the feed, with ordinates ξ_M and i_M, pole P_A of the stripping column, with ordinates ξ_A and $(i_A - q_B)$ and pole P_v of the enriching column with ordinates ξ_E and $(i_E - q_R)$ will be located on a straight line, the *main line* in Figure 110.

The same relation follows from the conception of the mixing line, by assuming mixture M to be composed of two constituents represented by points P_v and P_A.

In Figure 110 it has been assumed that the temperature of the mixture to be decomposed is below boiling temperature. Pole P_A may also be located below the abscissa, if the value of $i = i_A - q_B$ becomes negative. The orientation of the main line reveals the total heat balance of the rectifier.

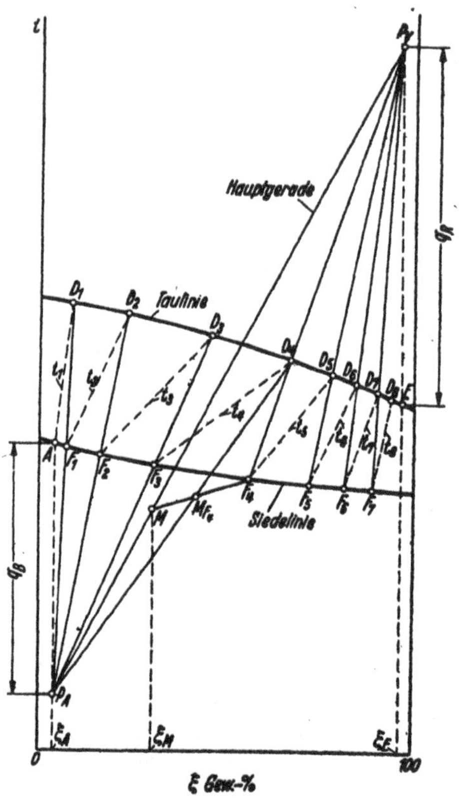

FIGURE 110

Determination of the number of theoretical plates for a continuous rectifier

Taulinie—Dewpoint curve
Siedelinie—Bubble-point curve
Hauptgerade—Characteristic (Main line)
Gew. %—Per cent by weight

If, for instance, point M and pole P_v (and with P_v the heat recycled per kilogram of distillate) are given, the operating heat required per kilogram of effluent can be read off the vertical through ξ_A. This simple and easily plotted relation, given by the i vs. ξ plot, is important from a thermodynamic viewpoint.

1. Finding the number of plates in a stripping column

The theoretical number of plates of a stripping column is found on the basis of the assumption that mixing of liquid, and interchange of heat and materials are perfect. If this is the case, the vapor rising from a plate must be in phase equilibrium with the reflux leaving the same plate.

We thus have the following conditions in the stripping column (Figure 111): The vapor from the still is in equilibrium with

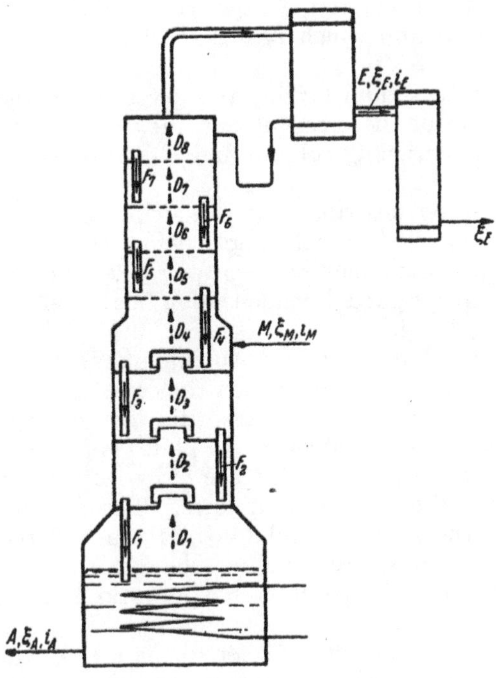

FIGURE 111

Continuous rectifier

Location of points determined in Figure 110

the batch. Therefore, the point D_1 of this vapor is located on the dewpoint line and on isothermal t_1 passing through point A of the effluent condition. Point F_1 of the reflux to the still is the intersection of cross-section line $P_A D_1$ with the boiling-point line.

The vapor reaching the second plate (Figure 111) is in equilibrium with effluent F_1. Its point D_2 is therefore located in the i vs. ξ plot, Figure 110, on isothermal t_2 passing through F_1. By means of the cross-section line $P_A D_2$ one obtains F_2 and, from the latter, point D_3 of the vapor in equilibrium. If D_3 is connected with P_A, point F_3 holds for the reflux leaving the top plate of the stripping column. Isothermal t_4, which yields intersection D_4 on the dew-point line passes through F_3. This indicates the condition of the vapor rising from the top plate of the stripping column which passes to the bottom plate of the enriching column.

Point F_4 of the liquid leaving the lowest enrichment plate is the intersection of the cross-section line, which passes through pole P_v of the enriching column and D_4, with the boiling-point line.

By the method described for the stripping column and explained by Figure 104, F_4 will supply points D_5, F_5, D_6, F_6, D_7, F_7 and D_8. The enrichment between points D_8 and E represents the enrichment by partial condensation in the reflux condenser; see section D, IV, 4.

The theoretical number of stripping plates is given by the number of points F_1, F_2, etc., which the cross-section lines of the stripping column supply in intersecting the boiling-point line. The theoretical number of enriching plates is given by the number of points F_4, F_5, etc., which the cross-sectional lines supply in intersecting the boiling-point lines. The actual number of plates, n, of the stripping column and that of the enriching column are obtained by dividing the theoretical number of plates n_t by the average interchange ratio s_M; see equation (132).

For numerical values of s_M, refer to the case of the enriching column in section D, IV, 2. Actual conditions in the stripping column and actual number of plates may be obtained as explained by Figure 105 for an enriching column.

The effluent from the bottom plate of the enriching column, having point F_4 in Figure 110, is mixed with feed M. Therefore, if, in the i vs. ξ plot in Figure 110 points F_4 and M on the one hand, and P_A and D_4 on the other, are connected by straight lines, intersection M_{F4} occurs, which represents the condition of the mixture of liquids F_4 and M, before they begin their interchange with the vapor.

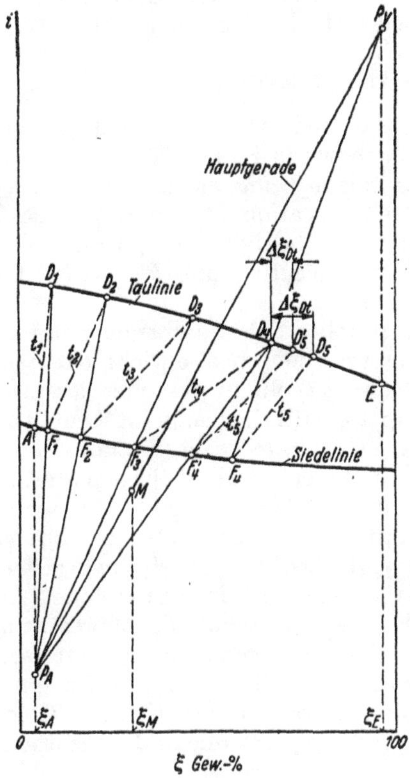

FIGURE 112

Conditions in a column with the feedpoint placed too high

Hauptgerade—Characteristic (Main line)
Taulinie—Dewpoint curve
Siedelinie—Bubble-point curve
Gew. %—Per cent by weight

If the mixture is fed at boiling temperature, its representing point M will, in the i vs. ξ plot, be located on the boiling-point line. Depending on the intensity of its curvature, mixing point M_{F4} will shift towards the wet-vapor zone. This means that vapor is liberated when the liquid feed and reflux are mixed. As a rule, however, the mass of this vapor is insignificant, because the mixing point is very near the boiling-point line. Furthermore, under usual conditions the feed will usually have a temperature slightly below the boiling point.

2. Location of the feedpoint

Assume that the feed is introduced at a point located one plate higher than shown in Figure 111. Thus mixture M would be mixed, other things being as in Figure 110, with reflux F_5 (Figure 111). The location of the reflux leaving the fourth plate (counting from the bottom upwards) is then given by intersection F_4' of connecting line P_AD_4 with the boiling-point line; see Figure 112.

Isothermal t_5' supplies the equilibrium point D_5' on the dewpoint line. Thus the theoretical enrichment $\Delta\xi_{Dt}'$ of the upper plate of the stripping column becomes less than enrichment $\Delta\xi_{Dt}$ of the same plate with the feedpoint located as in Figure 111. If the feedpoint were located several plates higher than in the latter case, the enrichment of all of these plates, located between the new feedpoint and that of Figure 111, would be less than in the case described in Figure 111. In that case, the upper stripping plates might become entirely ineffective.

However, if the feedpoint is located one plate below that considered in Figure 111, point F_3' representing the effluent from the third plate is defined as the intersection of cross-sectional line P_vD_3 with the boiling-point line (Figure 113), and the vapor in equilibrium with this liquid corresponds to a point located on the dewpoint line and on isothermal t_4' through F_3'.

In this case again, the theoretical enrichment $\Delta\xi_{Dt}'$ of the third plate is less than enrichment $\Delta\xi_{Dt}$, which would result with the feedpoint located as in Figure 111. As compared to a feedpoint located farther up or down, the latter will require, for the same separation effect of the unit, a smaller number of plates, or, with the same number of plates in stripping and enriching columns, a better separation of the mixture. The optimum condition exists if the point representing the vapor rising from the uppermost stripping plate is located on the main line P_AP_V.

3. Minimum reflux

The location of the poles for the minimum reflux is characterized by the fact that the number of plates of the enriching or

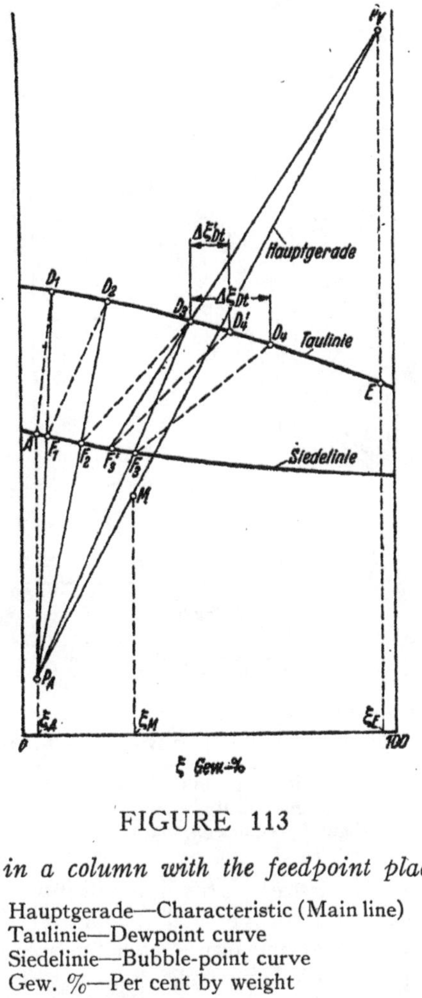

FIGURE 113

Conditions in a column with the feedpoint placed too low

Hauptgerade—Characteristic (Main line)
Taulinie—Dewpoint curve
Siedelinie—Bubble-point curve
Gew. %—Per cent by weight

stripping column is infinitely large. Increasing the minimum reflux by any small amount makes it finite.

If the mixture is introduced cold, as indicated by point M in Figure 114, the highest position of pole P_{Am} which corresponds to the minimum reflux, as well as the corresponding lowest position of pole P_{Vm} are determined by that isothermal t_m the straight extension downwards of which passes through

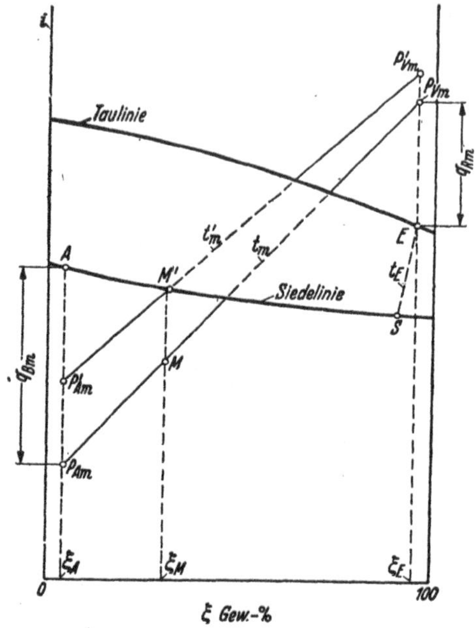

FIGURE 114

Minimum reboil heat and operating heat—Feed entering at ordinary temperature and at boiling temperature

Taulinie—Dewpoint curve
Siedelinie—Bubble-point curve
Gew. %—Per cent by weight

the point M of the mixture. This line will intersect the vertical through ξ_A at P_{Am} and the vertical through ξ_E in P_{Vm}. These two points indicate the minimum specific operating heat q_{Bm} and the minimum specific reflux heat q_{Rm}; see Figure 114.

If the feed is heated to boiling temperature, point M shifts to M'; see Figure 114, t_m' isothermal passes through M'; the straight extension of t_m' yields intersections P_{Am}' and P_{Vm}'. It is seen that preheating the feed to boiling temperature reduces the minimum specific operating heat, but will increase the specific minimum reflux heat.

The minimum reflux heat is usually determined in this way. However, an exception will take place if the mixture involves isothermals for values exceeding t_m or t_m' (see Figure 114) which intersect the vertical through ξ_E at points located above points P_{Vm} or P_{Vm}' in Figure 114.

As explained by Figure 107 for the case of the batch rectifier, that isothermal which yields the maximum numerical value for the minimum reflux heat on the vertical to abscissa ξ_E will be the controlling factor. This also holds for the continuous unit. If in the i vs. ξ plot this intersection is connected by a straight line to the point representing the mixture to be separated, this line will intersect the vertical through abscissa ξ_A at a point indicating the minimum specific operating heat. This may occur in the case of an ethanol-water mixture.

The actual recycled heat exceeds the minimum and is so selected that the sum of operating cost and investment (interest and amortization) becomes a minimum; see section E, III. q_R may equal 1.3 to 10 times q_{Rm}. In the alcohol industry, q_R usually equals 1.5 q_{Rm}.

In an i vs. ξ plot, q_R determines pole P_V of the enriching column. The main line yields pole P_A of the stripping column with specific operating heat q_B; see Figure 110.

In designing a unit, find the masses of distillate E and of effluent A from the mass of mixture M to be separated, by means of equations (141) and (142). The total heat of condensation (recycled heat) to be removed for providing the reflux then results from

$$Q_R = q_R \times E$$

and the operating heat from

$$Q_B = q_B \times A$$

In Figure 115, the specific minimum reflux heats, q_{Rm}, for distillate concentrations of 80 and 95 per cent by weight of alcohol, and the minimum specific operating heats for alcohol-free effluent ($\xi_R = 0$) are plotted against the concentration ξ_M of an ethanol-water mixture fed at boiling temperature.

For $\xi_E = 85$ there is a steady decrease of q_{Rm} with an abscissa intersection at $\xi_m = 78$ per cent. This concentration yields isothermal t_E through the distillation point E (in Figure 114) at intersection S with the boiling-point line. For $\xi_E = 95$, the minimum reflux heat stays constant over a large range of concentrations; when this value changes, it rapidly decreases to zero.

The horizontal course of q_{Rm} for $\xi_E = 95$ in Figure 115 is due

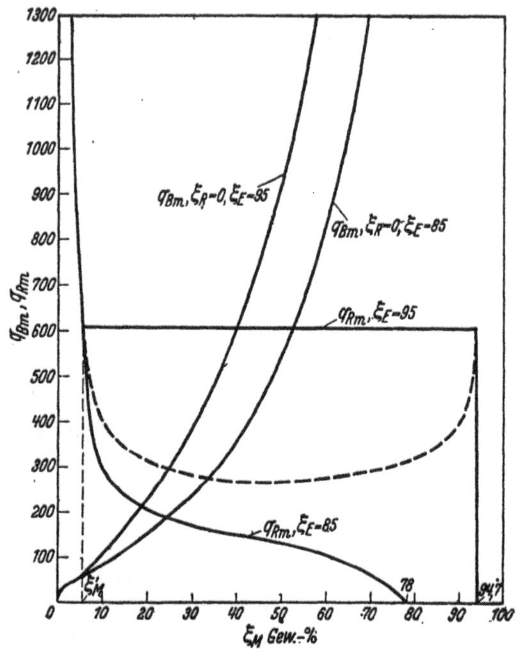

FIGURE 115

Specific minimum reboil heat and operating heat as functions of the concentration of the feed introduced at its boiling point (Ethanol-water)

Gew. %—Per cent by weight

to the fact that the number of plates of the enriching column is supposed to become finite if the minimum reflux heat is increased by any small amount. The location of pole P_{Vm} which indicates the minimum reflux is then found by the method outlined in Figure 107.

However, if the minimum reflux were determined directly, without considering the above conditions from the lowest position of the pole resulting as the intersection of a given isothermal and the vertical to the abscissa at ξ_E (Figure 106), the dotted line in Figure 115 would result.

The usual outline of the minimum operating heat plotted

against ξ_M is as shown in Figure 115 by the q_{Bm} curve for $\xi_R = 0$ and $\xi_E = 85$. As ξ_M increases, q_{Bm} increases and approaches the vertical asymptotically through abscissa value ξ_E. Basically, the same would hold for $\xi_E = 95$. However, at the feed concentration ξ_m', beginning at which the q_{Rm} line for $\xi_E = 95$ runs horizontally, the q_{Bm} line for $\xi_E = 95$ takes a steeper course than that for $\xi_E = 85$. This is due to the circumstance that the situation of the pole indicating the minimum reflux is not given by the lowest intersection of the wet-vapor isothermal with the vertical through ξ_E, but must rather be determined by the procedure explained in Figure 107.

It has to be admitted that in the example mentioned the determination of the minimum reflux from the i vs. ξ plot will not yield extremely precise results, unless plotting is done in a very elaborate manner. As a matter of fact, the isothermals for temperatures near the azeotropic boiling temperature intersect the vertical through ξ_E at a very acute angle. On the other hand, by use of the equilibrium curve isothermals have to be plotted in the i vs. ξ diagram which are very close together and hold for temperatures which may be 0.1°C or less apart. Drawing a tangent to the equilibrium curve takes less time and is more accurate; see sections D, V and E, II, Figures 64 and 79.

Numerical example. For the example from section E, V, point M, applying to the ethanol-water mixture introduced at 70°C, with $\xi_M = 10$ per cent weight of alcohol, has been shown in the i vs. ξ plot in Figure 116. The distillate concentration, ξ_E is 94 per cent of weight of alcohol and the specific reflux heat, q_R is 912 kcal per kilogram of product (reflux ratio, $v = 4$). Determine pole P_V of the enriching column. Point A with $\xi_A = 0$ gives the condition of the effluent. The vertical through A intersects the main line passing through P_V and M at pole P_A of the stripping column with the abscissa value, $i = i_A - q_B$. The vertical distance between P_A and A represents the operating heat required per kilogram of effluent. It is q_B which equals 172 kcal per kilogram.

Since mass of mixture M and mass of distillate E are in the ratio of the differences $\xi_E - \xi_A$ and $\xi_M - \xi_A$, the ratio of distances $\overline{P_VP_A}$ and $\overline{MP_A}$ will also represent the ratio of masses M and E (9.4 : 1). Further, the ratio of $\overline{P_AM}$ and $\overline{MP_V}$ is the same as that of the masses of distillate and effluent (1 : 8.4).

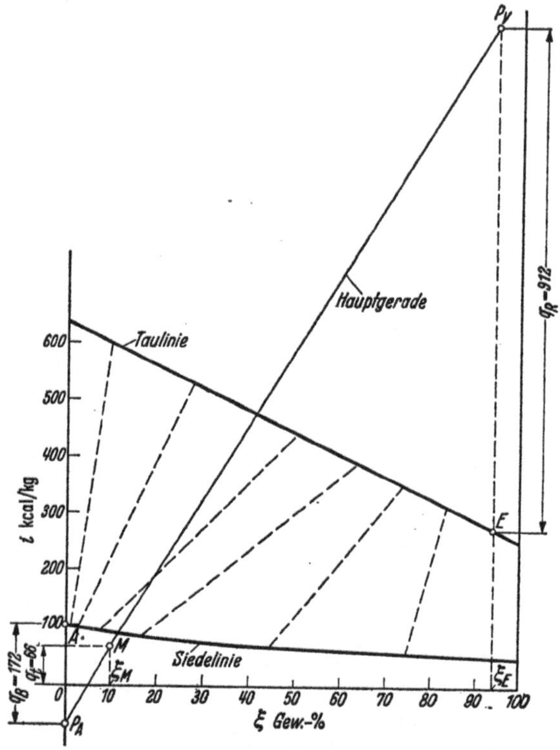

FIGURE 116

Relation between reboil heat, process heat and physical condition of an ethanol-water mixture to be separated

Hauptgerade—Characteristic (Main Line)
Taulinie—Dewpoint curve
Siedelinie—Bubble-point curve

G. SEPARATING MIXTURES CONTAINING MORE THAN TWO COMPONENTS

I. Eliminating small amounts of certain components in a mixture

Sometimes a mixture of liquids to be separated will contain certain components in so small an amount that a special rectifier to recover them is not needed. Then it will suffice to provide some simple measures by which secondary components can be separated. As a matter of fact, they will accumulate at certain points of the separating column.

Thus, Figure 117 shows the layout of a unit for recovering aldehydes and esters from a mixture of ethyl alcohol and water Since these substances boil at a lower temperature than the principal mixture, they will concentrate at the top of the enriching column. A few interchanger plates are provided for their recovery, and the reflux condenser is so hooked up that it will have an enriching action by partial condensation. Should a similar device prove insufficient to supply alcohol of the purity required, the spirits delivered by the rectifying column must be stripped of the last traces of aldehydes and esters in a final column.

As a rule, the rectifiers employed in the alcohol industry are equipped with a device for recovering another secondary product, which is fusel oil. This consists principally of amyl alcohol, and will be most concentrated at that point of the separating column where the mixture contains 16–23 mol per cent of alcohol. According to Figure 77, this concentration is encountered on the lowest plates of the enriching column. Therefore, the lowest plate, or the two or three lowest plates are equipped with cocks and piping to drain liquid; Figure 88. The liquid is cooled first, and mixed with washing water to permit complete separation of the fusel oil in a decanter. The fusel-free liquid is preheated and returned to the top section of the stripping column.

II. Separation of ideal ternary mixtures

1. Basic equations and determination of the number of stages

To investigate the changes of concentration in a rectifier handling ternary mixtures, the same procedures and basic equations will be applied which are used in the case of binary mixtures. However, instead of the equilibrium curves of a pair of substances, distillation lines in a triangular coordinate system are employed, in which, furthermore, lines of equal temperature for liquid and for vapor are indicated.

As an example, Figure 23 shows this mode of representation for the ideal ternary mixture benzol-toluol-*m*-xylol. If it is assumed that the molar heat of evaporation of the ternary mixture is independent of the composition, and that the heats of solution can be neglected, the concentration of each of the

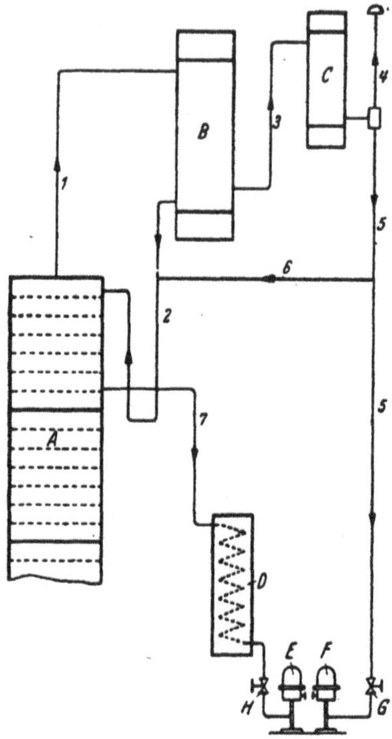

FIGURE 117

Layout for obtaining aldehyde-free spirits and recovering concentrated aldehydes

The vapors leaving rectifier *A* pass through line *1* into condenser *B*, where the bulk of the vapors is liquefied. The residual vapors rich in low boiler (aldehydes, esters) enter, together with non-condensables (mostly carbon dioxide and air) through line *3* into vent condenser *C*. Non-condensables are eliminated through line *4*. The remaining vapor, enriched with the above-mentioned materials, is liquefied in vent condenser *C*. This condensate passes through line *5* and control valve *G* to tester *F*. Any material not withdrawn in this fashion is recycled through *6*, together with the reflux from condenser *B*, and returns through line *2* to rectifier *A*. Highly purified alcohol is extracted as a boiling liquid from the column through line *7*. This extraction point is located several plates (say *4* or *5*) below the inlet connection for the reflux from condenser *B*. After cooling in *D*, this alcohol is withdrawn through control valve *H* and tailbox *E*. The supply of by-product may be adjusted by means of valve *G* in accordance with readings at tester *F* in such a way that the alcohol obtained at tailbox *E* has the proper purity as to presence of aldehydes and esters. The cooling water passes in series flow through product cooler *D*, vent condenser *C* and hot condenser *B*.

three materials in the enriching column can be represented by an enrichment line by using equations (56) and (60).

$$
\begin{aligned}
x_D &= \frac{v}{v+1} \times x_F + \frac{x_E}{v+1} \\
y_D &= \frac{v}{v+1} \times y_F + \frac{y_E}{v+1} \\
z_D &= \frac{v}{v+1} \times z_F + \frac{z_E}{v+1}
\end{aligned}
\tag{145}
$$

where v = molar reflux ratio, which is the ratio of the amount of liquid F returning from the reflux condenser to the column and the quantity of distillate E.

Let

x_F = concentration, in mol per cent, of low boiler in the liquid, in a horizontal section between two plates

y_F = same for the medium boiler

z_F = same for the high boiler

x_D, y_D, z_D = corresponding concentrations in the vapor in the same column section

x_E, y_E, z_E = corresponding concentrations, in mol per cent, in the product

On the basis of the assumptions made, the amounts of liquid and vapor passing through the enriching column, F mols per hour and D mols per hour, respectively, are constant. Thus, in a plot where the abscissa represents the concentration of a substance in the liquid phase, and the ordinate the concentration of the same material in the vapor phase, each of equations (145) represents a straight line which intersects the diagonal at a point, the abscissa value of which represents the concentration of the distillate. All of the lines have the same slope α against the abscissa, for which

$$\tan \alpha = \frac{v}{v+1} = \frac{F}{D} \tag{146}$$

The ordinate sections of the three enrichment lines are given by the last member of the right-hand sides of equations (145). Consequently, any point on the enrichment line will indicate the interrelation of the liquid and vapor concentrations in a horizontal section between two plates.

For the stripping column, which is located below the feedpoint

(Figure 119) of the ternary mixture, the interrelations are analogous to those demonstrated for the case of binary mixtures. Using the ′ sign for the values applying to the stripping column, the stripping line for the different substances, on the basis of relation 88, will be

$$
\begin{aligned}
x_D' &= \frac{F'}{F' - A} \times x_F' - \frac{A}{F' - A} \times x_A \\
y_D' &= \frac{F'}{F' - A} \times y_F' - \frac{A}{F' - A} \times y_A \\
z_D' &= \frac{F'}{F' - A} \times z_F' - \frac{A}{F' - A} \times z_A
\end{aligned}
\tag{147}
$$

where x_A, y_A and z_A represent the mol percent in concentrations of the three components of the effluent and A is the amount of effluent in mols per hour. The slope against the abscissa is determined by the masses of the liquid and vapor flowing through a horizontal section of the column. According to equations (147) this slope corresponds to each of the three lines.

$$\tan \alpha' = \frac{F'}{D'} \tag{146a}$$

The concentration of any substance in the liquid and vapor phases between two plates is indicated by a point on the stripping line of the same material. These lines will intersect the diagonal of the plot showing the liquid concentrations on the abscissa and the vapor concentrations on the ordinate at points, the abscissa values of which indicate concentrations x_A, y_A and z_A in mol per cent.

The intersection of the enrichment and stripping lines of each material is also determined by the equations holding for binary mixtures, (92), (96) and (97); see also Figure 73. If the feed has concentrations x_M, y_M and z_M and is at boiling temperature, the enrichment and stripping lines of the low boiler will intersect at x_M, those of the medium boiler at y_M and those of the high boiler at z_M. Furthermore

$$
\begin{aligned}
x_F + y_F + z_F &= 100 \\
x_D + y_D + z_D &= 100 \\
x_D' + y_D' + z_D' &= 100 \\
x_F' + y_F' + z_F' &= 100
\end{aligned}
\tag{148}
$$

If a mixture described by point P in the triangular plot is decomposed into two mixtures having points P_1 and P_2, these three points P, P_1 and P_2 will be located on a straight line. Whether one or several points refer to the liquid or to the vapor state of the mixture makes no difference. By means of this interrelation of the enrichment and the stripping line, and of the distillation lines of a ternary mixture, the concentration changes in a rectifying column can be determined and the number of plates found.

This procedure is demonstrated for the ideal ternary mixture benzol-toluol-*m*-xylol in Figure 118, for the case of a continuous rectifier. Good mixing of the liquid and perfect interchange between the phases on a plate have been assumed so that the vapor rising from a plate is in phase equilibrium with the liquid leaving the same plate.

The feed is of a composition corresponding to point M in Figure 118, and is assumed to enter at boiling temperature. The intersection of the enrichment and stripping lines for the low boiler is then given by the benzol concentration, x_M mol per cent of the mixture.

The two lines mentioned have been shown in Figure 118. In this x_F vs. x_D plot the ordinate is located in the extension of the toluol-*m*-xylol side of the triangle. The abscissa value, $x_F = 100$ mol per cent, equals the height of the triangle. It becomes thus possible to transfer the benzol concentrations from the triangular into the x_F–x_D plot by drawing lines parallel to the toluol-*m*-xylol side of the triangle.

If such parallels pass through points E of the distillate and A of the effluent, on the diagonal of the x_F vs. x_D plot, intersections E and C appear; the enrichment and the stripping lines pass through these points. Their common intersection follows from the horizontal through M. The enrichment line intersects point B on the ordinate; with $x_E = 95$ mol per cent of benzol and $v = 3.5$, the ordinate value of B yields an o of 21.1 mol per cent.

The point in the triangular plot representing the effluent is located on the liquid isothermal for 125°C. Point D_1 on the vapor isothermal for 125°C corresponds to the vapor liberated from this liquid; D_1 is also on the distillation line passing through point A (see Figure 23).

To make things plain, Figure 119 indicates the points corresponding to those indicated in Figure 118. To find point F_1 of the liquid leaving the bottom plate of the stripping column, vapor concentration x_D of point D_1 must be transposed as an rodinate into the x_F vs. x_D plot. Line D_1 1 I is drawn, since point

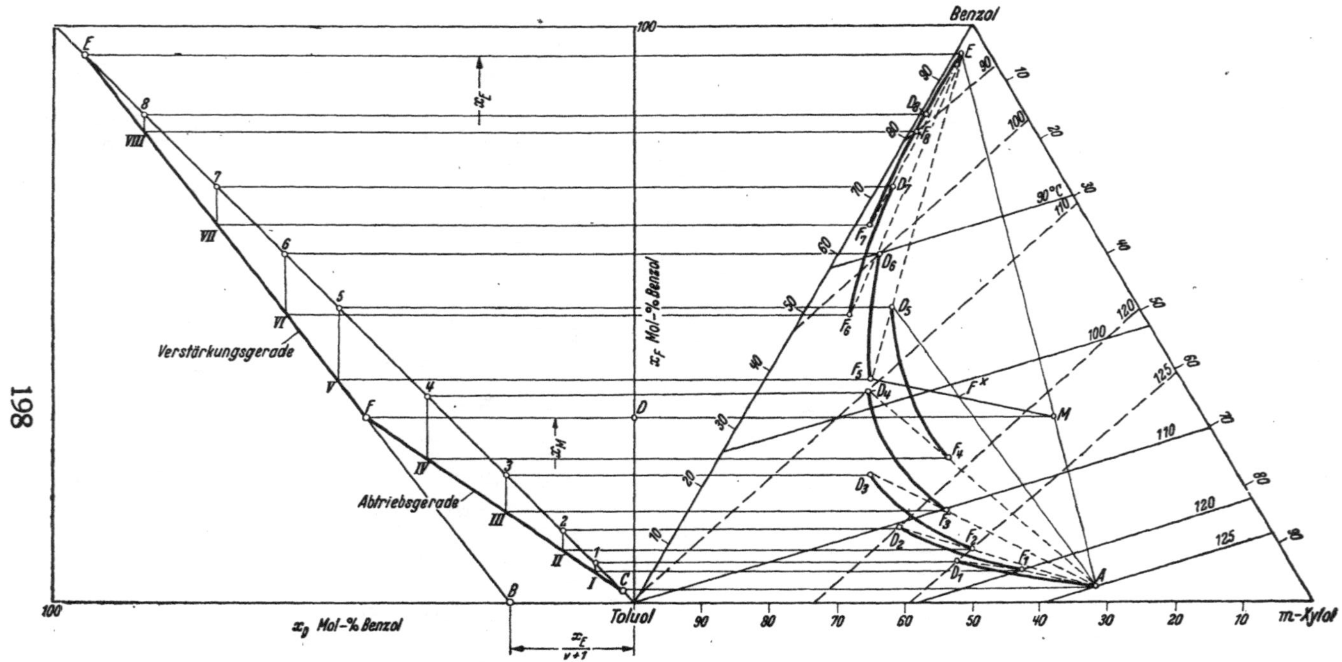

FIGURE 118

Separation of an ideal ternary mixture. Determination of the theoretical number of plates for a continuous rectifier

Versärkungsgerade—Enrichment line
Abtriebsgerade—Stripping line

I indicates the interrelation between the liquid and vapor contents in the horizontal section below the low rectifying plate. Then point F_1 is located on the horizontal through point I as well as on the connecting line D_1A. The vapor rising from the bottom plate has the composition of point D_2, located on a vapor isothermal for the temperature applying to the liquid isothermal through F_1. Furthermore, point D_2 must be located on the distilling line through F_1.

In the same way by which F_1 was determined from D_1, point F_2 for the liquid flowing from the penultimate plate is found. In this connection, line D_2 2 II is drawn, and the parallel to the toluol-xylol side of the triangle passing through II is made to intersect line D_2A. The vapor represented by point D_3 is in phase equilibrium with the liquid represented by point F_2. Point F_3 is found from point D_3 by means of line D_3 3 III on line D_3A. F_3 will yield vapor equilibrium point D_4 and the latter supplies point F_4 of the mixture leaving the fourth plate from the bottom.

The vapor of composition D_5 will be in equilibrium with this liquid. D_5 is the intersection of the distillation line through F_4 and through the vapor isothermal of the temperature determining the liquid isothermal through F_4.

Since the vapor corresponding to point D_5 enters the bottom plate of the enriching column, its benzol concentration x_D can be plotted in the x_F vs. x_D diagram as an ordinate by means of lines D_5 and 5V, to find the composition of the reflux leaving the bottom plate of the enriching column. Point V located on the enrichment line indicates the relation between the liquid and vapor concentrations below the bottom plate of the enriching column.

The parallel to the x_D ordinate through V yields point F_5 on line ED_5; vapor equilibrium point D_6 corresponds to F_5. These two points are located on the same distillation line and on isothermals for the same temperature. Point F_6, representing the liquid, can be located on line ED_6 by means of line D_6 6 VI; vapor equilibrium point D_7 corresponds to F_6.

Points F_7 and D_8 are found in the same fashion. The latter indicates the composition of the vapor flowing to the top plate of

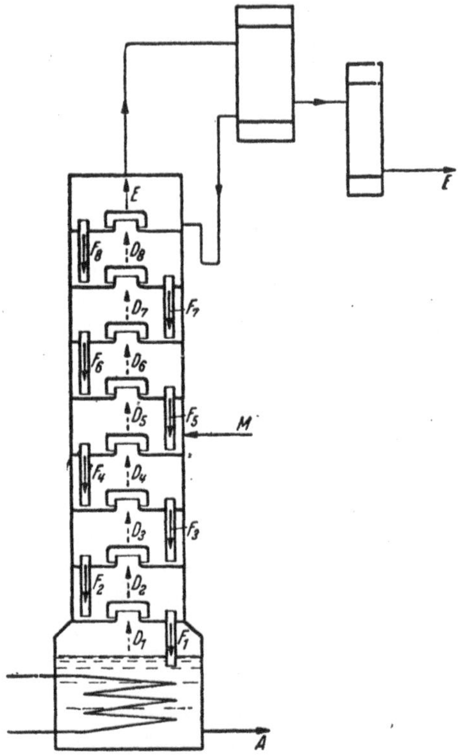

FIGURE 119

Location of the points determined from Figure 118 in the continuous rectifier

the enriching column which yields a liquid corresponding to point F_8; point F_8 is obtained from D_8 by means of line D_8 8 *VIII* and line ED_8. Vapor at condition E is in equilibrium with this liquid.

Since it has been assumed that no enrichment takes place in the reflux condenser, E will indicate the composition of the distillate. Thus, the theoretical number of plates equals that of the corner points *I*, *II*, *III*, etc., located on the enrichment and stripping lines. Dividing this figure by the mean enrichment ratio (see section D, IV 2) will give the actual number of plates.

The intersection $F^{\times}$ of lines MF_5 and AD_5 is shown in Figure 118. It indicates the condition of the liquid which is caused by mixing the reflux from the bottom plate of the enriching column, F_5, with the ternary mixture to be split up, M, before interchange takes place.

If all of the vapor entering the reflux condenser is liquefied and returns to the column, v becomes infinite. Then, the enrichment and stripping lines merge with the diagonal of the x_F vs. x_D plot, and points F_1, F_2, F_3, F_4, F_5 merge with points D_1, D_2, D_3, D_4, D_5 etc. The changes of the conditions in the column are dependent upon the course of the distillation line passing through A.

By proper selection of the reflux ratio and plate numbers of the two columns, different compositions of the distillate and effluent can be obtained. One limit in a case appears if point E coincides with the benzol corner, and point A is located on the side of the triangle connecting the toluol and the xylol corners. At the other extreme, point A is located in the xylol corner, and E falls on the benzol-toluol side of the triangle. While, in the first case, the effluent must be treated in another column for complete separation into the primary components, in the second case such a treatment would have to be applied to the distillate; see Figures 120a and 120b.

By means of the procedure explained in Figure 118, for the case of the continuous outfit, it is also possible to investigate conditions in the column of a batch-type unit, and to determine the number of plates in it. If, for instance, the composition in the still pot (Figure 43) is represented by point F_4 in Figure 118, the vapor liberated has a composition which establishes point D_5. Liquid of a composition corresponding to point F_5 flows to the still pot. Vapor liberated on the bottom plate, corresponding to point D_6, is in equilibrium with this liquid.

2. Finding the minimum reflux ratio

In the case of a ternary mixture, too, the two criteria for the minimum reflux ratio are that it yields an infinite number of plates, and that the number of plates becomes finite if the minimum reflux ratio is increased by any small amount.

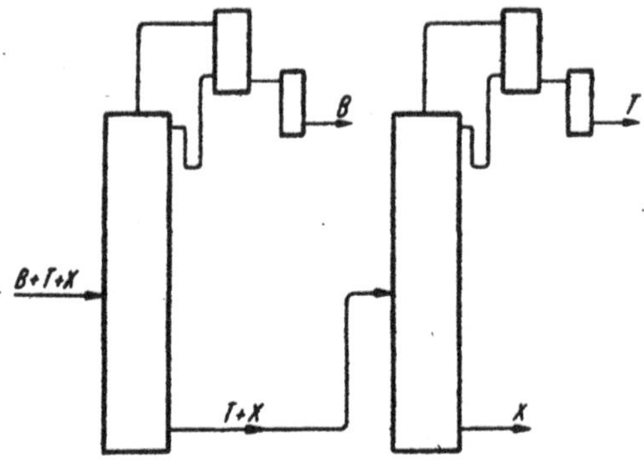

FIGURE 120a

Continuous rectifier for separating an ideal ternary mixture

The lowest-boiling fraction is recovered at the top of the ternary column.

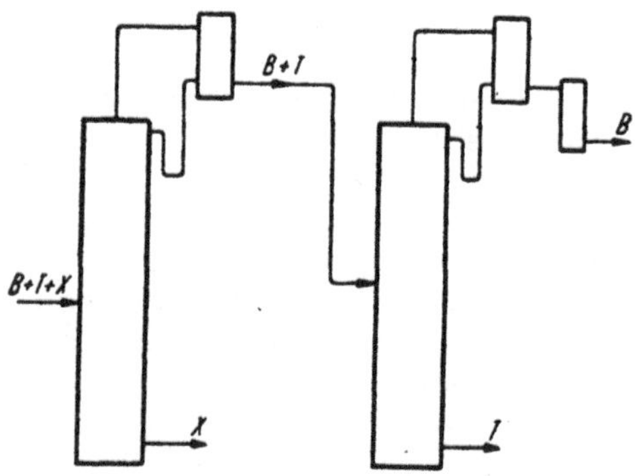

FIGURE 120b

Continuous rectifier for separating an ideal ternary mixture

The highest boiling fraction is recovered at the bottom of the ternary column.

Figure 121 for the ternary mixture benzol-toluol-xylol explains how the minimum reflux ratio is found for a batch-type unit consisting of still pot and enriching column.

Points E and B_F determine the compositions of product and liquid in the still pot. The still-pot vapor, represented by point B_D is in equilibrium with the liquid in the still pot.

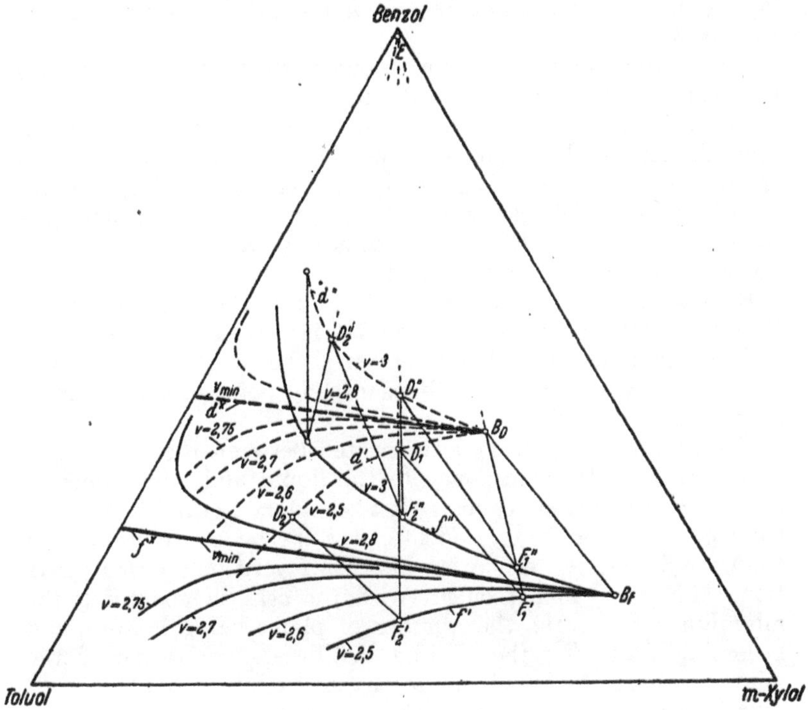

FIGURE 121

Separation of a ternary mixture

Determination of the minimum reflux ratio of a rectifier

The concentration of the reflux from the bottom rectifying plate into the still depends on the reflux ratio, as is seen from Figure 118. Point F_1' results from a reflux ratio v or 2.5, and point F_1'' from $v = 3$. Vapors corresponding to D_1' and D_1'' will be in equilibrium with the liquids corresponding to F_1' and F_1'', respectively. The bottom plate receives a liquid, the composition of which is given by point F_2' in the case where v is 2.5, and by F_2'' in the case where v is 3.

The vapor rising from the penultimate plate is in equilibrium with this liquid and is represented by points D_2' for $v = 2.5$ and D_2'' for $v = 3$. By connecting the points representing the liquid, one obtains liquid line f' for $v = 2.5$ and liquid line f'' for $v = 3$.

The points for the vapor conditions, with the reflux conditions in question, lie in vapor lines d' and d''.

Lines f' and d' extend, starting at B_F and B_D, in the direction towards the toluol-xylol side of the triangle, rather than towards the distillate point E. It follows that, with a v of 2.5, even an infinite number of plates could not provide a product corresponding to point E. This becomes, however, possible with a v of 3, as shown by lines f'' and d''.

Between f' and f'', and between d' and d'' there are still other lines for v-values between 2.5 and 3. It is seen that a v of 2.75 will not give the desired result even if the number of plates is infinite, while at $v = 2.8$ the lines already point in the direction of distillate point E.

The minimum reflux ratio is located between the last-named reflux values. It corresponds to the liquid and vapor lines $f^{\times}$ and $d^{\times}$ which may be called reflux boundary lines. They start from points B_F and B_D, and intersect the benzol-toluol side of the triangle without deviating in a downward or upward direction. They are in agreement with the criteria governing the minimum reflux ratio, the number of plates being infinite, and becoming finite for the desired distillate composition if the minimum reflux ratio is increased by any small amount.

It is seen that in this case the minimum reflux ratio cannot be found by any means other than a stepwise trial and error method.

III. Number of rectifying columns required to separate multicomponent mixtures and their hookup

It has been shown that two columns are required to decompose an ideal ternary mixture into its pure constituents. A complete separation of a binary mixture takes one rectifying column. As a general thing, the rectification of an ideal mixture of w components requires z columns where

$$z = w - 1 \tag{149}$$

to obtain the pure constituents. This relation also holds for two-stage rectification sets, if the two stages are considered as one unit. For instance, if it is desired to recover argon when decomposing liquid air, an additional rectifying column becomes necessary.[53] According to a procedure used by Linde, the second column received a mixture of argon and oxygen recovered from the lower part of the low pressure column of a two-stage unit (Figure 85) and which is separated into oxygen and argon.

Equation (149) will not hold in the case of a rectification of mixtures with an azeotropic point, as in the case of the manufacture of absolute ethyl alcohol from aqueous solutions by azeotropic processes. The principle of the latter has been explained in section B, III 6 d, and in the literature.[54,55]

Contrary to equation (149), a complete separation of that binary mixture employs several columns, as shown in Figure 122 for the Drawinol process. The ternary mixture resulting upon addition of an entrainer (i.e., stabilized trichlorethylene) to the alcohol-water mixture is at first rectified in the ternary column, from the still of which absolute alcohol is recovered. This still must be designed in such a way that leakages which would cause contamination of the product by the steam used for heating are prevented. The top product of this column forms two layers. The water is continuously extracted from the alcohol-water layer in a rectifying unit.

Some multicomponent mixtures exhibit an azeotropic point or behave as non-ideal mixtures. The decomposition of such mixtures is among the most difficult problems of distillation.

In many cases, the numbers of plates in the individual columns cannot be determined in advance. Designing similar plants requires pilot-plant experiments or data on existing similar installations. As an example, Figure 123 shows an installation for separating a mixture of acetone, ethyl alcohol, butyl alcohol and water.

IV. Rectifying ideal mixtures of more than three components

1. Separating key components

Computing the capacity of a unit for separating multicomponent mixtures is comparatively simple if such mixtures have the characteristics[56] of ideal solutions, i.e., if, with constant boiling temperature, the partial vapor pressure of each substance is proportional to its concentration in mol per cent in the liquid.

Petroleum and similar oils composed of hydrocarbons are mixtures of this kind. The computation becomes still simpler by introducing the concept of *key components*.[58] These are the

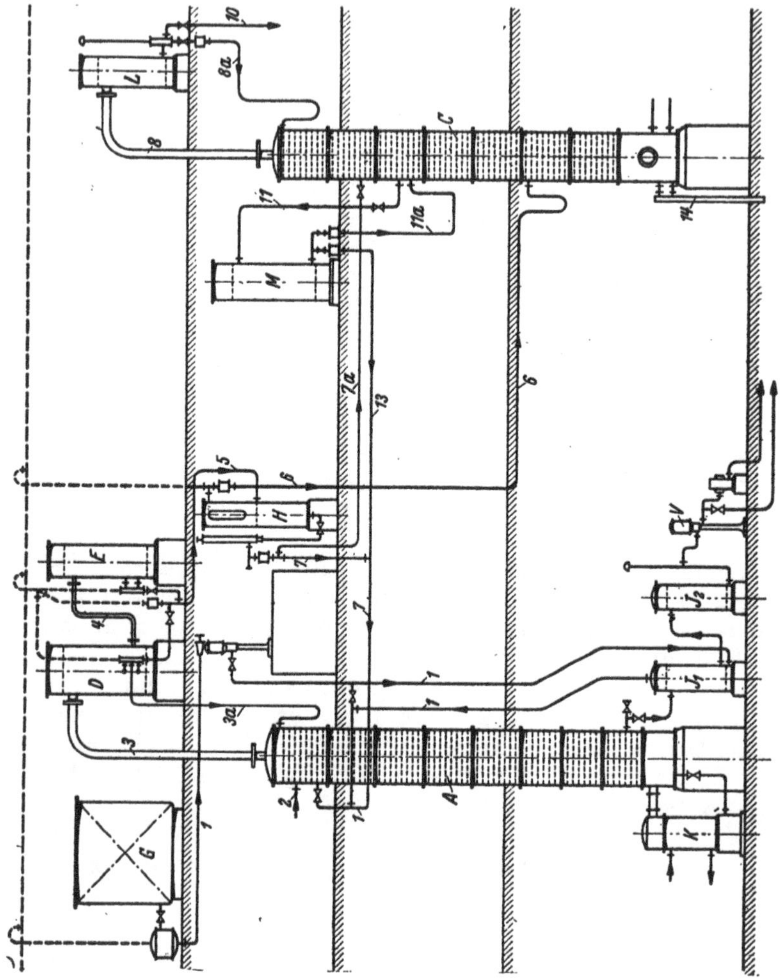

FIGURE 122

Alcohol dehydration by the Drawinol process

The crude spirits flow from tank *G* via line *1* and heat exchanger J_1, in which the absolute alcohol produced serves as a heating medium, to rectifying column *A*, which receives Drawinol (stabilized trichlorethylene) through connection *2*. The ternary vapor passes through line *3* to heat exchanger *D*. Remaining gases and vapors go through line *4* to heat exchanger *E*, the liquid from which flows to settling tank *H*.

Column *A* receives its reflux partly through line *3a* from condenser *D*, partly through line *7* from the lower layer in separator *H*, which contains little water. The remainder of this layer passes through line *7a* to the top section of column *C*.

The top layer from decanter *H* is rich in water, this passes through line *6* to the lower portion of column *C*. In it, the ethyl alcohol is enriched to at least 94 per cent; methyl alcohol free from ethyl alcohol and Drawinol are obtained as a top product. This ternary mixture flows through line *8* into heat exchanger *L*, from which line *8a* returns the reflux to the column, and line *10* drains the methyl alcohol and trichlorethylene.

The latter can be washed out upon addition of water; the trichlorethylene is passed to column *A*, and the methyl alcohol can be dehydrated completely in a third rectifier column.

The ethyl alcohol vapors flow, with small amounts of trichlorethylene through line *11* to condenser *M*, from which the condensate flows via line *11a* to column *C*, and via line *13* to column *A*.

Pure water is obtained from the lower part of column *C* and absolute alcohol from the lower part of column *A*. The alcohol is cooled in exchanger J_2 and flows to tailbox *V*. *K* is the heater for *A*. Dotted lines denote vents.

---

two components *X* and *Y* to be separated. *X* is assumed to be the low boiler.

The problem usually is to split up an ideal multicomponent mixture into two liquid mixtures in such a way that the distillate contains all of the material boiling lower than *X*, practically all of *X* and a small amount of *Y*, while the effluent contains all materials boiling higher than *Y*, as well as *Y* itself and a small amount of *X*.

Let

P_x and P_y = the boiling pressures of the pure key components, corresponding to a given boiling temperature, kg per sq cm.

x_F, y_F and x_D, y_D = their liquid and vapor concentrations, mol per cent

p = total pressure

then on the basis of equations (4), (9) and (9a) we have the following relations between the liquid and vapor compositions

$$x_D = \frac{P_x}{p} \times x_F \qquad y_D = \frac{P_y}{p} \times y_F \tag{150}$$

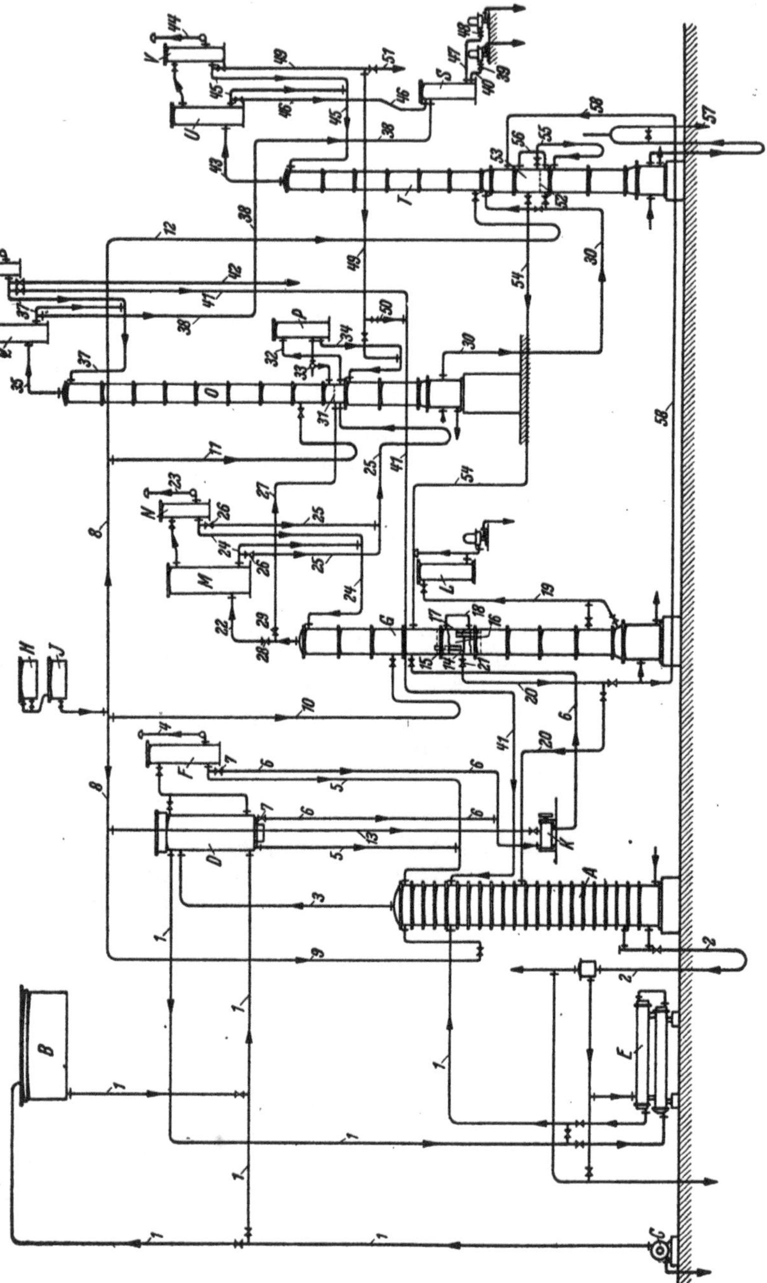

FIGURE 123

Layout of a rectifying plant for separating acetone, ethanol, butanol and water

The aqueous original mixture is fed into distilling column *A* from overhead tank *B* or by means of pump *C* from a container located on a lower level. The liquid first passes through condenser-heater *D*, which utilizes the heat of condensation of the crude distillate for preheating purposes. The liquid then passes through heat exchanger *E*, where the liquid leaving the lower part of distilling column *A*, stripped of acetone, ethanol and butanol (stillage) yields part of its heat to the crude mixture to be distilled.

The lower part of distilling column *A* is heated by direct steam. The upper part of the distilling column, acting as an enricher, is connected by line *3* with condenser-preheater *D* and vent-condenser *F*.

The vapor mixture leaving the column is liquified in condenser-preheater *D*. Non-condensed gases pass through line *4* of vent-condenser *F*. The gas-free condensate partly returns through lines *5* to column *A*, and partly passes through lines *6* to butanol column *G*. Valves *7* control the amounts of reflux and distillate.

Containers *H* and *J* serve to provide an alkaline solution for neutralizing the volatile acids by injecting either into the charge or distilling column *A*, or into intermedi te fractions free of carbonic acid. The alkaline solution is either introduced into the column directly, or made to react with distillates in special vessels. The alkaline solution passes to the columns by lines *8* to *12*, which are fitted with the necessary valves. Line *6*, for instance, contains a neutralizer device *K*. Lines *8* and *13* transfer the alkaline solution.

The mixture fed to butanol column *G* is decomposed in the column section above blind plate *14* into a distillate of acetone and alcohol, containing little butanol and water, and into a mixture of butanol and water, with traces of acetone and ethanol. The latter passes through line *15* into a separating space *16*, closed off by blind plate *14*. Here, the liquid separates into two layers.

The upper layer, rich in butanol, passes through overflow pipe *17* into the column section below the blind plate, which is equipped with indirect heating. Here, the butanol is dehydrated by distilling off the azeotropic butanol-water mixture. At the same time, any fractions of acetone and ethanol are driven out of the butanol.

The distillate vapors leave the lower column section below blind plate *14*, and return through line *18* to the column section embodying the separating chamber. The water-free butanol is drained from the foot of the column by line *19* and passes to cooler *L*. The lower layer of the liquid in the separator space passes through line *20* containing control valve *21*, back to distilling column *A*.

From the upper part of butanol column *G*, the distillate vapors pass via line *22* to condenser *M*, and from the latter partly to supplementary condenser *N* which is fitted with vent *23*. The mixture liquefied in condensers *M* and *N* partly returns via lines *24* to column *G*, and partly flows as condensate through pipes *25* and acetone column *O*. Valves *26* in line *25* control the flow.

Line *27* connected to pipe *22* facilitates the bypassing of condensers *M* and *N*. After closing valve *28* and opening valve *29*, the distillate vapors pass from column *G* via line *27* to column *O*. Here, rectification separates the acetone which passes upwards, while the acetone-free liquid is drained by line *30* from the lower part of the column which may be directly or indirectly heated.

To improve separation efficiency, acetone column *O* is fitted with an inter-condenser *P*. A blind plate *31* in column *O* forces the vapors to leave the column by line *32*, and to pass through intermediate condenser *P*. After separating a condensate rich in higher-boiling constituents, the vapors pass through line *33* and return to the column section above the blind plate in *O*. The fraction of distillate liquefied in *P* returns by pipe *34* to the section of column *O* located below the blind plate.

(The rectifying effect of intermediate condenser *P* might also be obtained by increasing the number of rectifying plates.)

The rectified acetone vapors leave column *O* and pass through line *35* into condenser *Q*. Supplementary condenser *R* fitted with vent *36* is connected to the latter. Condensers *Q* and *R* liquefy the acetone vapors.

Part of the condensate is returned to column *O* by lines *37*; from condenser *Q* part of it passes through line *38* to coil-type cooler *S*, which it leaves by line *40* fitted with control valve *39*.

Lines *41* facilitate the extraction of condensate from supplementary condenser *R*, which is returned to distilling column *A*. Line *42* will drain any condensate formed in *R*.

The remaining liquid, stripped of acetone, passes through line *30* into ethanol column *T* heated by direct or by indirect steam. The column yields concentrated ethanol upwards, and alcohol-free water downwards. The vapors of the ethanol rectificate leave column *T* via line *43*. They are liquefied in condenser *U* and supplementary condenser *V*, which is vented by *44*.

The condensate is partially returned via lines *45* to ethanol column *T*. From condenser *U*, part of the condensate is passed through line *46* to cooler *S*, the secondary coil of which supplies a highly concentrated ethanol via line *47* and valve *48*.

Line *49* connected to supplementary condenser *V* allows the return of condensate to acetone column *O*. By conecting line *50*, this fraction of distillate can be passed to distilling column *A*, via line *41*. If necessary, the condensate is drained by line *51*.

The butanol fraction introduced into ethanol column *T* with the remaining liquid, through line *30*, becomes more concentrated in the middle portion of the column, in which blind plate *52* provides a separation space *53* below. The liquid which accumulates here forms two layers. The upper one contains much butyl alcohol and is continuously drained by line *54*. A pump (not shown) transfers it to butanol column *G*. The lower layer, mostly composed of water, passes through line *55* to the dealcoholizing column below blind plate *52*. The distillate vapors of this section go through line *56* above the liquid level to separating space *53*. The spent water leaves the lower section of column *T* by line *57*. Should one intend to rework the lower liquid layer appearing in separating chamber *16* of butanol column *G*, by means of ethanol column *T*, a connection can be established between *16* and *T* by pipes *20* and *58* fitted with valves.

By division

$$\frac{x_D}{y_D} = \frac{P_x}{P_y} \times \frac{x_F}{y_F} = d \times \frac{x_F}{y_F} \tag{151}$$

The ratio of the boiling pressures of the pure key components

$$d = \frac{P_x}{P_y} \tag{152}$$

is called the relative volatility, and is assumed to be constant for the range of temperature variations encountered in the column.

The concentrations x_{Dn} and y_{Dn}, x_{Fn} and y_{Fn} are in the horizontal section above a given plate of the column. The concentrations $x_{D(n+1)}$, $y_{D(n+1)}$, $x_{F(n+1)}$ and $y_{F(n+1)}$ are in the corresponding section for the next plate above.

If the vapor rising from a plate is in phase equilibrium with the liquid leaving the same plate, which is the case for conditions of perfect interchange and liquid mixing, equation (151) yields

$$\frac{x_{D(n+1)}}{y_{D(n+1)}} = d \times \frac{x_{Fn}}{y_{Fn}} \tag{151a}$$

With these relations, the more important data for the two limiting conditions can be found; these are the minimum number of plates and the minimum reflux ratio. The former is concomitant with a complete reflux, $E = 0$ and $v = \infty$. In this case, basic equation (57) for the two key components assumes the forms

$$D \times x_{D(n+1)} = F \times x_{F(n+1)}$$

and (153)

$$D \times y_{D(n+1)} = F \times y_{F(n+1)}$$

from which follows

$$\frac{x_{D(n+1)}}{y_{D(n+1)}} = \frac{x_{F(n+1)}}{y_{F(n+1)}} \tag{154}$$

From equation (151a) and (154)

$$\frac{x_{F(n+1)}}{y_{F(n+1)}} = d \times \frac{x_{Fn}}{y_{Fn}} \tag{155}$$

If the concentrations in the section above the bottom plate are indicated by subscript 1, those above the penultimate plate by subscript 2, etc., and those above the top plate by subscript m, equation (155) yields

$$\frac{x_{F2}}{y_{F2}} = d \times \frac{x_{F1}}{y_{F1}}$$

$$\frac{x_{F3}}{y_{F3}} = d \times \frac{x_{F2}}{y_{F2}} = d^2 \times \frac{x_{F1}}{y_{F1}}$$

$$\frac{x_{Fm}}{y_{Fm}} = d^{(m-1)} \times \frac{x_{F1}}{y_{F1}} \tag{156}$$

From this, the theoretical minimum number of plates becomes:

$$m = \frac{\log \dfrac{x_{Fm} \times y_{F1}}{y_{Fm} \times x_{F1}}}{\log d} + 1 \tag{156a}$$

With the minimum reflux ratio v_M, even an infinite number of interchanger plates would achieve no enrichment

$$\begin{aligned} x_{F1} &= x_{F2} = x_{F3} = \cdots x_{Fn} = x_{F(n+1)} \\ y_{F1} &= y_{F2} = y_{F3} = \cdots y_{Fn} = y_{F(n+1)} \end{aligned} \tag{157}$$

According to equation (60) we must have, with subscript n applying to any desired plate

$$\begin{aligned} x_{Dn} &= \frac{v_M}{v_M + 1} \times x_{Fn} + \frac{x_{Fm}}{v_M + 1} \\ y_{Dn} &= \frac{v_M}{v_M + 1} \times y_{Fn} + \frac{y_{Fm}}{v_M + 1} \end{aligned} \tag{157a}$$

From this, and from equations (157) and (151) there follows

$$\frac{x_{Dn}}{y_{Dn}} = \frac{v_M \times x_{Fn} + x_{Fm}}{v_M \times y_{Fn} + y_{Fm}} = d \times \frac{x_{F(n-1)}}{y_{F(n-1)}} = d \times \frac{x_{Fn}}{y_{Fn}}$$

from which follows the minimum reflux ratio

$$v_M = \frac{1}{d - 1}\left(\frac{x_{Fm}}{x_{Fn}} - d \times \frac{y_{Fm}}{y_{Fn}}\right) \tag{157b}$$

Numerical example. A petroleum contains 5 mol per cent of n-heptane (C_7H_{16}) and 5 mol per cent of the next higher (3°C) boiler, methyl cyclohexane (C_7H_{14}). The distillate stands for half of the mass of the original mixture. The other half forms the effluent. The distillate is supposed to contain all materials boiling lower than heptane, and 90 mol per cent of the heptane together with 10 mol per cent of the methyl cyclohexane. The rest of the materials present go into the effluent. The relative volatility of the two key components, d, is 1.14.

For the enriching column

$$\frac{x_{F1}}{y_{F1}} = \frac{1}{1}$$

and

$$\frac{x_{Fm}}{y_{Fm}} = \frac{9}{1}$$

so equation (156a) will yield a minimum plate number

$$m_V = \frac{\log \dfrac{9 \times 1}{1 \times 1}}{\log 1.14} + 1 = 18$$

For the stripping column

$$\frac{x_{F1}}{y_{F1}} = \frac{1}{9} \quad \text{and} \quad \frac{x_{Fm}}{y_{Fm}} = \frac{5}{5}$$

which yields a minimum plate number for the stripping column

$$m_A = \frac{\log \dfrac{5 \times 9}{5 \times 1}}{\log 1.14} + 1 = 18$$

To find the minimum reflux ratio, insert in equation (157b)

$$x_{Fm} = 9 \text{ mol per cent}$$
$$y_{Fm} = 1$$
$$x_{Fn} = 5$$
$$y_{Fn} = 5 \text{ mol per cent}$$

since 100 mols of feed, containing 5 mols of heptane, will yield 50 mols of distillate with

$5 \times \frac{90}{100} = 4.5$ mols (corresponding to $x_{Fm} = 9$ mol per cent heptane)

$$v_M = \frac{1}{1.14 - 1}\left(\frac{9}{5} - 1.14\,\frac{1}{5}\right) = 11.$$

2. Finding the Variations of Conditions in the Column and the Number of Plates

It will again be assumed that the molar heat of evaporation is independent of the composition of the mixture, and that no heats of solution occur in the interchanging column. From this it follows that the masses of liquid and vapor, F and D (mols per hour), remain constant. If the concentrations x_F, y_F, z_F, etc., in the refluxed liquid, and x_D, y_D, z_D, etc., in the vapor exist between two plates, we can write an equation for any of the various materials, of the type of relation (145).

This means that, in a plot applying to the individual substance considered, where the concentration of that substance in the liquid is given on the abscissa, and its concentration in the vapor is given on the ordinate, the interrelation of the vapor and the liquid concentration is described by a straight line.

Thus an enrichment line can be given for each substance.[58,59] All of the enrichment lines run parallel and have a slope against the abscissa which is given by the ratio F/D; see equations (145). Each line intersects the diagonal of the plot in a point, the abscissa value of which corresponds to the concentration of the material considered in the product.

The enrichment line for a component with the distillate concentration x_E is shown in Figure 124. Intersection B of the enrichment line with the ordinate is given by ordinate section o in accordance with equation (61), as in the case of binary and ternary mixtures. Again, the reflux ratio v represents the ratio between the masses of reflux F and distillate E.

Analogous considerations will apply for the stripping column, located below the feedpoint. A stripping line for each of the substances can be found from equations (147). In the plot, in accordance with equation (146a), each of these lines has a slope against the abscissa as determined by ratio F'/D' of the mass of liquid and the mass of vapor.

The location of the intersection of each stripping line with the diagonal of the plot is dependent upon the concentration of the material in the effluent, as in the case of binary or ternary mixtures; see equation (89). Figure 124 shows the stripping line of a substance having concentration x_A in the effluent.

Also, the intersection of the enrichment and the stripping lines is determined by the concentration of the substance in the feed and its temperature, according to equations (96), (97) and (101), as in the case of binary mixtures. If, for instance, the mixture is fed at the boiling temperature, enrichment and stripping lines will intersect at a point, the abscissa value of which is given in the plot by the concentration of the substance in the feed.

For ideal mixtures there is another indication, besides the enrichment and stripping lines, which is useful in determining the composition of the vapor liberated when a liquid of known concentration boils. According to equation (150), the con-

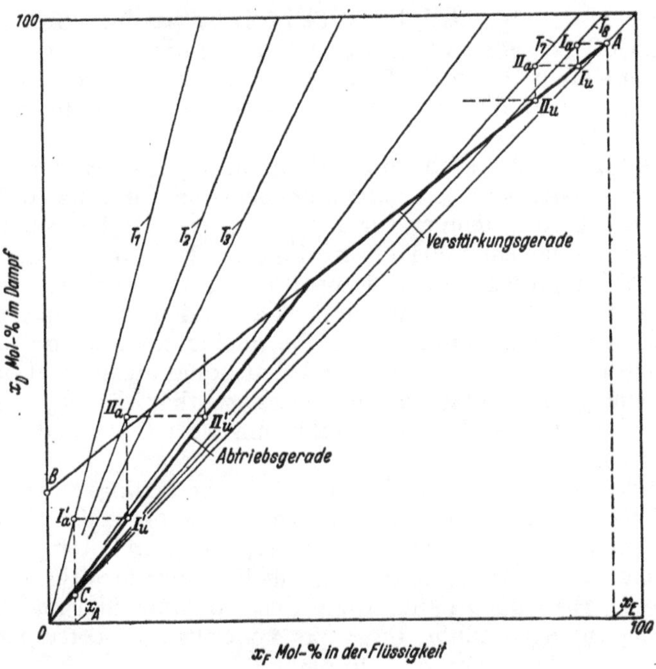

FIGURE 124

Ideal mixture of more than three components

Determination of the theoretical number of plates for a continuous rectifier

Mol % in der Flüssigkeit—Mol per cent in liquid
Mol % im Dampf—Mol per cent in vapor
Verstärkungsgerade—Enrichment line
Abtriebsgerade—Stripping line

centration of a substance in a vapor can be computed from the concentration of that substance in the liquid phase, its boiling pressure in a pure condition at the boiling temperature given, and the total pressure under which the mixture evaporates. Thus, for a given boiling temperature, the vapor concentration is directly proportional to the liquid concent·ation.

Therefore, in a plot showing the liquid concentration of a component on the abscissa, and the vapor concentration on the ordinate, the equilibrium conditions for possible boiling temperatures are represented by straight lines passing through the origin of the plot and having a slope against the abscissa dependent upon the ratio of the boiling pressure of the pure substance at the temperature for which the line holds, to the total distilling pressure.

In Figure 124, such equilibrium lines are given for one of the components of a multicomponent mixture, for different absolute boiling temperatures T_1, T_2, etc. If the reflux condenser has no enriching effect (Figure 52), point A in Figure 124 will represent the concentration of the distillate, of the reflux to the topmost enriching plate, and of the vapor rising from this plate. Assume that the boiling temperature on the top enriching plate is known, and that liquid mixing and perfect interchange take place on all plates, so that the vapor rising from any plate will be in equilibrium with the liquid leaving the same plate.

Under the conditions of Figure 124, the boiling temperature at the top plate of the enriching column is T_8; a definite equilibrium line, shown in the figure, corresponds to this temperature. Equilibrium point I_a, located on this line, indicates the relation between the vapor rising from the top plate and the liquid leaving the same plate; the vapor concentration corresponding to point A also corresponds to point I_a.

The abscissa value of point I_a intersects the enrichment line at point I_u. Its ordinate defines the concentration of the component in the vapor reaching the top plate of the enriching column, rising up from the second plate below the first one. This vapor is supposed to have been liberated by the liquid leaving the second plate. To find the concentration (in this vapor) of the substance considered in the plot, the ordinate value of point I_u must intersect the equilibrium line for the boiling temperature applying to the second plate.

This temperature can be obtained only from actual operating results. It can be found only by trial and error, and by employing a third relation, according to which the sum of the concentrations of the individual substances in the liquid leaving a plate must yield 100 mol per cent.

In this connection, the enrichment line for each component must be rendered in its plot, and the variation of the conditions must be found in the manner shown. Point II_a in Figure 124 is found this way and point II_u on the enrichment line follows from it. It indicates the relation between liquid and vapor concentrations in a horizontal section below the second plate.

The parallel to the abscissa, passing through this point, supplies an equilibrium point on the equilibrium line applying to the boiling temperature at the third plate. This point indicates the concentration in the reflux leaving the third plate.

The corresponding temperature is found in the same way as that for the second plate. In this fashion, concentrations and temperatures in the enriching column can be determined, step by step, from plate to plate.

With liquid concentrations below that indicated by the intersection of the enrichment and the stripping lines, one must pass to the stripping line; the lower corner points of the sequence of steps will then be located on the stripping line.

One may also take the composition of the effluent as a basis to determine changes of concentration and temperature in the rectifying column. If its boiling temperature is T_1 and its concentration in the substance under concentration x_A, the concentration of a substance in a vapor rising upward in a distilling unit can be indicated, in accordance with Figure 124, by equilibrium point I_a' located on the equilibrium line for T_1 and having abscissa value x_A. The parallel to the abscissa through I_a' intersects the stripping line at I_u'. The latter gives the interrelation of vapor and liquid concentrations in the horizontal section below the bottom plate of the stripping column.

Since the vapor rising from this plate will be in equilibrium with the liquid on the plate and, consequently, with the liquid leaving it, its concentration in the individual substance considered may be found from an equilibrium point II_a', which in Figure 124 is located on the vertical through I_u' and on the equilibrium line for the boiling temperature at the first plate.

The latter, however, can be found only by trial and error from the requirement that the sum of the concentrations of the individual substances in the vapor rising from a plate must equal 100 mol per cent.

The parallel to the abscissa through point II_a' thus located will supply intersection II_u' on the stripping line. Point II_u' indicates the relation between vapor and liquid concentrations in the section below the second plate of the stripping column.

From this, an equilibrium point III_a' can be found by the same method which yielded II_a' from I_a', etc. With liquid concentrations exceeding the abscissa value obtaining from the intersection of enrichment and stripping lines, one passes over to the enrichment line.

Regardless of the composition on which the determination of the number of plates has been based, whether that of the distillate or that of the effluent, the theoretical number of rectifying plates should be equal in both cases, and correspond to the number of corner points on the enrichment and the stripping lines, I_u, II_u, III_u, . . . , I_u', II_u', III_u', ...

One may also use the two point A and C in Figure 124 as a start for drawing the sequence of steps, and find in each case the corner points between the starting points and the intersections with the two lines. Dividing the number found by the mean enrichment ratio s_M (see section D, VI, 2) yields the actual number of plates.

Numerical example. Find the theoretical number of plates and the changes in conditions in a rectifying column for decomposing a primary mixture having a temperature, T, of 294°K and containing 26 mol per cent of methane and permanent gases, 9 mol per cent ethane, 25 mol per cent propane, 17 mol per cent butane, 11 mol per cent pentane and 12 mol per cent higher hydrocarbons behaving like heptane.

On the top plate of the enriching column there is a temperature of 294°K, and the effluent leaves the stripping column at 413.5°K. Distillate and effluent are intended to contain quantities and concentrations as follows:

Substance	Product		Effluent	
	Mols	Mol %	Mols	Mol %
Methane	26	42.9	0	0
Ethane	9	14.9	0	0
Propane	24.96	41.2	0.04	0.1
Butane	0.61	1	16.39	41.6
Pentane	0	0	11	27.9
Heptane	0	0	12	30.4
Total	60.57	100	39.43	100

By means of these data, the condenser pressure can be computed first, since, if one neglects the pressure losses in the column, the total pressure is the same as the condenser pressure.

Furthermore, the solubility of methane and of the permanent

gases in the refluxed liquid is assumed to be negligible. Let the concentrations of ethane, propane and butane in the liquid equal x_F, y_F and z_F mol per cent. Their boiling pressures, at a temperature of 294°K, are 39.4, 9.1 and 2.28 absolute atmospheres.

On the basis of equation (150), if p_1 = the sum of the partial pressures of these three constituents in the distillate

$$39.4 \times x_F = 14.9 \times p_1$$
$$9.1 \times y_F = 41.2 \times p_1$$
$$2.28 \times z_F = 1 \times p_1$$

and

$$x_F + y_F + z_F = 100$$

from which

$$p_1 = 18.7 \text{ absolute atmospheres}$$

Let p_2 = partial pressure of methane and the permanent gases. The total pressure follows from

$$18.7 + p_2 = p$$

and

$$\frac{p_2}{p} \times 100 = 42.9:$$

$$p = 32.8 \text{ absolute atmospheres}$$

Using the procedure explained by Figure 124, the variation in the butane concentration is shown in Figure 125. The composition of the effluent yields the starting point. Its butane concentration of 41.6 mol per cent, supplies intersection C of the stripping line with the diagonal. The vertical to the abscissa or the value of 41.6 mol per cent intersects the equilibrium line for T = 413.5°K at a point, the ordinate value of which indicates the butane concentration in the vapor which enters the bottom plate of the stripping column.

The parallel to the abscissa passing through this ordinate value yields intersection I_u' on the stripping line. It indicates the interrelation between the liquid and vapor concentration of butane in the horizontal section below the bottom stripping plate, while points II_u', III_u', IV_u' and V_u' below the second, third, fourth and fifth plate of the stripping column, and points I_u, II_u, III_u and IV_u follow for the sections below the corresponding interchanger plates of the enriching column. The enrichment line intersects the diagonal at the abscissa value of 1 mol per cent. Thus, one obtains a theoretical number of

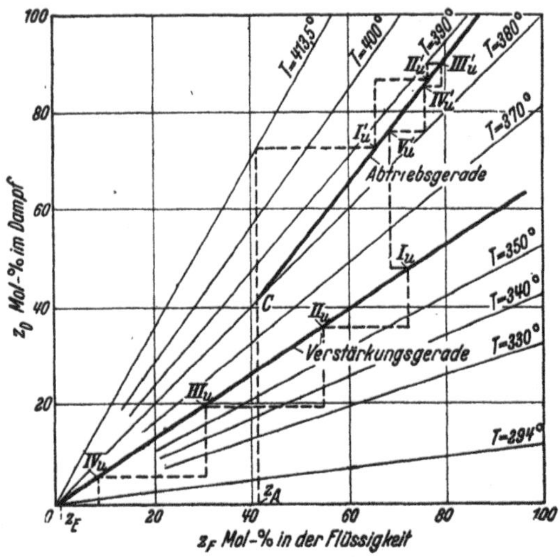

FIGURE 125

Variation of the butane concentration in the column

Mol % in der Flüssigkeit—Mol per cent in liquid
Mol % im Dampf—Mol per cent in vapor
Verstärkungsgerade—Enrichment line
Abtriebsgerade—Stripping line

plates of 5 for the stripping column, and of 4 for the enriching column.

In the same fashion, the variation in the propane concentration can be found, as shown by Figure 126. In this case, one starts with the distillate composition. The propane concentration of 41.2 mol per cent in the distillate yields intersection *A* of the enrichment line with the diagonal. From this the equilibrium point located on the equilibrium line for $T =$ 294°K follows; this in turn supplies point IV_u on the enrichment line. The latter indicates the interrelation between liquid and vapor concentrations in the section below the top plate of the enriching column.

Analogous considerations hold for points III_u, II_u and I_u, as well as for points V_u', IV_u', III_u', II_u' and I_u'. To find the latter, the part of the plot applying to a low propane concentration must be drawn on a larger scale, as in Figure 75.

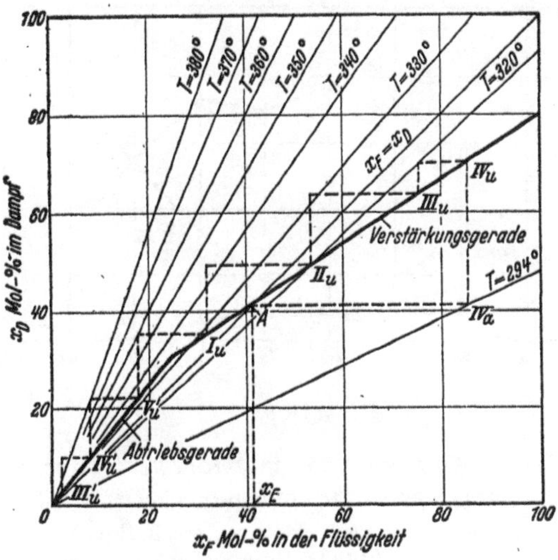

FIGURE 126

Variation of the propane concentration in the column

Mol % in der Flüssigkeit—Mol per cent in liquid
Mol % im Dampf—Mol per cent in vapor
Verstärkungsgerade—Enrichment line
Abtriebsgerade—Stripping line

The variation in the composition of any individual substance can be found in the way explained. The result is plotted in Figure 127, where the concentration of a substance is given as in the effluent (*A*), in the liquid leaving the bottom plate of the stripping column (I_u'), at other points, and in the distillate (*E*). The temperatures at the same points are shown in Figure 128.

3. Sidestream recovery

So far, it has been assumed that all of the reflux supplied by the reflux condenser to the top plate of an enriching column passes across one plate after another continuing its downward path. This is the average condition.

Where, however, at some point of the column, some liquid is withdrawn laterally as a sidestream, this will exert a certain

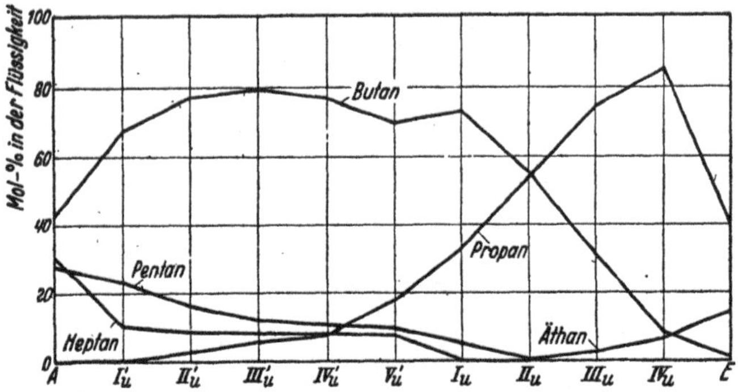

FIGURE 127

Changes in the composition of the individual components of an ideal mixture in the continuous rectifier

Mol % in der Flüssigkeit—Mol per cent in liquid
Butan—Butane
Propan—Propane
Pentan—Pentane
Heptan—Heptane
Äthan—Ethane

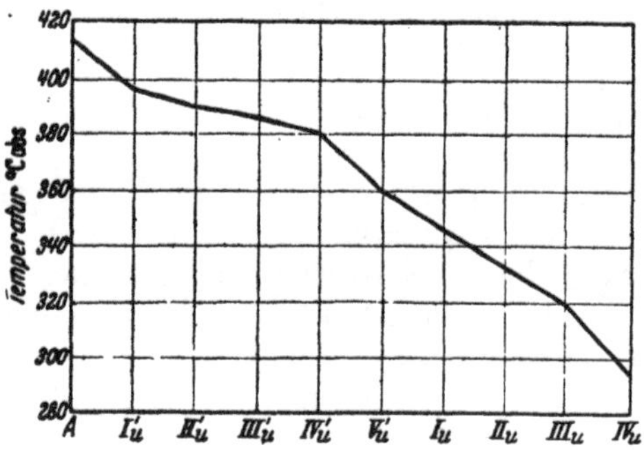

FIGURE 128

Temperature changes in the separation of an ideal mixture by means of a continuous rectifier

Temperatur, °C abs.—Temperature, degrees Kelvin

influence, not only on the mass of effluent but also on the efficiency of the rectifying plates, since the plates are located below the extraction point. This effect on the rectification efficiency is due to a change in the location of the enrichment line.

In the following discussion, a constant molar heat of evaporation for the mixture will again be assumed. If the mass of vapor leaving the top plate of the enrichment column = D mols per hour and the mass of liquid F_0 mols per hour returns as reflux to the same plate, the slope of enrichment line v_0 for the upper portion of the enriching column is given by the enrichment ratio F_0/D. This line intersects the diagonal of the x_F vs. x_D plot at A (Figure 129), at an abscissa value equal to the

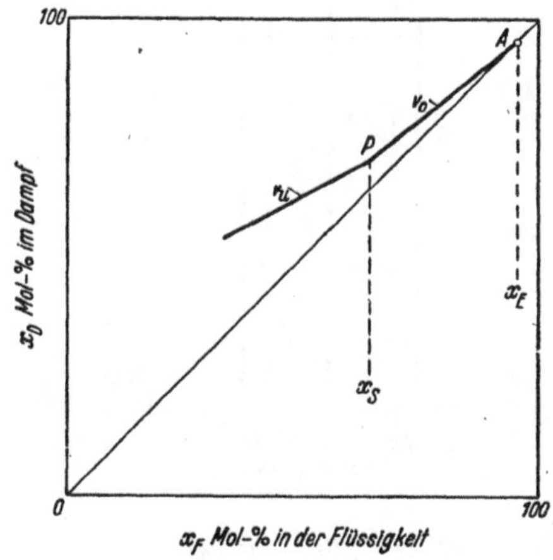

FIGURE 129

Location of the enrichment lines in the case of side streams withdrawn from different plates

Mol % in der Flüssigkeit—Mol per cent in liquid
Mol % im Dampf—Mol per cent in vapor

concentration in the distillate of the material for which this plot holds.

At point s_u or s_0 a sidestream equal to S mols per hour is withdrawn from the reflux exit plate B (Figure 130); the concentration in the sidestream of an individual component is x_S mol per cent. Then, only a mass of liquid $F_u = F_0 - S$ will pass across the rectifying plates below. The corresponding enrichment line v_u then has a slope against the abscissa of F_u/D. Both mass equations and Figure 130 reveal that the two enrichment lines will intersect x_S at point P with the abscissa value (Figure 129). If sidestream S is extracted at s_u, P may be considered as a point located on line v_0; this point indicates the relation between x_{Du} (see Figure 130) and $x_{Fu} = x_S$.

If, however, mass S is extracted at s_0, the top plate of the interchanger column below rectifying plate B will receive only a mass of $F_0 - S$, and point P may be considered to be located on the enrichment line v_u. Since concentrations x_{Du} and x_{Fu}

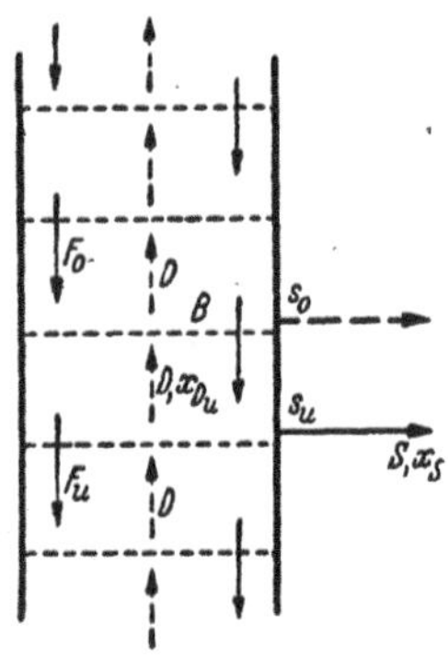

FIGURE 130

Diagrammatic section through column and symbols used in side-stream arrangement

are independent of the location of the extraction point (whether at s_0 or at s_u), $x_{Fu} = x_S$ will indicate intersection P.

If, furthermore, the sidestream is extracted from the interchanger column at some other point below plate B (Figure 130), line v_u in Figure 129 will be extended by another one, the slope of which can be computed as outlined. It will be less than that of line v_u. The intersection of this third line with v_u follows from the concentration of the second sidestream.

Because of these extractions, the amount of reflux will be reduced in the lower part of the rectifying column. Nevertheless, the mass in mols per hour of the vapor ascending stays the same. For this reason, the diameter of the column should be the same over the entire height.

Figure 131 shows the layout of a petroleum refinery as an

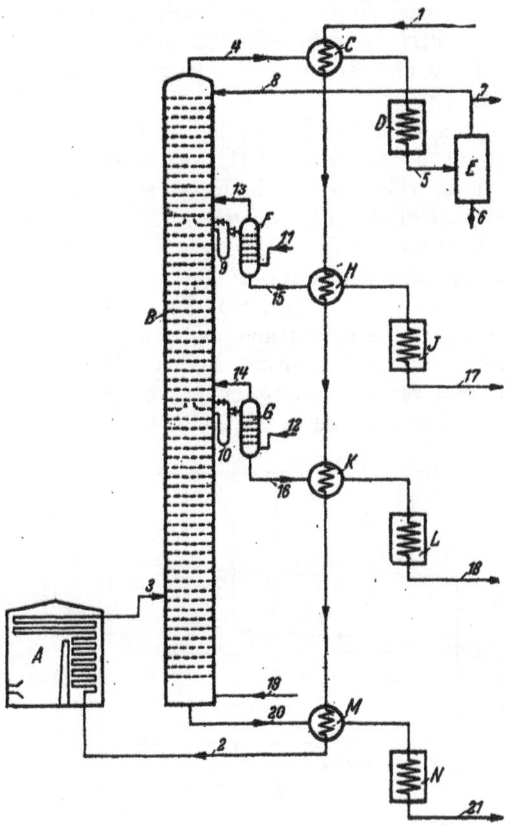

FIGURE 131

Oil refinery layout

The crude petroleum entering through *1* is heated in exchangers *C*, *H*, *K* and *M*, and passed to furnace *A* by line *2*. Here, the oil is heated to such a point that it enters rectifying column *B* as a liquid-vapor mixture through line *3*. The lowest boilers pass through line *4* into heat exchanger *C* and condense. The condensate is cooled in *D* and passed by line *5* to decanter *E* from which the water is removed by line *6*, while the lowest boiler (gasoline) is withdrawn by line *7*, and the reflux passed to rectifying column *B* by line *8*. Extraction lines *9* and *10* permit of withdrawing sidestreams. Their low-boiling components are stripped by injecting steam through *11* and *12*, which passes on to the column by *13* and *14*.

The sidestreams go through *15* and *16* to heat exchangers *H* and *K*, are cooled in *J* and *L* and removed through *17* and *18*.

Line *19* introduces superheated steam into the lower section of the column. The effluent from column *B* is drained by line *20* after having been cooled in exchangers *M* and *N*.

example of a rectifying column with sidestream extractions. Before the oil mixture enters the column, it is heated to such a point that the feed becomes a mixture of liquid and vapor; see section E, VIII, 4.

The stripping column located below the feedpoint is heated by direct superheated steam. If the temperature of the latter exceeds that of the residue leaving, its temperature will be lowered and part of the oil will be evaporated. Thus, a mixture of steam and oil vapors flows through the stripping column.

The steam causes a reduction of the oil vapor pressure throughout the entire rectifying column, and lowers the boiling temperature; this lowering is sometimes necessary to obtain a product of a given quality and characteristics. By way of example, Figure 132 renders the boiling curves of products obtained in an installation like the one in Figure 131.

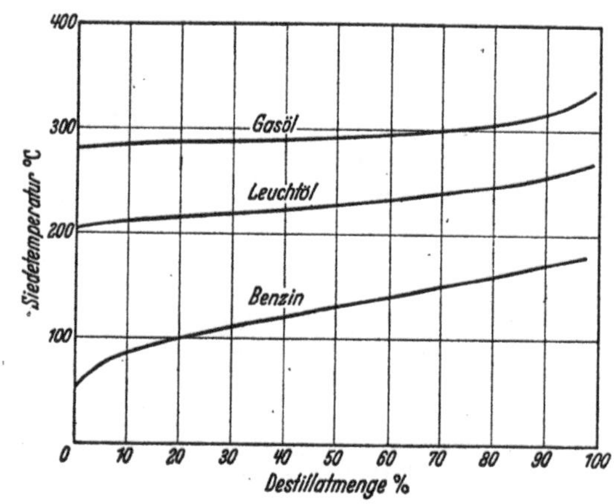

FIGURE 132

Distillation curves for three fractionated refinery products

Siedetemperatur—Boiling temperature
Destillatmenge—Quantity of distillate (Volume per cent over)
Gasöl—Gas oil
Leuchtöl—Kerosene
Benzin—Gasoline

Some oils are rectified by applying both superheated steam and vacuum. The steam remains superheated throughout the entire column. For a given quantity in mols per hour its partial pressure stays the same. It may be computed from equation (4).

Recent continuous plants for separating tars also have layouts analogous to that in Figure 131.

H. DETERMINING THE DIMENSIONS OF RECTIFYING COLUMNS WITH EXCHANGER PLATES; PLATE EFFICIENCY

I. Cap-type and tunnel-type plates

1. Fundamental equations for determining the column diameter

Among the principal points to be considered in the computation of a rectifying unit are the number of plates and the cross-section of the column. Columns, in practically every case, have a circular section. The sectional area is calculated like that of a pipeline from the mass flowing through it per unit of time and the flow velocity of the material. Consequently, the column section results as the ratio of the mass of vapor flowing and the vapor velocity in the free column section.

If the former is given as mols per hour = D, and the vapor velocity in meters per second = w_D, the area of the column section in square meters becomes

$$f = D \times \frac{22.4}{w_D \times 3600} \times \frac{273 + t}{273} \times \frac{760}{p} \qquad (158)$$

where t = average temperature, °C, and p = average pressure in mm Hg absolute. If the molar heat of evaporation is constant, the mass of vapor D flowing through the column may also be expressed by the mass of the product E in mols per hour and the reflux ratio v:

$$D = E(v + 1) \qquad (159)$$

To find the diameter of the stripping column, replace the value of D in equation (158) with the value of the mass of the vapor flowing through this section. Since the temperature variation in the column is usually small as compared with the level of the absolute temperature, it suffices to write an average t value in the case of the rectifying as well as in that of the stripping column. The same holds for p in the case of rectifying columns, as the pressure loss in the streaming vapor is low if compared with the absolute pressure level.

a. Selection of the superficial vapor velocity

Due to their wide load range and lack of sensitivity to contamination and incrustation, cap-type plates are usually preferred. Since heats and materials are handled on any interchanging plate, it might be thought opportune to use general equations for finding the best vapor velocity. Experience has shown, however, that such general equations neglect certain important factors, so that it is necessary to obtain reliable data by tests.

A yardstick for plate efficiency is the enrichment ratio, also called interchange ratio (see section D, IV, 2.) This is the ratio of the actual enrichment on a plate in low boiler b mol per cent, to a theoretical enrichment, a_F mol per cent. The latter would result with perfect mixing of the liquid and perfect interchange. Both enrichments may be determined from the liquid concentrations above and below a plate (Figure 59). The larger the numerical value of the ratio s of the two enrichment figures, the higher the efficiency of the rectifying plate.

Figure 133, based on tests performed by Kirschbaum, shows

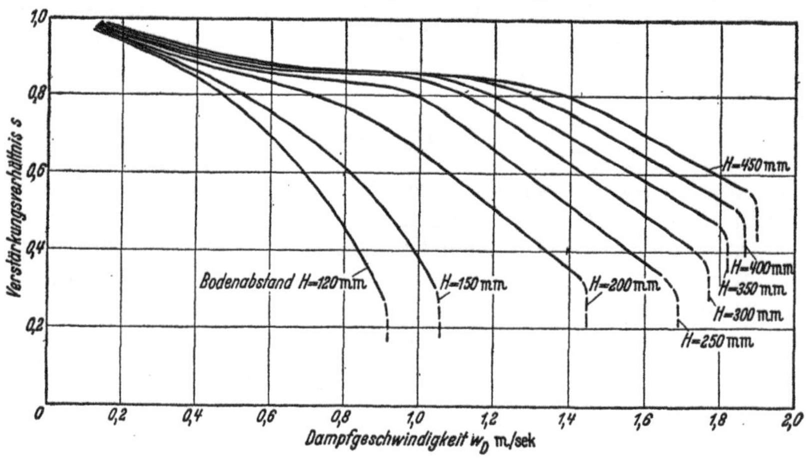

FIGURE 133

Enrichment ratio s and vapor velocity w_D with different plate spacings

Mixture of ethanol and water; reflux ratio v = infinity; final plate concentration x_F = 50 mol per cent, at atmospheric pressure.

Dampfgeschwindigkeit—Vapor velocity
Verstärkungsverhältnis—Enrichment ratio
Bodenabstand—Plate spacing

enrichment ratios of rectifying plates plotted against the load (superficial vapor velocity) for different plate spacings. The graph holds for a mixture of ethyl alcohol and water and a final plate concentration of $x_{Fu} = x_F = 50$ mol per cent alcohol.

The reflux ratio of $v = \infty$ (no recovery of distillate) was selected because only in this way can the method outlined in Figure 59 and text be applied without involving a separate determination of the entrainment from plate to plate.[60] If the moisture in the vapor were considered, the amount of reflux would be different for each plate, as entrainment varies from plate to plate. With $v = \infty$, however, the rectification line, corresponding to the variable liquid recovery from the plate, will merge with the diagonal in the x_D vs. x_F plot. Furthermore as the enrichment line for $v = \infty$ will also merge with the diagonal mentioned, the points C and E in Figure 59, following from the final plate concentrations, will also merge with the diagonal.

Both experiment and computation show, by the way, that in the case of the mixture considered the enrichment ratio is almost independent of the reflux ratio; see also section H, IV.

With $v = \infty$ and for the usual reflux ratios required to separate the mixture considered, the enrichment ratio was practically constant.

When the efficiency of the plates in a column is determined by measuring concentrations, the final concentration of any given plate will equal the initial concentration of the next plate below. Figure 134 shows the plates used in finding the results shown in Figure 133, as well as the sampling connection for concentration measurements. The design of the plate must be stated in similar cases, since it may influence the rectifying effect; see section H, I, e.

Figure 133 indicates that for all plate spacings the enrichment ratio decreases as vapor velocity increases. This decrease is due to the reduction in the contact-time of the two phases on the interchanging plate. However, the decrease itself becomes less as the spacing increases. With larger spacings, as $H = 400$ mm, the enrichment ratio will even stay constant over a larger range of velocities.

This is caused by increased interchange of heat and materials occurring with higher velocities, as well as by an increased area of contact between phases. Constancy of efficiency over a wide range of loads is a particular advantage of the cap-type plate.

If, with a virtually constant enrichment ratio, the range of loads is exceeded, efficiency will suffer least with large plate spacings; see Figure 133. For small plate spacings, as for instance H = 120 mm, the enrichment ratio decreases steadily and intensely with increasing loads, despite the previously mentioned improvement in the interchange figures and surfaces.

Obviously, an unfavorable item superposes itself on the favorable influences mentioned, namely, entrainment of liquid by the vapor. This becomes so effective with small plate spacings because it is not only the pressure of the steam acting on the slower liquid particle that counts, but also the velocity with which the liquid drops and lamellae are projected upwards from the plate with the vapor passing through the passages provided at high speed.

This in turn causes a sudden decrease in the enrichment ratio for larger plate spacings and increasing vapor velocity, once a certain load value has been exceeded. As a matter of fact, under the conditions existing in the case of Figure 133, a vapor space forms between the plates above the plate which contains a suspension of liquid drops, which is followed by another vapor space over it containing a thinner suspension of liquid drops, a horizontal separating plane being developed between these two spaces. As the velocity increases, the distance of this separating plane from the plate increases, until eventually the plane reaches the next plate above. As the velocity further increases, the enrichment ratio diminishes up to a certain upper load limit, at which the column becomes inoperable and the enrichment ratio practically zero.

This load limit is created by a condition where the liquid can no longer, or not to a sufficient extent, be drained from the plates via the weirs and downspouts. It should be kept in mind that the mass of liquid to be drained from the plates is given not only by the liquid from the reflux condenser, but also by liquid carried over by the vapor. Liquid will then build up between the plates, and eventually return from the top plate to the reflux condenser. Two circumstances may be responsible for this. First, there is a limit to what the downspouts can handle. Since liquid will not flow into a downspout in the same fashion as it flows across a weir, but will rather have the properties of a foam, drops or lamellae, a weir of a sufficient capacity must surround the downspout; see Figure 134. It may also happen that, with a certain velocity, the pressure loss in the vapor in passing through the interchanging plate becomes so

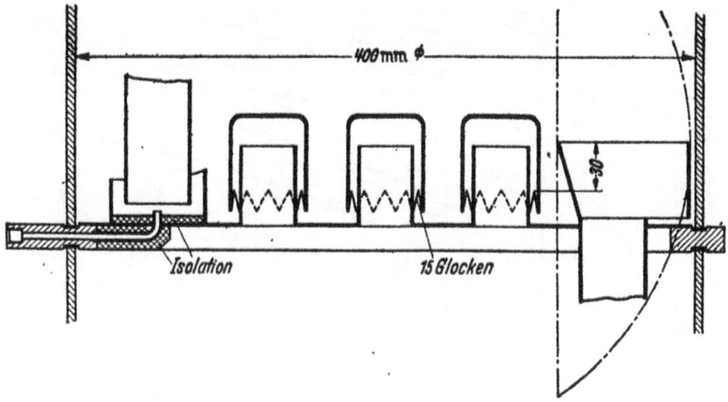

FIGURE 134

Bubble plate used in experiments

Isolation—Insulation
15 Glocken—15 caps

high that, due to the excess pressure in the vapor space below a plate (as compared to that above that plate), the liquid in the downspout is forced upwards again, causing the reflux condenser to be flooded.

In Figure 133, the upper load limit imposed by the liability of choking is indicated by dotted ends on curves. It is located at higher load ranges than the region where s begins to diminish under increasing velocity and for larger plate spacings. Since the latter phenomenon decides on the maximum operating load permissible, the weir area was made ample, after preliminary tests; Figure 133. Considering a certain margin of safety, under the conditions of Figure 133, and with $s = 0.83$, for a plate spacing of $H = 250$ mm, a vapor velocity w_D of 0.8 m per second, and for a spacing of $H = 150$ mm a velocity w_D of 0.4 m per second is found.

The influence of the plate spacing on the enrichment ratio becomes evident from Figure 135, which was derived from Figure 133. At low loads, the efficiency of the plates does not vary much for different spacings. At a certain load, according to Figure 135 with $w_D = 1$ m per second the influence of the plate spacing on the s value is most outspoken.

At this velocity, entrainment varies most with plate spacing.

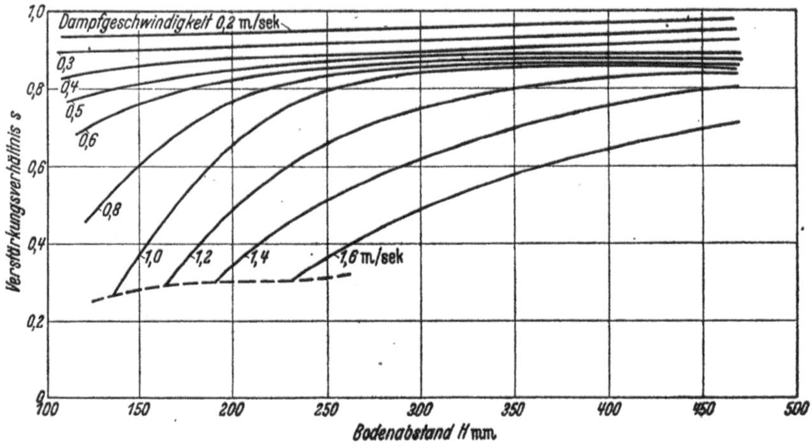

FIGURE 135

Enrichment ratio plotted against plate spacing, for various vapor velocities

Ethanol-water; reflux ratio $v = \infty$; final plate concentration x_F = 50 mol per cent

Bodenabstand—Plate spacing
Verstärkungsverhältnis—Enrichment ratio
Dampfgeschwindigkeit—Vapor velocity

With very high loads, s increases rather regularly with H. The dotted line in Figure 135 indicates those operating conditions where the upper load limit is reached.

Figure 133 demonstrates the influence of vapor velocity and plate spacing on the rectifying effect of interchanger plates.

The author has conducted further tests, with the same equipment that was used to obtain the data for Figure 133, to find other influences to be considered. In this case, however, the absolute operating pressures in the column were 50, 100, 250, 500 and 760 mm Hg. In this way, a wide variation of the physical properties of both the liquid and the vapor appeared.[61]

The vapor density γ_D kilograms per cubic meter, which is the most important value in the computation of the column diameter, varied within two limiting values in a ratio of 1 : 13. Extrapolating beyond these limits would make the test results even more significant than becomes evident from the variation of the physical properties alone.

The interdependence of the enrichment ratio and the load at different pressures, and consequently different vapor densities is shown in Figure 136. Generally speaking, it follows that the vapor velocity may be greater the smaller the absolute pressure p and the specific weight of the vapor.

With $s = 0.7$, $p = 760$ mm Hg absolute ($\gamma_D = 1.254$ kg per cubic meter), w_D becomes 0.9 m per second and with $p = 50$ mm Hg absolute ($\gamma_D = 0.1$ kg per cubic meter), w_D becomes 2.5 m per second. The density figures hold for vapors in phase equilibrium with the liquid leaving the plates and having a concentration of 50 mol per cent of ethyl alcohol.

For greater or smaller plate spacings than $H = 200$ mm (the value used in the tests), velocities were correspondingly lower or higher.

A conspicuous fact is that with low vapor velocities, at which liquid entrainment by vapor can no longer exist, the enrichment ratio is practically constant at all pressures. According to Figure 136, its value for a w_D of 0.4 m per second would be located between $s = 0.8$ and $s = 0.9$. This means that, barring entrainment, the physical properties of the mixture to be separated have only a small influence on rectifying-plate efficiency. Other tests, as conducted with benzol and toluol, lead to the same conclusion.[62] These comparative tests do not take into account secondary factors, as a leakage of joints between nozzle and plate, which may occur with small velocities and particularly with low absolute pressures, and may impair efficiency. Uniform results, as plotted in Figure 136, were obtained only after these joints between nozzles and plate had been tightened perfectly.

In Figure 136, with a w_D of 0.4 m per second, there are still small variations in the enrichment ratios corresponding to different pressures. The small reduction of s which resulted when the pressure was lowered from 760 to 550 mm Hg, can be explained by a reduced tendency of the mixture to foam, as the thickness of the foam layer will diminish with higher vacua. Obviously this influence of the foaming intensity is not appreciable; if the pressure goes down further, s increases again to a limited extent, although the conductivities of both phases as well as the surface tensions of the liquid against its vapor diminish. However, the amount of liquid flowing across a plate becomes much less with constant w_D when p becomes 50 after having been as high as 500 mm Hg. This means that the duration of the stay of the liquid on the plate increases to a

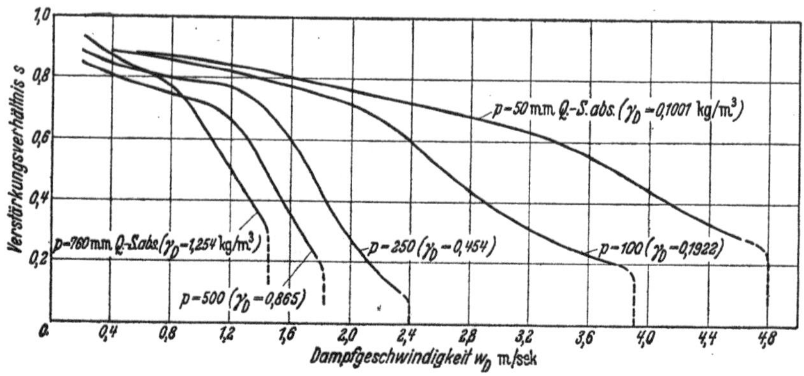

FIGURE 136

Enrichment ratio as a function of vapor velocity at various pressures

Plate spacing 200 mm; ethanol-water; reflux ratio $v = \infty$; final plate concentration $x_F = 50$ mol per cent

Dampfgeschwindigkeit—Vapor velocity
Verstärkungsverhältnis—Enrichment ratio
mm Q.-S. abs.—mm Hg absolute pressure

multiple of its original value. Nevertheless, the increase in s is negligible. It would seem that these secondary influences counteract themselves to at least some extent.

What may be stated with certainty from the data in Figures 133 and 136, however, is that *plate spacing H, vapor velocity* w_D and *density* γ_D *of the vapor* are the *predominant influences governing the efficiency* of rectifying plates. These three data are therefore essential in the computation of the column diameter.

Different vapor densities mean different maximum velocities; while (according to Figure 136) a rectifying column at 760 mm Hg absolute pressure may be operated with a vapor velocity w_D of almost 1.5 m per second, flooding of the reflux condenser with an operating pressure of 50 mm Hg will take place only when the velocity w_D is as high as 4.8 m per second.

While Figure 136 will demonstrate the influence of vapor velocity on rectification with different vapor densities, it fails, however, to show the influence of the throughput upon the efficiency of an interchanger column with different vapor pressures and identical dimensions. Figure 136 shows, for constant plate spacings, curves giving the interrelation of enrichment and load as functions of different γ_D values. These lines change their location if throughput m mol per $m^2 \times hr$ is plotted on the abscissa instead of vapor velocity w_D meters per second.

m is the number of mols of vapor flowing per hour and per square meter of free column area. Figure 137 shows that

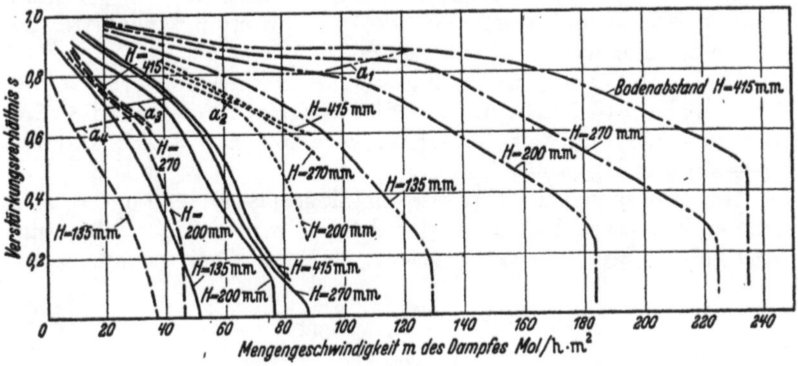

FIGURE 137

The enrichment ratio as a function of the mass velocity of the vapor, at different plate spacings and absolute pressures (vapor densities)

Test liquid ethanol-water; plates as shown in Figure 134

——— p = 50 mm Hg abs., γ_D = 0.1001 kg per cu m
- - - - p = 100 mm Hg abs., γ_D = 0.1922 kg per cu m
. . . . . . p = 250 mm Hg abs., γ_D = 0.454 kg per cu m
- . - . - p = 760 mm Hg abs., γ_D = 1.254 kg per cu m
Mengengeschwindigkeit des Dampfes—Mass velocity of vapor
Verstärkungsverhältnis—Enrichment ratio
Bodenabstand—Plate spacing

smaller vapor densities will also yield smaller throughputs for any given plate spacing and constant rectification efficiency, in spite of increased vapor velocity.

In this figure, the data for p = 500 mm Hg have been omitted for the sake of clarity, while the rest of the curves apply again to the same pressures and plate spacings as before. It is seen that with increasing mass velocity of the vapor the enrichment ratio decrease is more pronounced, the lower the vapor density and the smaller the plate spacing.

Figure 137 also indicates that with plate spacings exceeding 200 mm, increased plate spacing will mean a percentage load increase for constant efficiency which is more pronounced the higher the vapor density. In the case of plate spacings below 200 mm, the reverse holds true.

The throughput, which is expressed as kilograms per hour and square meter, is computed from the mass velocity m given in mol per hour and square meter by multiplying the latter by the molecular weight of the vapors.

Under the conditions of Figure 137, this yields, for $p = 760$ mm Hg, a μ of 36.26; for $p = 500$ mm Hg, a μ value of 36.56; for $p = 250$ mm, $\mu = 36.95$; for $p = 100$, $\mu = 37.02$; and for $p = 50$ mm Hg, $\mu = 37.1$.

If the mass velocity expressed in kilograms per hour and square meter is divided by the vapor density expressed in kilograms per cubic meter, the vapor velocity in meters per second results

$$w_D = \frac{m \times \mu}{\gamma_D \times 3600} \tag{160}$$

If, however, a column section were computed on the basis of a certain mass velocity and a corresponding vapor velocity from Figure 137, the enrichment ratio being assumed as constant, economy considerations would be neglected. The total manufacturing price of a column depends not only on the column diameter, but also on the plate number, and thus on the enrichment ratio.

If $s = 0.8$ and $H = 135$, Figure 137 will give, with $\gamma_D = 0.1$, a mass velocity m of 2, and with $\gamma_D = 1.254$, $m = 60$. If, other things remaining unchanged, s is lowered to 0.7, in accordance with Figure 137 the load for $\gamma_D = 0.1$ will increase by 300 per cent, and for $\gamma_D = 1.254$ by 32 per cent only.

With $H = 200$ mm, under these conditions, the output increases are 85 and 32 per cent. It is, furthermore, conspicuous that only with high vapor densities, such as $\gamma_D = 1.254$, efficiency stays practically constant over a given range of loads. If the throughput is increased beyond the upper limit of this range, rectifying-plate efficiency decreases.

For $\gamma_D = 1.254$ it would be advisable to let the permissible load coincide with the upper limit of the load range mentioned; curve a_1 in Figure 137. Thus, for instance, a value of $H = 200$ would correspond to a value of $m = 90$. Considering the relation shown between load and enrichment ratio at different pressures, it would further seem purposeful to indicate the load permissible for low densities by lines a_2, a_3 and a_4.

The permissible load follows, in this way, from the intersection of the lines a_1, a_2, a_3 and a_4 with the curves $s = f(m)$ holding for different spacings and densities. The load is expressed as the mass velocity m of the vapor. From the angle of economy, rectifying plates working vapor mixtures of low density are allowed to operate at a lower enrichment ratio, and plates are slightly more numerous than in the case of high density vapors. However, so far these interrelations between design and economy have not been elucidated sufficiently.

However, defining curves a_1, a_2, a_3 and a_4 as in Figure 137 will be of help in the selection of the load. If the superficial vapor velocity w_D, expressed as meters per second, which can be computed from equation (160), is plotted against the density on a logarithmic scale, straight lines result for certain plate spacings. These conform to the equation

$$w_D = C \times \gamma_D^n \tag{161}$$

where, for any value of the plate spacing, C is a constant and n a constant exponent. These have been shown in Figure 138 as functions of the plate-spacing value. The individual points were obtained from the tests conducted by the author. Extrapolation is possible for plate spacings exceeding 415 mm, the largest used in the experiments.

One experimental result, applying to $H = 135$ and $\gamma_D = 0.1001$ did not fit in with equation (161).

In Figure 139, the vapor velocity allowable for load m is plotted against the vapor density for different plate spacing, the C and n values having been obtained from Figure 138. We thus obtain: For a plate spacing of

$$\begin{array}{ll} H = 500 \text{ mm}, & w_D = 1.14 \times \gamma_D^{-0.465} \\ 400 & 1.1 \times \gamma_D^{-0.47} \\ 300 & 1.02 \times \gamma_D^{-0.49} \\ 200 & 0.82 \times \gamma_D^{-0.545} \\ 150 & 0.62 \times \gamma_D^{-0.49} \; * \\ 135 & 0.54 \times \gamma_D^{-0.425} \; ** \end{array} \tag{161a}$$

* and ** for γ_D values exceeding 0.2 kg per cubic meter.

In Figure 139, which is helpful in the computing of column diameters, lines covered by tests are heavy, extrapolated lengths are dotted. Since the extreme values of the densities investi-

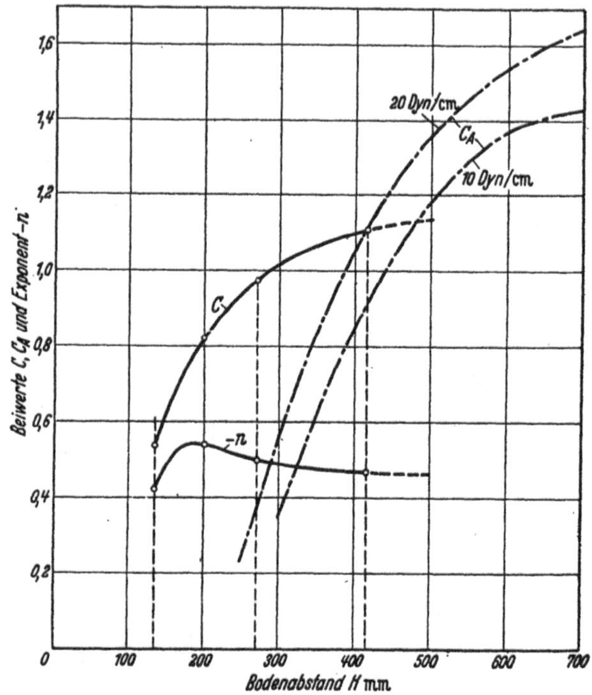

FIGURE 138

Coefficients C and C_A, and exponent $-n$ plotted against plate spacing

Bodenabstand—Plate spacing
Beiwerte und Exponent—Coefficients and exponent

gated are as 1 : 13, it would seem permissible to extrapolate to a higher density, the highest density still tested being as 1 : 3. Within narrower limits the extrapolation for very small densities takes place. If, for instance, in the case of a rectification of oils, oil vapor under vacuum has a density of $\gamma_D = 0.26$ kg per cubic meter, with a plate spacing of 500 mm the graph would indicate a permissible vapor velocity of $w_D = 2.2$ m per second. If, however, in atmospheric rectification the vapor density increases to 3.5 kg per cubic meter, for $H = 500$ mm the w_D value becomes 0.65 m per second.

When Figure 139 is used, it should be kept in mind that it

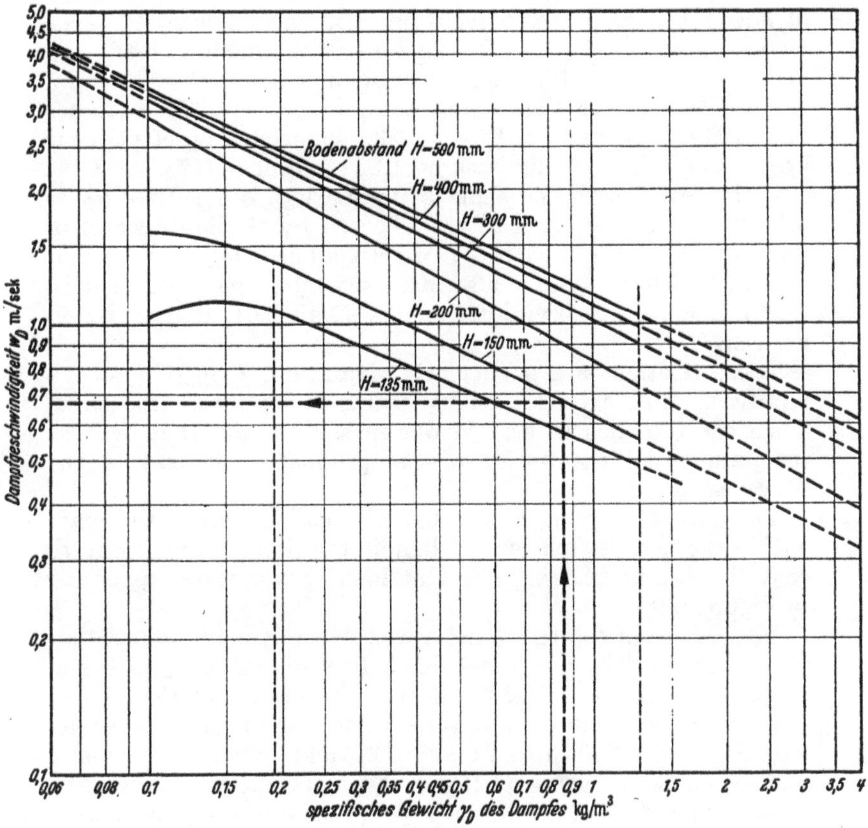

FIGURE 139

Permissible (computed) vapor velocities for different vapor densities and plate spacings (Kirschbaum's experiments)

Spezifisches Gewicht des Dampfes—Vapor density, kilograms per cubic meter
Dampfgeschwindigkeit, m /sek—Vapor velocity, meters per second
H = Bodenabstand—Plàte spacing

Note: For a vapor density of 0.1 kg per cu m, the enrichment ratio (s) is approximately 16 per cent smaller than for a vapor density of 1.25 kg per cu m

was derived from Figure 138. Thus, the enrichment ratio in Figure 139 is about 16 per cent smaller with γ_D = 0.1 kg per cubic meter than with γ_D = 1.254 kg per cubic meter.

Suitable use of Figures 138 and 139 permits the prediction of how the behavior of a rectifying column will vary if vapor velocities vary. If, for instance, the permissible load and velocity with $\gamma_D = 1.254$ kg per cubic meter and $H = 415$ mm, which according to Figure 137 amounts to $m = 125$ mols per hour and square meter with $s = 0.88$, is exceeded by 50 per cent up to $m = 187$, s will diminish by 18.2 per cent to 0.72.

If, however, with the same plate spacing, but with a $\gamma_D = 0.1922$ the permissible load as indicated by the intersection of line a_3 in Figure 137 with the corresponding $s = f(m)$ curve at $s = 0.73$ with $m = 43$, is also increased by 50 per cent, that is, up to $m = 64.5$, s diminishes from 0.73 to 0.41, that is, by 44 per cent.

These relations are particularly significant if, in designing a rectifying unit, a certain safety factor for overload is desired. Then the unit has to be laid out in such a way that even at maximum load the purity of the product is assured to the degree required.

The data underlying Figure 139 will hold as long as the space within the weir is sufficient to handle the liquid going down, so that there is no choking effect at loads below those appearing in Figure 139.

In the case of too restricted a weir space it may happen that the reflux condenser becomes flooded even with a w_D of 0.4 m per second, for $\gamma_D = 1.254$ and $H = 200$, and that the column can no longer achieve the desired separating effect. Nevertheless, Figure 139 indicates a w_D of 0.81, and actually the column may be operated even with higher loads, its efficiency decreasing merely as indicated from the $s = f(w_D)$ and $s = f(m)$ curves in Figures 133 and 137.

As a rule, it will not suffice to simply provide a circular tube as a downspout,[63] without an enlarged opening, since the liquid will not flow downwards as a coherent mass, but may foam intensely, and be in a highly turbulent condition. The weir must be able to destroy the foam; therefore ample dimensions of the weir edge and liquid path are required.

The total length of the weir edge must considerably exceed that of the downspout circumference. In the case of the plate in Figure 134, the weir edge is straight; this design is suitable for small to medium column diameters (1.5 m), in which case the liquid on a plate usually moves in a direction opposite to that of the liquids on the plates above and below. This condition also exists in the case of the plates in Figure 134, which were used in gathering the data on Figures 135 to 139.

Measurements were also taken on rectifying and absorbing columns which are stated to have been operated at maximum capacity with satisfactory production, and in which oils, alcohol-water mixtures and recovered materials were handled at different pressures. Nothing regarding the constructional details of these plates has been stated by the authors;[64] but the data found have been employed in finding constants C' to be used in the theoretically developed equation

$$k = C'\sqrt{\gamma_D(\gamma_F - \gamma_D)} \qquad (162)$$

where k = throughput in kilograms per hour and square meter, γ_D = vapor density and γ_F = liquid density, in kilograms per cubic meter. In the figures in parentheses in equation (162), γ_D may always be neglected by comparison with γ_F. If, further, γ_F is assumed to be 8100 kg per cubic meter, in correspondence with the data underlying Figures 135–139, dividing both sides of equation (162) by γ_D will yield

$$w_D = C_A \times \gamma_D{}^{-0.5} \qquad (163)$$

where w_D is again expressed in meters per second. The C_A values have been given in Figure 138 as functions of the plate spacing. The one line for C_A holds for a surface tension of the liquid against the vapor of 20 dynes per square centimeter, the other for 10 dynes per square centimeter.

The surface tension may have some influence on the dispersion of the liquid. However, tests and observations by the author have shown it to be so small that it is negligible as compared with those of the vapor density and velocity. The influences of tendency to foaming and of the keeping-qualities of the foam formed also superpose themselves; the author's C values comprise these factors.

According to tests run by the author, the surface tension of alcohol is within the limits applying for the two C_A curves in Figure 138.[65] The surface tension of water is higher.

Even if the density of the liquid actually has as large an influence on the load-carrying capacity of interchanger plates as assumed in equation (162), still this influence is hardly significant, since most γ_F values are of the same order, and because in equation (163) the γ_D value figures in its -0.5th power.

If one assumes that the extent of entrainment depends mainly on the density and velocity of the vapor, equation (161) becomes of more general applicability. Further experiments will have to decide whether secondary influences of other kinds are still present.

Referring to Figure 138, it is seen that the data obtained by the author fit in satisfactorily with the outline of the C_A curve for a plate spacing of 400 mm. For larger spacings, the C_A line yields larger, and for spacings below 400 mm, smaller vapor velocities than does Figure 139. Due to its practical significance, the latter discrepancy should be discussed.

Even for a plate spacing of H = 250 mm and a vapor density of γ_D = 1.3 kg per cubic meter, as encountered in the rectification of ethyl alcohol, the C_A line will yield, by equation (163), a permissible vapor velocity of w_D = 0.21 m per second for a surface tension of 20 dynes per centimeter. For a surface tension of 10 dynes per centimeter, w_D becomes even smaller.

It is, however, a generally known fact that alcohol rectifying columns, even with plate spacings of 150 mm, operate perfectly well with vapor velocities of 0.4–0.5 m per second. With this spacing, discrepancy with the measurements mentioned becomes still more pronounced.

Figure 139 yields a maximum velocity of w_D = 0.54 m per second for H = 150 mm. This figure corresponds to actual conditions; furthermore, all of the data in Figure 139 were based on a large number of systematic tests. Figure 134 shows that in the case of the plates investigated the depth of immersion was 30 mm as measured between the top point of the serrations and the weir edge.

The circumstance that the C_A line yields much greater plate spacings from Figure 138 than from Figure 139 might be explained by the possibility that Souders Jr. and colleague investigated vacuum columns having large plate spacings. On account of the low pressure-loss in the vapor, the highest point of the cap edge may be at the level of the weir edge. It has, however, never been proved that, in that case, entrainment is less and that spacings may be greater than in the case of lower cap edges, so that for this reason also Figure 139 offers reliable gu dance.

Example. The numerical example in section E, V, yielded, for a rectifying column handling an ethanol-water mixture, with a reflux ratio v of 4, a vapor flow of D = 17.36 mols per hour. Find the diameter of the rectifying column for a constant plate spacing H of 200 mm, with a safety margin of 25 per cent against overloads.

The maximum density of the vapors, $\gamma_D = 1.39$ kg per cubic meter, occurs in the top section of the column, where alcohol concentration comes up to 94 per cent by weight or 86 mol per cent. From Figure 139, with $\gamma_D = 1.39$ kg per cubic meter and $H = 200$ mm, w_D is found to be 0.68 m/sec. Considering the margin of safety specified, vapor velocity becomes 0.545 m per second. (For a spacing of 150 mm, a velocity of 0.42 m per second would have resulted.)

With an average temperature of 85°C and an average pressure of 760 + 40 mm Hg absolute within the column, equation (158) will yield a section of

$$f = 17.36 \frac{22.4}{3600 \times 0.545} \times \frac{273 + 85}{273} \times \frac{760}{800} = 0.247 \text{ sq m}$$

A diameter d_v of 560 mm corresponds to this section.

In the lower section of the rectifying column, the vapor density falls to a value γ_D of 0.785, according to Figure 77. Therefore, on the basis of Figure 139, with the same vapor velocity and considering an overload margin, a plate spacing smaller than 200 mm, namely 150 mm, can be selected. Usually, plate spacings are made uniform throughout a column. On the basis of the data following from Figure 139, however, it would seem purposeful to use larger plate spacings for high vapor-densities in those columns where vapor density is subject to intense fluctuations.

In the example, the vapor mass flowing through the stripping column is 20.81 mcls per hour. For an average temperature of 96°C in the column, and an average pressure of 815 mm Hg, with the plate spacing H of 200 mm a diameter of the stripping column d_A of 670 mm results.

b. Selection of the plate spacing

Figure 133 shows that with otherwise unchanged conditions and with a constant enrichment ratio the velocity of the vapor may be higher the larger the plate spacing. Consequently, with a given total efficiency and output of a rectifying column, plate spacing must increase as the column diameter decreases.

For the purpose of comparison, let us assume that for finding the purchase price of a column the conditions of Figure 133 and an enrichment ratio s of 0.83 are taken as a basis. If the vapor velocity which that enrichment ratio determines for any given plate spacing is exceeded, the efficiency of the interchanging plates is reduced appreciably.

For a plate spacing of 300 mm, a column diameter *d* of 700 mm is assumed. From this, Figure 133 will yield column diameters for different plate spacings and velocities corresponding to an enrichment ratio of 0.83.

The purchase cost of similar rectifying columns, working at atmospheric pressure and fitted with 20 plates, made of iron and/or copper, was found as an average from estimates given by several fabricators.

These averages have been given in Figure 140, where the price per plate, including the shell belonging to each plate, is

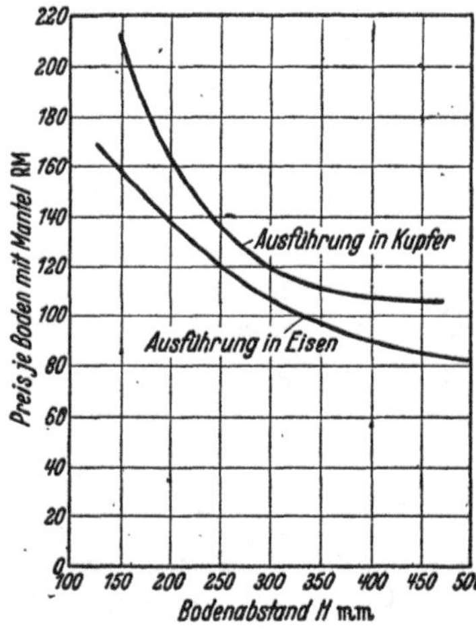

FIGURE 140

First cost of columns as a function of plate spacing

Bodenabstand—Plate spacing
Preis je Boden mit Mantel, RM — Reichsmark Price per plate including shell, ($1.00 = RM2.50)
Ausführung in Kupfer—Equipment made of copper
Ausführung in Eisen—Equipment made of steel

plotted against the plate spacing. It is seen that investment decreases with increasing plate spacing.

It should be kept in mind that this investment does not comprise the building, and that the convenience of operation also depends on the plate spacing and consequently on the column height. Also, open-air columns are usually assembled from several lengths, and the shell is provided with manholes to facilitate assembling and upkeep. Such items have not been considered in Figure 140.

The building cost plays a prominent part in the determination of economical plate spacing; it works towards a reduction of this spacing. Many housed rectifying columns have spacings of 150 mm, while, in the case of rectifying equipment operating in the open, spacings of 600 mm and more may be encountered.

c. *Liquid distribution and phase motion on cap-type plates; influence of entrained liquid on rectification efficiency*

The effect of liquid entrained in the vapor on the efficiency of interchanging plates may be conceived when the motions of the phases on a plate are observed. As a rule, a transparent liquid layer forms immediately on top of the level plate of a cap-type section; on top of this there is foam (Figure 141), or in the case of non-foaming mixtures, a turbulent layer.

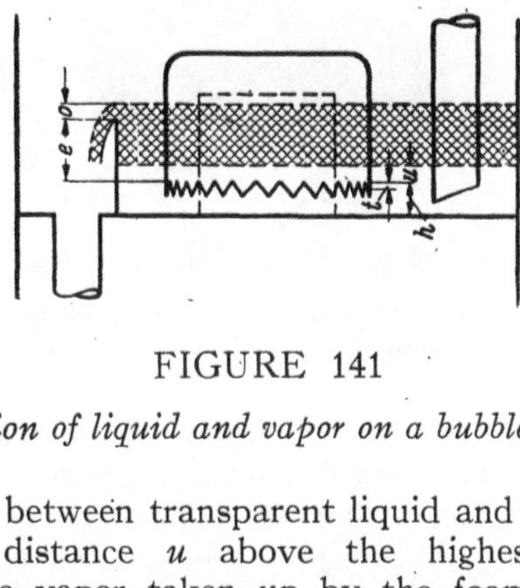

FIGURE 141

Distribution of liquid and vapor on a bubble-cap plate

The interface between transparent liquid and foam is located at a certain distance u above the highest point of the serrations. The vapor taken up by the foam will occupy a certain space between the serrations, the lowest point being situated below the top point of the serrations; distance t.

For instance, in the case of an ethanol-water mixture, the distance $u + t$ was 12 mm, with vapor velocities of 0.1–0.5 m per second based on the free column section, and with alcohol concentrations of 20–60 mol per cent. With this mixture, the foam layer is always present as long as the immersion depth

e (Figure 141) exceeds u. If e is less than u, the foam layer disappears, even in the case of a foaming mixture, and dimension o may even become negative. In the latter case, the liquid is projected over the weir in the shape of drops or lamellae.

Between the serrations, the vapor advances in the form of thin streams, as long as (in the case of the mixture mentioned and with a density of the vapor of 1.3 kg per cubic meter) the superficial vapor velocity exceeds 0.1 m per second. At low loads, the streams appear intermittently, and the vapor then advances in the shape of bubbles through the transparent liquid layer.

The conditions explained by means of Figure 141 become apparent in Figure 142, which was photographed from a column

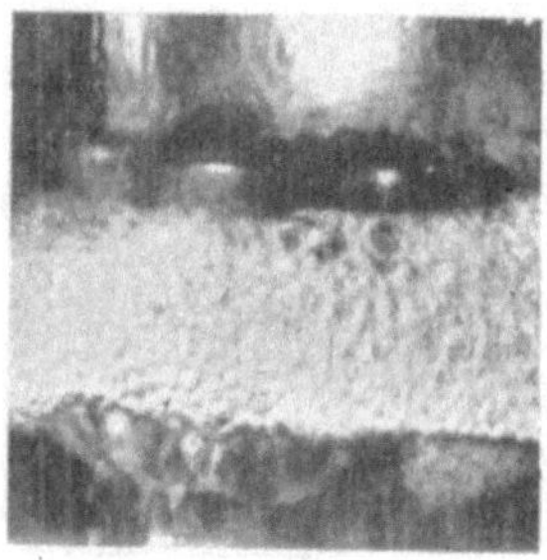

FIGURE 142

Fluid motion on a bubble-cap plate

Vapor velocity w_D = 0.3 m per second; liquid used was an ethanol-water mixture containing 50 mol per cent alcohol; atmospheric pressure

fitted with glass shell sections which handled the mixture mentioned. Figure 142 applies in the case of cap-type plates. With a slotted cap the flow is similar to that occurring in the case of a serrated plate; Figure 143. At the load value applying in the case of the last two figures, the foam layer projects but little liquid. With larger loads, however, a layer comprising drops and lamellae develops on top of the foam layer. This layer is bulkier the higher the load and the greater the vapor density. It may increase to a point where it fills all of the space between two plates, so that considerable entrainment results. Conditions at high loads are shown in Figure 144. The circumstance, that the liquid on the lower plates is projected upwards to a lesser

FIGURE 143

Fluid motion through slotted bubble caps

Vapor velocity w_D = 0.15m per second; ethanol-water with 50 mol per cent alcohol; atmospheric pressure

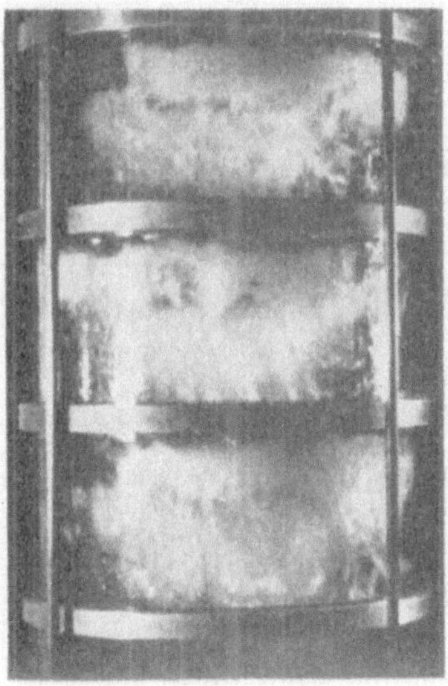

FIGURE 144

Fluid distribution in a rectifying column

Vapor velocity w_D = 1.5 m per second; ethanol-water; column diameter 400 mm, plate spacing 200 mm; atmospheric pressure

extent, is due to the lower density of the vapor, as compared with that in the upper region of the rectifying column.

An attempt will be made to determine the effect of entrainment by calculation, the reduction of the enrichment ratio being computed in proportion to the amount of liquid carried over by vapor.

Deductions and results are simplified by assuming complete reflux, that is, a reflux ratio v of ∞.

In Figure 145, x_{Du} = concentration of low boiler in the mixture fed to the plate considered, and x_D = concentration of low boiler

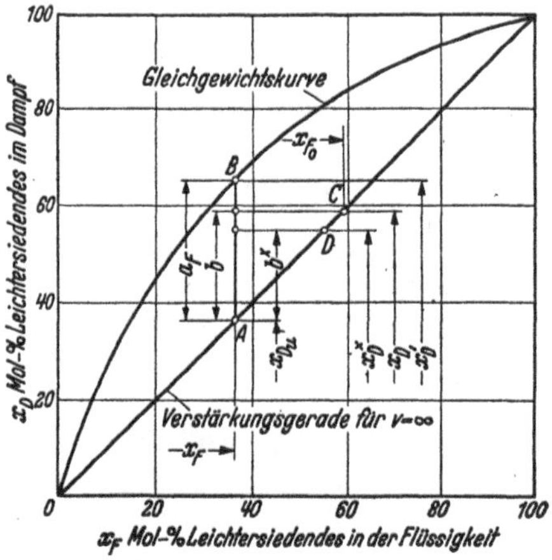

FIGURE 145

Determination of the enrichment ratio (s) when the reflux ratio (v) = ∝

Mol % Leichtersiedendes in der Flüssigkeit—Mol per cent of low boiler in liquid
Mol % Leichtersiedendes im Dampf—Mol per cent of low boiler in vapor
Gleichgewichtskurve—Equilibrium curve
Verstärkungsgerade für $v = \infty$—Enrichment line for the case of reflux ratio $(v) = \infty$

in the vapor ascending from the same plate, barring eventual entrainment. Amount b then corresponds to the enrichment in the case that the vapor does not carry any liquid, and that interchange is not perfect. If, however, entrainment takes place, the vapor-liquid mixture ascending from the plate will have an average concentration of x_D^* mol per cent of low boiler, which exceeds x_{Du} by the true enrichment b^*.

The enrichment line has, in the case where a vapor-liquid mixture ascends from plate to plate, the same significance as in the case of pure dry vapor. If entrainment takes place, the average concentration of the vapor-liquid mixture takes the place of that of pure dry vapor.

According to Figure 145, a liquid concentration x_F of the liquid flowing from a plate corresponds to a concentration x_{Du} of the vapor flowing to that plate. Again let us assume perfect mixing of the liquid, so that liquid concentration x_F can be considered to be uniform over the entire plate. Then, the liquid particles carried by the ascending vapor also have a concentration of x_F mol per cent.

Let f mol per cent = liquid concentration in the liquid-vapor mixture travelling upwards from the plate. The average concentration of the vapor-liquid mixture is, then:

$$x_D^* = \left(\frac{100 - f}{100}\right) x_D + \frac{f}{100} x_F \tag{164}$$

For $v = \infty$, $x_{Du} = x_F$; hence

$$b^* = x_D^* - x_F \tag{165}$$

and

$$\begin{aligned} x_D^* - x_F &= \left(\frac{100 - f}{100}\right) x_D + \frac{f}{100} x_F - x_F \\ &= \left(1 - \frac{f}{100}\right) (x_D - x_F) \end{aligned} \tag{166}$$

From this follows

$$b^* = \left(1 - \frac{f}{100}\right) b \tag{167}$$

If the vapor entrains liquid, the enrichment ratio becomes $s^* = b^*/a_F$, while in the case of perfectly dry vapor it is $s = b/a_F$. The decrease (in per cent) in enrichment due to entrainment amounts to

$$e = \frac{s - s^*}{s} 100 = \frac{b - b^*}{b} 100 \tag{168}$$

If the b^* value from equation (167) is substituted into equation (168), we obtain

$$e = f \tag{169}$$

This means that the decrease in the enrichment ratio (and therefore in efficiency) in the case of complete reflux (reflux ratio $v = \infty$) equals the share of the liquid in the vapor-liquid mixture travelling upwards. If, for instance, the vapor-liquid mixture released by a plate contains 20 mol per cent of liquid, the enrichment ratio will be 20 per cent lower than in a case where the vapor entrains no liquid.

Equation (169) will hold for any mixture and concentration, as long as the reflux ratio is infinite. With smaller reflux ratios, the loss in plate efficiency is greater than with $v = \infty$. Liquid entrainment increases the quantity of reflux, so that, with a finite value of the reflux ratio liquid, entrainment will depress the enrichment line in the x_F—x_D plot.

For finite v-values the decrease in the enrichment ratio can be computed by means of the same basic equations which have served above. For instance, in the case of an ethanol-water mixture and with an enrichment ratio $v = 3$ it follows that with a liquid concentration of 30 mol per cent and a liquid concentration in the vapor-liquid mixture of 20 mol per cent, the decrease due to entrainment is 40 per cent. As the alcohol concentration increases, the decrease in the enrichment ratio becomes still more noticeable.

This proves that an entrainment of liquid in the vapor causes an appreciable reduction of plate efficiency. Many experiments have been conducted to determine the extent of entrainment.[66,67,68,69,70] Figure 146 illustrates the results of entrainment tests on a cap-type plate column.[71] The column diameter was 458 mm, and the plates were fitted with 10 slotted caps of 76-mm diameter. The depth of immersion (distance between the top point of the vapor slots and edge of weir) was 25 mm. The reflux ratio was $v = \infty$. The lines apply to different plate spacings and to two pairs of phases. The vapor velocity plotted on the abscissa is based on the free column section.

It is seen that with smaller plate spacings larger amounts of liquid were entrained by the vapor. All known tests indicate the deciding influence of vapor velocity w_D and of vapor density γ_D on the amount of liquid entrained, although individual measurements vary.

It has been suggested to plot the amount of liquid entrained against $w_D\sqrt{\gamma_D}$, instead of against the vapor velocity itself. However, this does not make available test results more consistent.

It is easy to conceive that plate efficiency may be improved by eliminating entrained liquid between plates. This might be achieved by baffles, such as shown in Figure 147. The baffles are arranged at two different levels and are so distributed

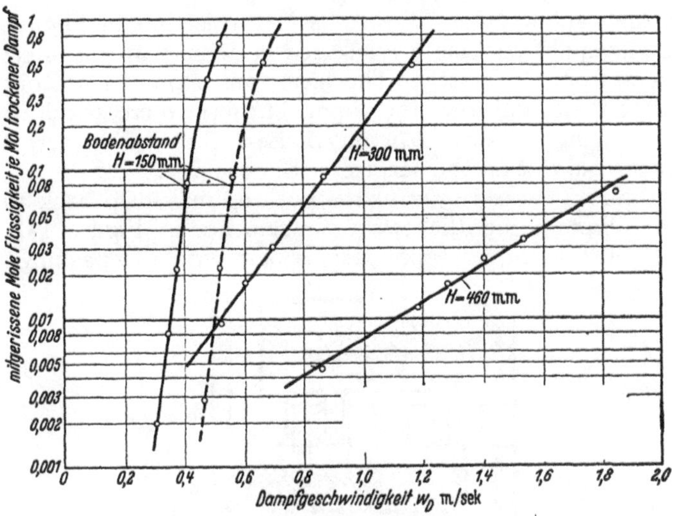

FIGURE 146

Entrainment plotted against vapor velocity for various plate spacings

——— Ethanol-water mixture; terminal concentration $x_F = 35$ mol per cent

- - - - Pure water

Dampfgeschwindigkeit, m /sek—Vapor velocity, meters per second

Mitgerissene Mole Flüssigkeit je Mol trockener Dampf—Mols of liquid entrained per mol dry vapor

Bodenabstand—Plate spacing

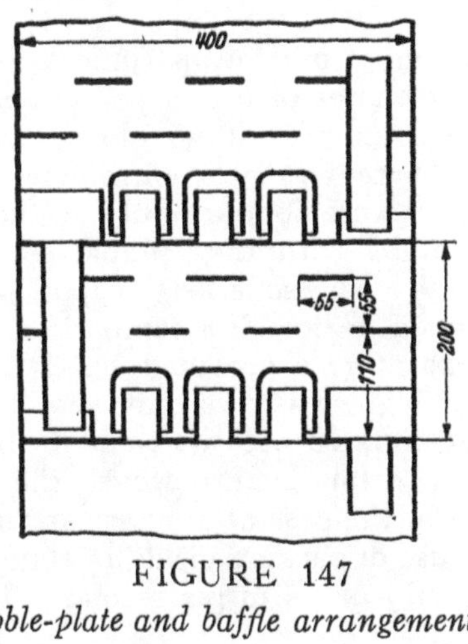

FIGURE 147

Combined bubble-plate and baffle arrangement in a column

that they are staggered. Another setup is that in Figure 148, where layers of packing are inserted between plates at a certain distance from the latter. The latter are supposed not only to separate the liquid from the vapor, but also to create additional interchanging surface between phases, the liquid carried up being spread out on the surface of the packing.

The liquid carried over to the packing may drip back to the plate. Tests using the two arrangements in Figure 147 and 148 show that liquid separation is best with the packing type

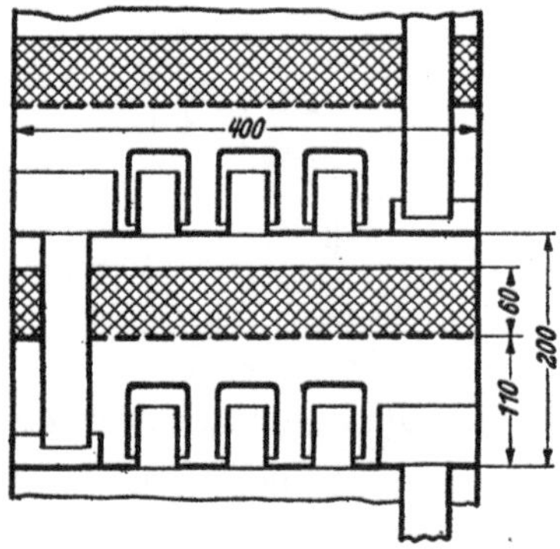

FIGURE 148

Packing layers arranged between bubble plates in a column

of device. By means of optical instruments it was found that even with vapor velocities of 0.1 m per second and less, there is always liquid entrainment if no obstructions are provided between plates. There was also entrainment in the case of baffles. Pure vapors were obtained only by the use of packings.

The results obtained with these liquid separation tests are given in Figure 149. The superficial vapor velocity is plotted against the enrichment ratio. For comparison, conditions in the case where no separator is provided have also been plotted. It is seen that for high velocities, baffles will increase the enrichment ratio, as compared with its value in the case of no separation. Beyond a certain load, however, the enrichment ratio diminishes, just as in the case of ordinary arrangements without baffles. Only the use of packing yields an almost constant effect up to a vapor velocity of 1.4 m per second. This load limit is

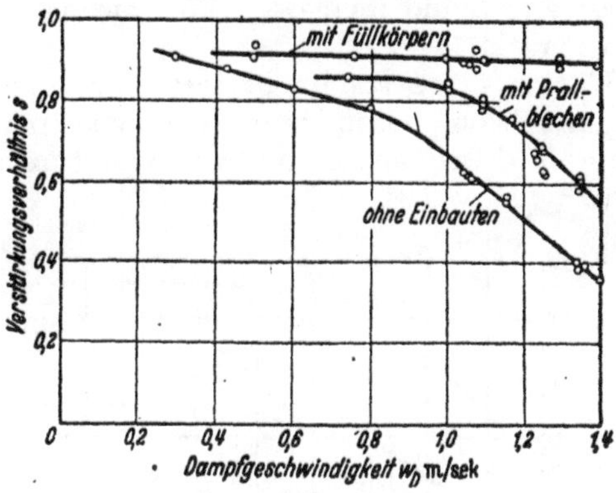

FIGURE 149

Enrichment ratios of interchanger plates, with and without internals added, as a function of vapor velocity

Ethanol-water; reflux ratio $v = \infty$; final plate concentration $x_F = 50$ mol per cent; plate spacing $H = 200$ mm; saddles of 25 mm size; packing height 60 mm; atmospheric operation

Dampfgeschwindigkeit—Vapor velocity
Verstärkungsverhältnis—Enrichment ratio
Mit Füllkörpern—With packing
Mit Prallblechen—With baffles
Ohne Einbauten—Without internals

imposed by the capacities of weir and downspout. The upper load limit may, however, also be imposed by the packing dimensions, since beyond a certain vapor velocity liquid will be entrained even through the layer of packing.

In the tests on which Figure 149 is based, 25-mm Raschig rings and porcelain saddles of the same size were employed.

d. Influence of the depth of immersion on the rectifying efficiency of cap-type plates

The depth of immersion, *e*, in Figure 141, is the vertical distance between the top point of the serrations of the caps, or the edge of the caps, and the edge of the weir. It has a certain influence on plate efficiency.

As the depth of immersion increases, with other conditions remaining unchanged, both the area of contact and the contact-time of vapor and liquid increase. This means an increase in the enrichment ratio.

Figure 150 shows the results of tests investigating the influence of immersion depth upon cap-type plate-efficiency. They were obtained with an ethanol-water mixture and the plate design shown.[72]

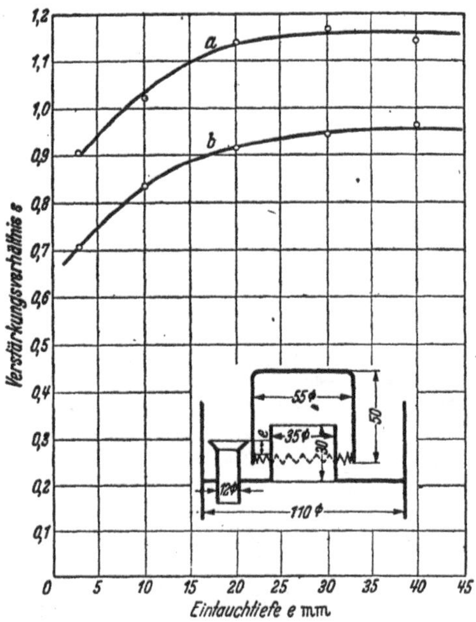

FIGURE 150

Submergence e plotted against enrichment ratio

Serrated bubble cap; ethanol-water; atmospheric operation.
Final plate concentration x_F = 55 mol per cent alcohol; reflux ratio = 1.
a. Enrichment ratio for 0.1 m per second vapor velocity
b. Enrichment ratio for 0.5 m per second vapor velocity
Eintauchtiefe—Submergence depth
Verstärkungsverhältnis—Enrichment ratio

It is seen that for small depths of immersion the enrichment ratio at first increases appreciably as *e* increases. This increase becomes less and less, and, beginning with a depth of about 30 mm the ratio remains constant. This holds for low and high vapor-velocities, and also for concentrations other than those on which Figure 150 is based. Thus, increasing the depth of immersion beyond 30 mm would not cause any further increase in plate efficiency and merely augment the pressure loss, which is undesirable.

With large column diameters, small depths of immersion should be avoided not only to prevent efficiency decrease, but also in view of a possible disturbance of the horizontal level, which would interfere with even vapor distribution.

From Figure 150 it would follow that even with $e = 0$, the enrichment ratio will not become zero. This is explained by the fact that even then the vapor would still pass through the liquid. To keep pressure loss down as far as possible, e is often made zero in vacuum rectifying equipment. Due to the comparatively low enrichment ratio existing in this case, the number of plates has to be made larger. In this way, however, the reduction in pressure loss may again be counterbalanced by the larger plate number.

In the case of sieve plates, tests showed the influence of the depth of immersion to be comparable to that in cap-type plates according to Figure 150. With sieve plates, the depth of immersion is the vertical distance between the level plate and the weir edge. Under the conditions in Figure 150, the depth of immersion reaches its optimum at 25 mm.

e. Influence of cap size and number on the rectification ratio

The rectifying efficiency of any plate depends on the size of the contact surface between phases. Therefore, the number of caps on a plate will have an effect on the enrichment ratio. As the number of caps increases, and, consequently, the size of the caps decreases, the total periphery of caps wetted increases. To obtain a high enrichment ratio, it would thus seem purposeful to have the highest possible number of caps. On the other hand, this means a reduction of the vapor passages in the plate, which may not be desirable in view of possible incrustation and choking. This consideration is mostly non-existent in rectifying columns. Nevertheless, fabrication cost limits the reduction in cap size. Furthermore, it might be doubted whether an increasing number of caps is always tantamount to improved plate efficiency, considering the flow of the liquid.

To clarify this point, the author has run tests with the plates shown in Figure 134 which are equipped with 15 caps each.

Also, plates of the type shown in Figure 151 were employed; these have 4 caps each. Plates like those in Figure 152 were also used, each of these is fitted with one tunnel-cap only, arranged at a right angle to the liquid flow.

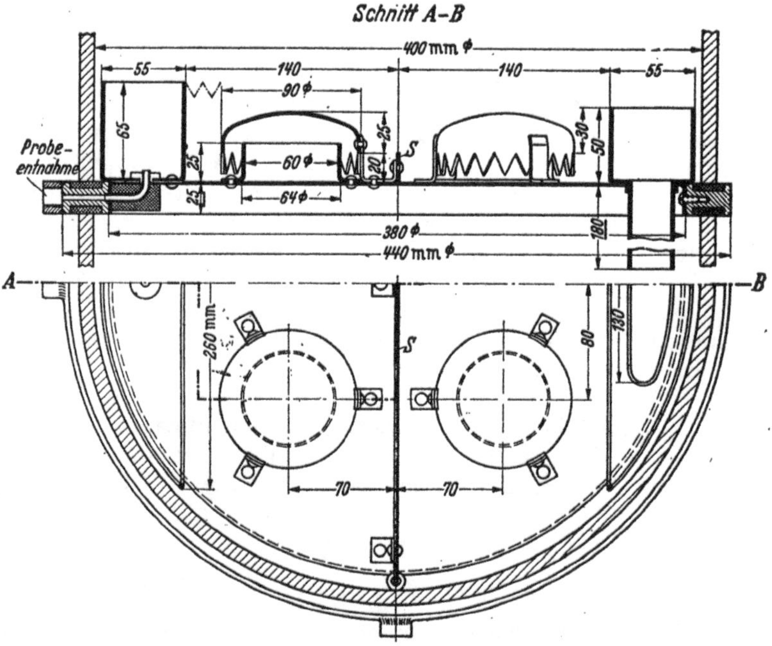

FIGURE 151

Experimental rectifying plate with four bubble caps and central weir S

Schnitt A-B—Section A-B
Probeentnahme—Sampling connection

Test results under equal conditions are plotted in Figure 153, with the superficial vapor velocity as abscissa and the enrichment ratio as the ordinate.

What is surprising is that the plate with 4 caps (curve *b*) with vapor velocities exceeding 0.5 m per second yields better results than the plate with 15 caps (curve *a*). The effect of the plate with one single cap, the nozzle having the outline h_1 of Figure 152, (curve *c*) is, on the whole not much different from that of the

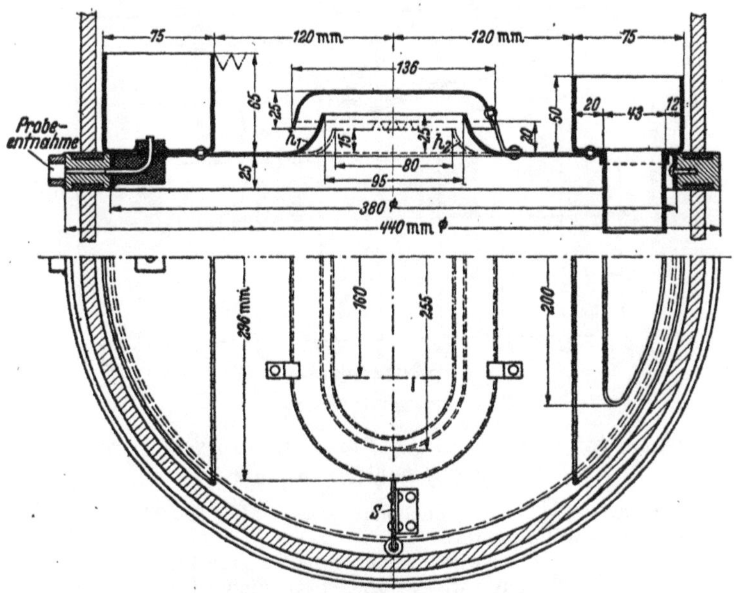

FIGURE 152

Experimental rectifying plate with one bubble cap and weir S

Probeentnahme—Sampling connection

plate with 15 caps. For an average velocity of 0.5 m per second, efficiencies are even identical.

Note, however, partition *S* in Figure 152. This prevents any liquid from flowing along the plate without participating in the interchange. If *S* were omitted, a large portion of the liquid would simply go the way of least resistance.

With a vapor velocity of 1 m per second, the enrichment ratio in the case of the plate having 4 caps is 23 per cent above that of the 15-cap plate. This demonstrates that it is not the number of caps alone that is important, but that other factors, such as flow and distribution of the liquid, are essential.

Curve *d* in Figure 153 refers to the cap-plate (shown in Figure 152), with the nozzle having outline h_2. The upper edge of the vapor nozzle is thus at the same level as the lowest points of the serrations, while in construction h_1 the edge of the vapor nozzle is located 5 mm above the top point of the serrations.

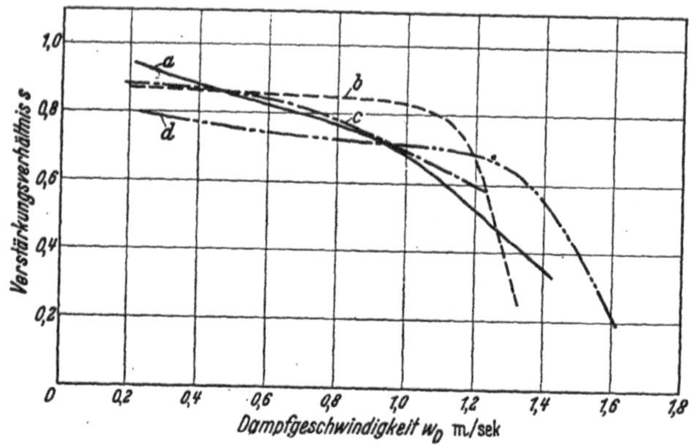

FIGURE 153

Enrichment ratio as a function of load (vapor velocity) for different types of bubble-cap plates

Spacing of plates $H = 200$ mm; final plate concentration 50 mol per cent alcohol in the ethanol-water mixture

a. Plate with 15 bubble caps; Figure 134
b. Plate with 4 bubble cap ; Figure 151
c. Plate with 1 bubble cap; Figure 152
d. Plate with 1 bubble cap; Figure 152

Dampfgeschwindigkeit, m /sek—Vapor velocity, meters per second
Verstärkungsverhältnis—Enrichment ratio

The enrichment ratio in curve *d* in Figure 153 is lower than that in the others, because part of the liquid can flow off the plate as shown in Figure 154, due to the low level of the vapor nozzle and to oscillating movements of the liquid on the plate.

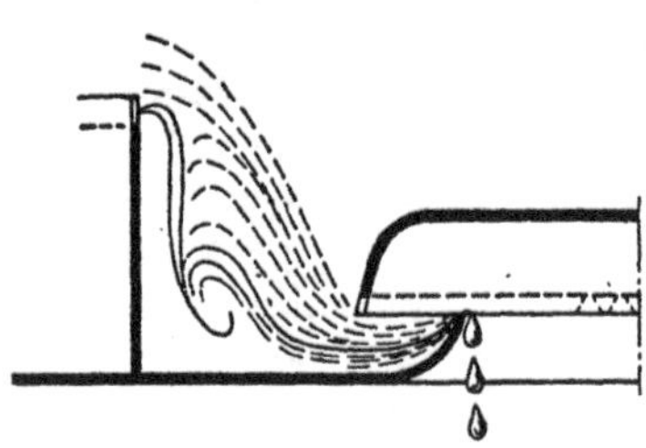

FIGURE 154

Flow conditions for the bubble cap shown in Figure 152

Low-style vapor nozzle; vapor velocity $w_D = 0.5$ m per second; Ethanol-water; final plate concentration 50 mol per cent alcohol; atmospheric operation

The reason why the efficiency of the plate with only one cap with a high nozzle, curve *c*, approximates that of the 15-cap plate, curve *a*, is mainly in the provision of the partition *S*; Figure 152. Considering the fact that the long tunnel-cap was located at right angles to the direction of liquid flow, one might suppose that the partition would not have much of an additional effect. However, comparative tests with plates of the type in Figure 152, with partition on the one hand, and no partition on the other, show definitely that even in this case the enrichment ratio for vapor velocities between 0.6 and 1 m per second was improved by 17 per cent if the partition was installed. With larger column diameters it has been confirmed by experience that columns with partitions will yield a much more uniform liquid distribution than designs featuring no such partitions.

f. Distance between caps and between cap and plate

To arrange the caps for best efficiency, knowledge of the phase movement and liquid distribution on the plates is required. The author conducted tests employing a column with glass sections, and containing rectifying plates as shown in Figure 134.

Figure 155 shows a scale sketch of one of the caps, and presents tests results, the horizontal passage depth l and passage height t having been plotted against the vapor velocity w_D. The serrations of the caps were 15 mm high, the lowest point being located 7 mm above the level plate. The flow conditions shown diagramatically in Figure 155 will occur as a general thing whenever a certain velocity is exceeded. The vapor then appears from under the caps in the shape of thin streams in the liquid, which, in the case of the mixture concerned (ethyl alcohol-water) is transformed into foam above a certain level.

The length l (= horizontal depth of passage) is the deciding fact in the determination of the pitch of the caps. It varies with the load, as shown by curve l in Figure 155. Under the conditions in Figure 155, a pitch of 50–60 mm between caps would be permissible and recommended with high loads for which the circular bubbling zones around the caps would just contact each other. With small vapor velocities, a pitch smaller than 50–60 mm would be advantageous.

To find a suitable distance of the caps from the plate, passage height t should be taken into account. It increases with increasing velocity. In the case of a vapor velocity w_D of 0.7

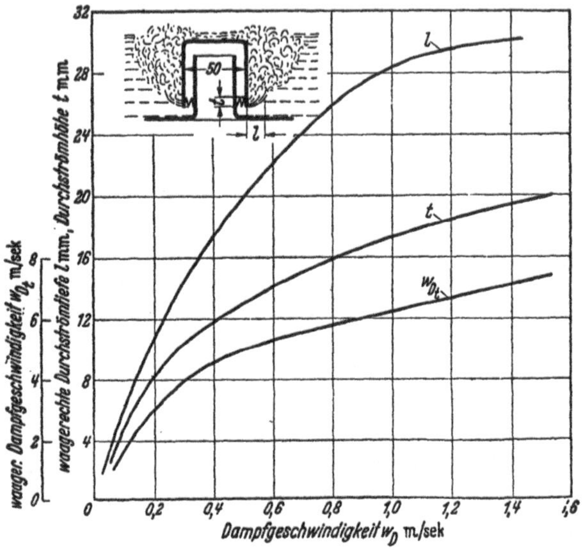

FIGURE 155

Horizontal extension of passage (l), *depth of passage* (t) *and vapor velocity in the horizontal* (w_{Dt}) *plotted against load* (*vapor velocity* w_D)

Ethanol-water; final plate concentration x_F = 50 mol per cent; reflux ratio infinite
Dampfgeschwindigkeit, m /sek—Vapor velocity, meters per second
Waager. Dampfgeschwindigkeit—Vapor velocity in the horizontal
Waagerechte Durchströmtiefe—Horizontal extension of passage
Durchströmhöhe—Depth of passage

per second, the lower limit of the passage height will just reach the lower tips of the serrations.

At higher loads, the vapor moves all around the ends of the serrations. The passage height t depends primarily on the vapor density, and thus on the physical properties of the mixture. As a rule, t increases as density increases, and vice versa.

The data in Figure 155 will hold for atmospheric pressure. With an absolute pressure of 50 mm Hg in the column, the area between serrations is taken up by the vapor only if the vapor velocity reaches w_D = 2.6 m per second, other conditions remaining the same.

From the height of passage and the load, a velocity w_{Dt} may be derived, which is encountered in a horizontal direction between cap serrations, or, in the case of high loads, in the section between the serrations and the ring area immediately below. This velocity w_{Dt} has also been plotted in Figure 155. At low loads it increases most for any given increase in load. As a general thing, values l and t are also influenced by the outlines of nozzles and caps.

Figures 156 and 157 indicate the contours of the vapor streams

FIGURE 156

Fluid flow around bubble cap

Air passes through water; air velocity through nozzle 3 m per second

formed below the caps and the state of movement on the plates. These pictures were obtained with cut-away models, air flowing through water.

The first of the two figures shows the movement for an air velocity of 3 m per second in the nozzle, the second one holds for a velocity of 8 m per second. With the latter, the limit of the gas passage-area below the cap is located at a lower level, and the horizontal passage depth is correspondingly greater.

In both cases, foam forms alongside and above the cap, and liquid drops and lamellae are projected above the foam layer. In the case of non-foaming mixtures, a turbulent layer takes the place of the foam layer.

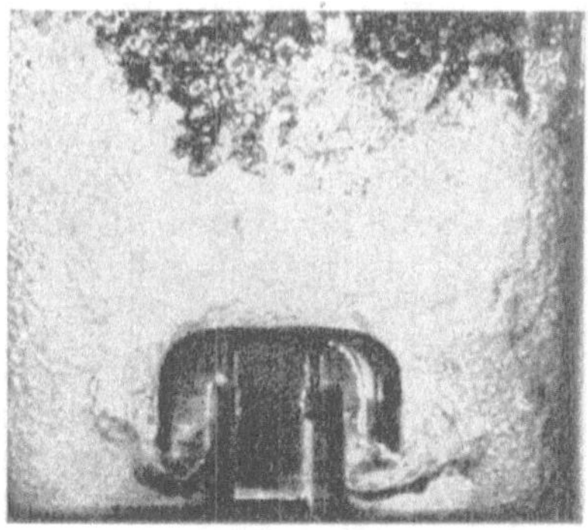

FIGURE 157

Fluid flow around bubble cap

Air passes through water; air velocity through nozzle 8 m per second

Figures 156 and 157 show that it would be purposeless to bring the vapor nozzle up to a level which is appreciably above that of the cap edge. This would yield only an unnecessary increase of the pressure loss in the vapor.

g. *Sloping plates*

To estimate operating results of rectifying plates, it may be interesting to know the influence exerted by an eventual deviation of the plates from a level situation. For this purpose, the author conducted a series of tests with a column fitted with plates of the type shown in Figure 134. The inclination was 1 : 100, one time in the direction of liquid flow, and another time at a right angle to the direction of this flow. In Figure 158, the enrichment ratio has been plotted against the vapor velocity for the two different directions of inclination, as well as for a horizontal plate system with a spacing of 415 mm.

The scattering of the experimental points is due to actual changes in efficiency, which are caused, especially with larger loads, by changes in the direction of the liquid. Only with low vapor velocities will the vapor always pass through the liquid on one side. This condition is shown in Figure 159, which holds for a vapor velocity of 0.15 m per second and a liquid concentration of 50 mol per cent of ethyl alcohol.

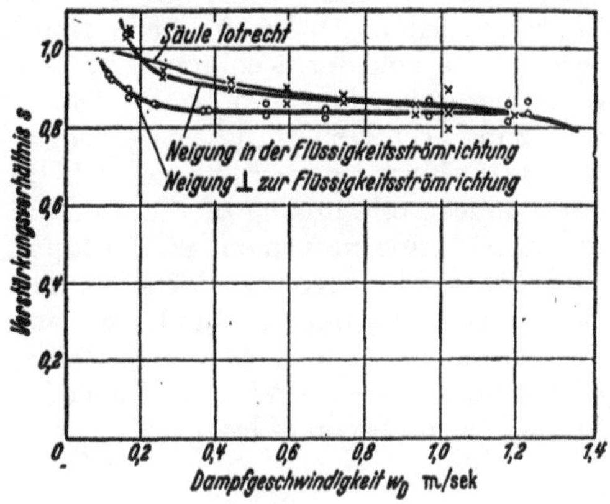

FIGURE 158

Effect of slope of plate upon the enrichment ratio

Plate pitched 1 per cent; ethanol-water; reflux ratio = infinity; plate-terminal concentration x_F = 50 mol per cent; plate spacing H = 415 mm; column diameter 400 mm.
Dampfgeschwindigkeit, m /sek—Vapor velocity, meters per second
Verstärkungsverhältnis—Enrichment ratio
Säule lotrecht—Column in perpendicular position
Neigung in der Flüssigkeitsströmrichtung—Inclination in direction of liquid flow
Neigung ⊥ zur Flüssigkeitsströmrichtung—Inclination at right angle to direction of liquid flow

FIGURE 159

Flow conditions on a sloping rectifier plate

Vapor velocity 0.15 m per second; final plate concentration x_F = 50 mol per cent alcohol; slope 1:100 at right angle to inherent direction of flow.

What is remarkable is that, according to Figure 158, the efficiency is not reduced much by an inclination of the plate. However, these results hold for a column diameter of 400 mm; with larger diameters conditions may be different.

On the other hand, the results may be explained as follows: If at low velocity the vapor penetrates through the liquid only at certain spots, undesirable mixing of the liquid is more easily avoided, and an oscillating movement of the liquid which may be observed in the case of horizontal plates can no longer occur.

Furthermore, in columns out of plumb, a foam layer forms at the points where the vapor breaks through; this will not occur with strictly horizontal plates. With high vapor velocities, the influence of an inclination becomes less.

II. Sieve plates

On perforated (sieve-type) plates, the distribution of liquid and vapor becomes analogous to that on cap-type plates; Figure 160 shows a case of foaming mixtures. The vapor flows in the

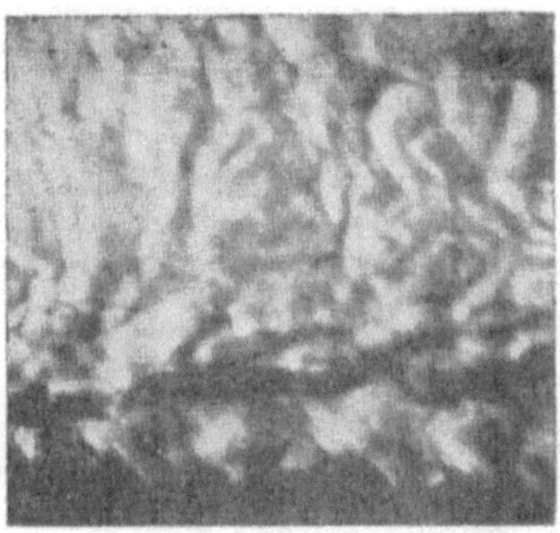

FIGURE 160

Fluid motion above a sieve plate 400 mm in diameter

Ethanol-water mixture with 50 mol per cent alcohol in the liquid phase; diameter of perforations 2 mm, pitch 7 mm; vapor velocity w_D = 0.3 m per second.

shape of *pipes* from the openings into the liquid layer. At their upper ends the ends expand so that they contact each other. Beginning at this point, the liquid of the mixture used in these tests is dispersed into foam. Figure 160 distinctly shows the interface between the transparent liquid and the foam above. The former has, under the conditions shown in Figure 160, a thickness of about 7 mm. This thickness decreases as the vapor velocity increases, and as the diameter of the bubbles becomes smaller.

General conditions being equal, and within a given range of load, the enrichment ratio and therefore the efficiency are, in the case of a sieve-type plate, superior to those of a cap-type plate. This is due to the fact that, with a sieve plate, a coherent bubbling or foam layer is formed over its entire area, which means a larger area of contact between the two phases than would result in a plate type column, where the exchanging area is reduced by the space occupied by the caps.

Figure 161 shows the results of tests run by the author, which

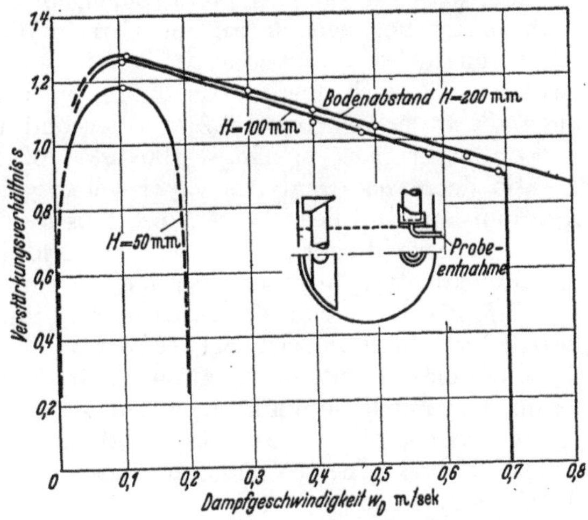

FIGURE 161

Enrichment ratio plotted against vapor velocity for different plate spacings

(Ethanol-water)

Final plate concentration 50 mol per cent; reflux ratio = 3. Column diameter 110 mm; 2 mm-perforations with 7 mm pitch.

Dampfgeschwindigkeit, m/sek—Vapor velocity, meters per second
Verstärkungsverhältnis—Enrichment ratio
Bodenabstand—Plate spacing
Probeentnahme—Sampling connection

explain the interrelation of enrichment ratio s of a sieve-type plate and the steam velocity w_D based on the free column section, for different plate spacings H.

A scale sketch of one of the seven plates used in the tests in the same figure, is shown. These plates were interconnected by glass shells, to permit observation of the phenomena within. The distance from the edge of the weir to the plate was 25 mm. With a vapor velocity of 0.1 m per second best efficiency was obtained, for all plate spacings, with a column diameter of 110 mm.

Efficiency decreases rapidly if the load diminishes below its optimum value, because the liquid then drips through the perforations. If vapor velocity increases beyond 0.1 m per second, the height of the foam layer increases. Despite the fact that the foam layer is always broken at the edge of the weir, it reaches a height sufficing, with the smallest plate spacing of 50 mm, to entirely fill the space between plates, and entrain liquid from plate to plate through the perforations.

For this reason, the enrichment ratio drops, with minimum plate spacing and a vapor velocity of 0.2 m per second, to practically zero. For larger plate spacings, 100 and 200 mm, the enrichment ratio decreases evenly as vapor velocity increases, once the optimum load has been exceeded. This decrease corresponds to the decrease in contact-time of vapor and liquid.

Liquid entrainment plays, in this connection, only a secondary part, since, differently from plates with caps, the liquid is not thrown upwards by a high velocity between caps. Due to the bursting of the bubbles, a state of agitation results immediately above the foam, and but little liquid is carried over.

With a plate spacing of H = 100 mm, and beginning with a vapor velocity of 0.7 m per second, the plates empty themselves, and efficiency is lowered appreciably. In the case of a sieve-type plate, at this velocity the pressure loss in the vapor has become so high that the excess pressure in the space below a plate, over the pressure in the space above, will make the liquid in the downcomers go upwards, until eventually it enters and floods the condenser from the top plate.

With a plate spacing of 200 mm, and under the conditions in Figure 161, this disturbance occurs only with vapor velocities of 0.8 m per second. A load where the plate efficiency decreases intensely due to liquid entrainment, as in the case of plate-type columns, is thus never reached. This low level of the limiting maximum velocity is a disadvantage of sieve-type plates.

Analogous conditions hold for low loads with larger column diameters. With a diameter of 400 mm, and with a perfectly horizontal plate, the liquid will rain through with a vapor velocity of 0.3 m per second; see Figure 162. Thus, below this

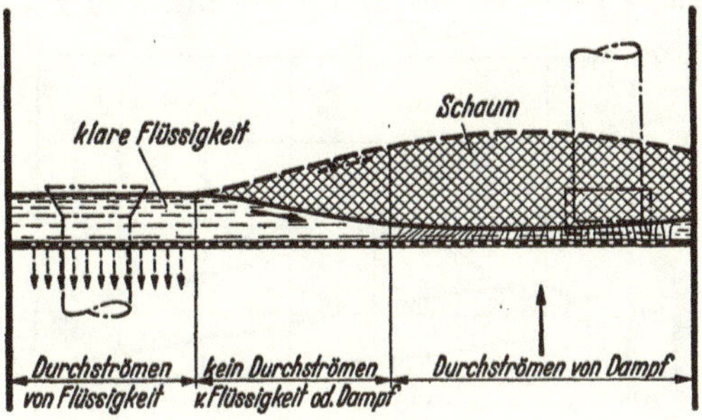

FIGURE 162

Layers of liquid and foam on a sieve plate 400 mm in diameter, operating at subnormal load

(Ethanol-water)

Klare Flüssigkeit—Clear liquid
Schaum—Foam
Durchströmen von Flüssigkeit—Liquid passes through
Kein Durchströmen v. Flüssigkeit od. Dampf—No passage of either liquid or vapor
Durchströmen von Dampf—Vapor passes upwards

load the plate efficiency decreases as the load diminishes. As shown in Figure 162, on the one side of the plate, vapor flows through the holes, forming a foam layer. Then, in the central region of the column, there is an area where there is neither a vapor nor a liquid flow.

In that part of the plate which is opposite to the region where the vapor passes through, liquid flows through the holes, and above these holes there is a perfectly transparent liquid layer. The latter is less thick in the region where there is no flow, and above this liquid the foam layer increases towards the region where vapor passes through.

Within the foam and transparent liquid layers there is a circulating motion, as indicated by arrows and by the direction of the vapor streams within the liquid.

With column diameters exceeding 400 mm, and especially with improperly leveled sieve-type plates, the liquid may begin to rain through even though the vapor velocity exceeds 0.3 m per second. Good leveling and workmanship are particularly important in the case of sieve-type plates. Large-diameter plates require proper stiffening and suitable staying.

A comparison between test curves *a* and *b* in Figure 163 will

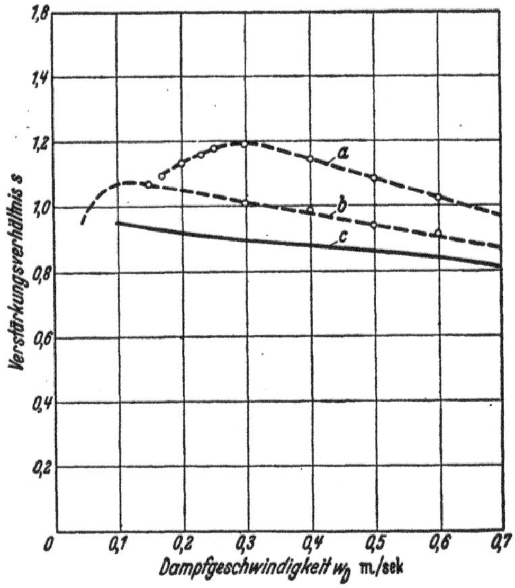

FIGURE 163

Enrichment ratio as a function of vapor velocity

Ethanol-water; final plate concentration 50 mol per cent alcohol; plate spacing 200 mm; reflux ratio $v = \infty$.

a. Sieve plate; column diameter 400 mm.
b. Sieve plate; column diameter 110 mm.
c. Bubble plate; column diameter 400 mm.
Dampfgeschwindigkeit—Vapor velocity
Verstärkungsverhältnis—Enrichment ratio

show that the capacity of sieve-type plates to carry low loads diminishes as the diameter increases, while the upper load limit, imposed by pressure loss in the vapor, stays almost constant, other factors remaining equal.

Within a load range of $w_D = 0.3$ and 0.8 m per second for a diameter of 400 mm and a spacing of $H = 200$ mm, (Figure 163), the larger diameter means, however, a better plate efficiency.

This fact may be due to the circumstance that, the larger the column diameter, the more liquid will flow through a unit of area per unit of time, other things being equal. The quantities passing through the column increase as the second power of the diameter, and the free area only as the first power, as long as the other dimensions involved remain unchanged.

For larger column diameters, this will increasingly prevent the mixing of liquid on the plates, which is always more or less present, and the mutual motion of the two phases becomes more and more of the cross-current type which is favorable to interchange.

With very large diameters, another influence, which may considerably impair the rectifying effect, may be superposed on the favorable influence just discussed. Then, the liquid no longer passes uniformly across the plate, but may have an increased velocity in some spots, and no motion at all in others. This circumstance must be kept in mind in the case of large column diameters; long parallel inlet and outlet weirs may be the answer; see Figure 151.

The importance of the area of contact between phases which results on a sieve-type plate will depend on the physical characteristics of the mixture, load, diameter of holes and their pitch. Where centers are at the apices of equilateral triangles as usual, the ratio ϵ = Pitch/Diameter is the deciding factor for the interchange efficiency. Where ϵ is large, a large, transparent and coherent liquid layer, which takes no part in the material interchange, will form on the plate. With small ϵ values, however, the vapor streams over the holes will merge, so that the vapor will hardly contact the liquid.

In the first case, there is a large pressure loss in the vapor. Tests [73] on sieve plates having holes of 1 to 2.5 mm diameter and varying pitch, yielded an optimum ratio of $\epsilon = 3.8$. In industrial practice, usual diameters are 2, 2.5, and 3 mm, to which correspond pitch values of 7, 9 and 10 mm. However, smaller diameters, down to 0.8 mm, as well as diameters larger than 3 mm are occasionally encountered.

One may conceive the idea of creating a large area of contact between vapor and liquid by making the vapor stream atomize the liquid on the plates. This may be achieved by letting the liquid flow from a level plate without forcing it over a weir. The upper edge of the downflow pipe then merges with the plate level, and the liquid will spread in a thin layer on the plate and will be atomized by the vapor flowing through the holes in a direction vertical to the path of the liquid.

Tests, conducted by Kirschbaum, with sieve-type plates having holes 2 mm wide and a 7-mm pitch indicate, however, that the atomization effect obtainable is not sufficiently perfect to obtain a rectifying action comparable to that which can be reached when a weir, which permits the formation of a foam layer, is provided.

The author also made certain tests in which corrugated plates with holes in the sides of the corrugations were employed, so that a sort of cross-current flow between liquid and vapor resulted. These gave results which were no better than for ordinary plates with weirs. It would thus seem that the latter design is the preferable one, as far as sieve-type plates are concerned.

III. Comparison of cap-type plates and sieve-type plates

1. Efficiency and incrustation

The advantage of the sieve-type plate, as compared with the cap-type plate, resides in its lower cost of fabrication, and in the higher interchange efficiency within its load range. Figure 163 contains curve *c*, which shows the efficiency of the cap-type plate in Figure 134 as a function of the load. It is seen that with a vapor velocity $w_D = 0.5$ m per second, a sieve-type plate (curve *a*) yields a better (25 per cent higher) interchange ratio than a cap-type plate, other conditions being equal.

However, the limitation of efficiency by the load range gives a considerable disadvantage to the sieve plate.

Cap-type plates have good efficiencies at vapor velocities of 0.1 m per second but at this vapor velocity the liquid would fall down through the holes of a sieve-type plate with a column diameter of 400 mm; see Figure 162. This raining-through of liquid is the more intense the larger the column diameter, and the more the plate is out of a true horizontal position.

Upper load limits also differ due to the high pressure loss inherent in sieve plates, so that, at a certain vapor velocity, the liquid will no longer flow from the plate and the plate becomes inoperable. This load range is higher in the case of the cap-type plate which is, therefore, less sensitive to load fluctuations.

The main disadvantage of sieve plates is their great tendency to clog up and to become incrusted. There are many instances where, apart from the solids contained in the material worked, crystallization takes place within the column, due to changes of temperature and concentration. These crystals may tend to obstruct the passages more and more, despite high vapor velocities.

Experience, as well as experiments by the author, demonstrate that the forces of attraction between a metal wall and the crystal particles being formed, or between the crystals present and those being formed, are much greater than the forces in the flowing vapor or liquid. When the holes are obstructed the pressure loss in the vapor increases until the return pipes can no longer handle the liquid.

Therefore, sieve plates are unsuitable for rectifying contaminating or incrusting liquids. They are rarely used, if at all, for rectifying columns; they may be satisfactory for stripping columns in special uses, such as the manufacturing of liquid air.

Incrustation in rectifying columns may be counteracted by suitable design. The incrustations should be given sufficient space to settle, so that passages for vapor and liquid remain available. Even with larger column diameters, only one cap may be provided, with the result that passages are ample, whether the plate be clean or contaminated. Even cap-type or tunnel-type plates may be clogged, if their vapor passages are insufficient. This is shown, for instance, by tests of ring-tunnel plates; Figure 164 and 165: Here, passages are narrow

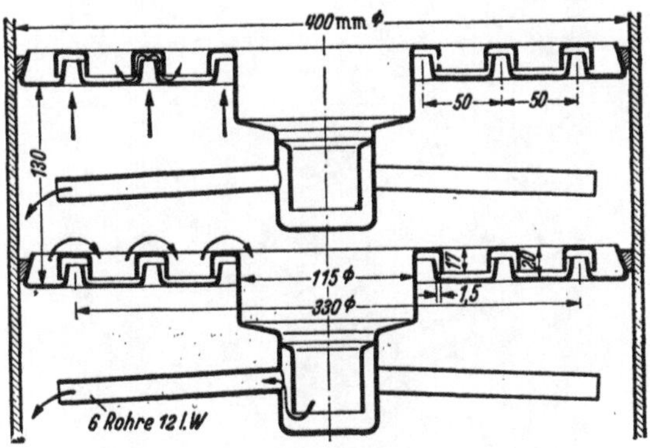

FIGURE 164

Plates with annular tunnels, Kühni type

6 Rohre 12 l.W.—6 tubes, 12 mm inside diameter

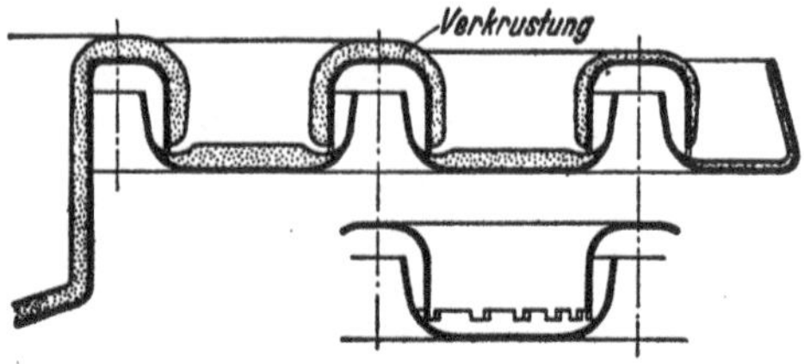

FIGURE 165

Scaling formed during six operating hours

Detail shows design of elements

Verkrustung—Scaling

and vapor velocities are high. During the tests, 200 kg per hour of water with a hardness of 17 g CaO per 100,000 g of water, were fed to the top plate of a stripping column at a temperature of 15°C. The superficial steam velocity below the top plate amounted to 0.3 m per second. As Figure 165 shows, the upper plates were, after 7 hr of operation, so heavily incrusted that the liquid accumulated on the plates because of the high pressure loss in the steam. Figure 166 shows the state of incrus-

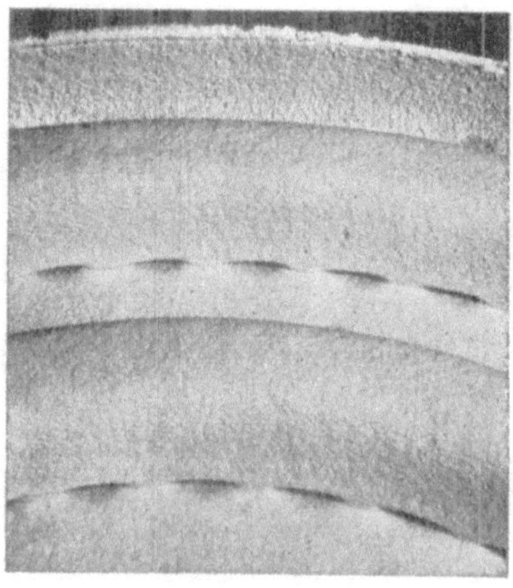

FIGURE 166

Incrustation formed during four hours of operation

tation reached after 4 hours of operation. It is seen that a constricted vapor passage was still available. According to these tests, even a vapor flowing at a rate of 50 m per second cannot dislodge the crystals formed.

However, in the case of cap-type or tunnel-type plates it is possible to arrange the caps or tunnels at a certain distance from the plate surface, and to make passages so ample that such plates will work satisfactorily whether or not some incrustation is present. Sieve plates cannot fulfill this requirement.

Even where there is no tendency to incrustation or contamination, holes in sieve-type plates will wear out in the course of a few years of operation, depending on the materials of construction and the physical properties of the substances worked.

2. Pressure loss caused by rectifying plates

Sieve-type plates on the one hand, and cap-type plates on the other, are different not only as far as their load-carrying capacity is concerned, and their sensitivity to incrustation and clogging, but also in their pressure losses. The pressure loss of a plate is equal to the pressure-difference between the vapor spaces above and below a plate. Since this pressure loss will prevent the flowing of the reflux if it is large enough, it is also connected with the load value.

The total pressure loss is composed of two individual losses. One of these is a static pressure caused by the fact that with both sieve and cap plates the vapor will enter the liquid above the plate in a spot above which there is a certain head of either liquid or liquid and foam. The static pressure is proportional to the height of this layer. The other individual loss comprises the impact, friction and contraction involved in the flow of vapor through passages. This *streaming* loss increases with the vapor density and approximately as the square of the velocity. In the case of small vapor velocities the static pressure loss is critical; with high vapor velocities the streaming loss is more important, especially with high vapor densities.

Generally speaking, the pressure loss depends on such properties of the liquid as tendency to foaming, surface tension, viscosity and density. Besides the main individual losses mentioned, there are additional losses, such as those due to the necessity of destroying the cohesion between the bubbles of a foam layer, after the surface tension has been overcome, and those occasioned by agitation which makes drops and lamellae jump upwards from the plate, to which they subsequently fall back. This obstructs vapor passage to a certain extent and creates an impact loss. As a consequence, plate spacing also has some influence on the total pressure loss.

The interdependence of pressure loss and load is different in the case of sieve plates on the one hand, and cap plates on the other. This is due to two facts. First, the thicknesses of the transparent liquid and foam and turbulence layers, which are critical in the case of low loads, differ. Second, the ratio of the column section F and the minimum vapor passage f differs, and this is critical in the case of higher vapor velocities.

The ratio F/f will be called ϕ. In the case of the sieve plate, f equals the sum of the sections of the holes. In the case of a cap-type plate, it must be determined which of the sections f_1, f_2, f_3 and f_4 is the most constricted one; see Figure 167.

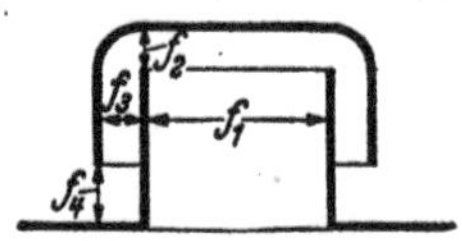

FIGURE 167

Vapor passages of bubble caps

f_1 is the section of the vapor nozzle, f_2 that between cap ceiling and edge of nozzle, f_3 the annular passage between nozzle and cap, and f_4 the area between cap edge and plate.

The smallest area is the deciding factor in regard to the pressure loss. Therefore, f in this case equals the sum of all of these minimum values for one plate. Generally speaking, the higher ϕ values apply to sieve plates. For a plate with holes of 2 mm diameter and 7 mm pitch, and a distance of 30 mm between edge of weir and surface of plate and a plate diameter of 400 mm, ϕ would equal 21.

In Figure 168, line *a* shows the interdependence of pressure loss and vapor velocity for this plate; the absolute pressure in the column is 760 mm Hg, vapor density 1.254 kg per cubic meter and the plate is charged with an ethanol-water mixture containing approximately 50 mol per cent of alcohol. All the curves in Figure 168 hold for this liquid concentration.

The outline of curve *a* indicates that the pressure loss is due

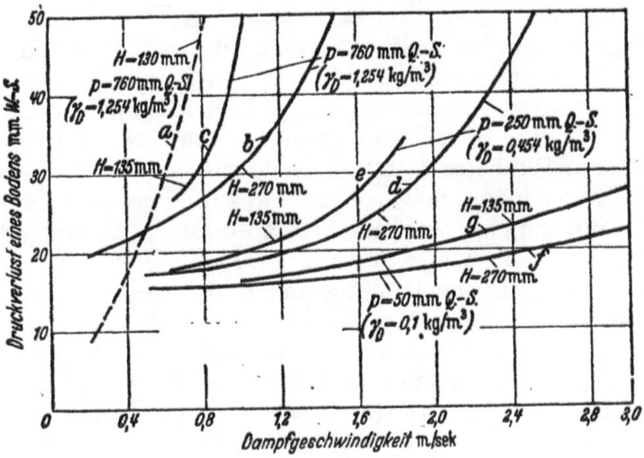

FIGURE 168

Pressure drops over plates plotted against load at various vapor pressures (densities)

(Ethanol-water mixture)

——— Bubble-cap plates according to Figure 134

– – – – Perforated plates (Sieve plates)

Dampfgeschwindigkeit, m/sek—Vapor velocity, meters per second

Druckverlust eines Bodens, mm WS—Pressure drop per plate, millimeters water

mm Q.-S.—mm Hg

principally to the streaming losses. The ϕ ratio in the case of the cap-type plate is much smaller than in that of the sieve-type plate; its average is around 10. The smaller ϕ is, the larger is the share of the static pressure in the total pressure loss. The larger ϕ becomes, the more will pressure loss increase, in the case of high vapor velocity and increasing load.

In the case of the cap-type plate shown in Figure 134, ϕ equals 9.2. Curve *b* in Figure 168 shows the pressure loss of this cap-type plate as a function of the load for a plate spacing of 270 mm, while curve *c* holds for a spacing of 135 mm. In both cases the pressure in the column is 760 mm Hg; for the alcohol-water mixture used, a vapor density of 1.254 kg per cubic meter corresponds to this column pressure.

It is seen that, for very small loads, the pressure loss of a cap-type plate exceeds that of a sieve plate. This is due to the fact that, under the conditions obtaining in Figure 168, a high foam layer accumulates on the sieve plate and the liquid layer to be penetrated by the vapor is only a few millimeters thick, while in the case of the cap-type plate the layer of transparent liquid is much thicker with the same load.

With larger loads, however, the pressure loss of a sieve plate becomes much higher than that of a cap-type plate. A comparison of curves *b* and *c* indicates that the pressure loss decreases as plate spacing increases, due to entrainment of liquid by the vapor. This decrease is larger the higher the vapor velocity and as shown by the curves *de* and *fg* the higher the vapor density becomes. Vapor densities of 0.454 and 0.1001 kg per cubic meter, respectively, correspond to these two pairs of curves, which also apply for plate spacings of 135 and 270 mm.

Curves *c* to *f* show that, with low velocities, the static pressure controls the total pressure loss. The static pressure will not depend much on the load. For a specific weight of the vapor of 0.1001 kg per cubic meter the pressure is nearly constant (16 mm water), up to a vapor velocity of about 1 m per second. For higher vapor densities, the streaming losses become more important, other factors remaining constant. Generally speaking, their influence becomes more noticeable the more pressure increases with the load.

Reducing the depth of immersion (distance of the top point of the cap serration, or of the straight cap edge from the weir edge; *e* in Figure 141) reduces the pressure loss. With a vapor density of 1.254 kg per cubic meter (according to experiments by the author) the pressure loss, for a vapor velocity of 0.5 m per second and a depth of immersion of 10 mm, amounts to only 57 per cent of the loss in the case of an immersion depth of 30 mm, other circumstances being equal. For a vapor velocity of 0.2 m per second, the pressure loss with the smaller immersion depth amounts to 62 per cent of that with the larger immersion depth.

IV. Influence of the direction of flow of the phases on the rectification effect of a plate; liquid mixing, vapor mixing, counterflow and parallel flow

1. General remarks

Assume that the reflux liquid and the vapor flowing to a rectifying plate remain for a certain time in a vessel impermeable to heat, and are then subjected to heat and material exchange for an infinite length of time. At the end of that time, the liquid and the vapor above it will be in phase equilibrium with each other.

On the basis of this fact, it may readily be assumed that, in determining the number of plates of a rectifier the perfect plate would yield a vapor rising upwards, containing $x_{D_o}{}^I$ mol per cent of low boiler, in equilibrium with the liquid leaving the same plate, and containing according to Figure 169, a concentration of x_{Fa} mol per cent of low boiler in a binary mixture.

$$x_{D_o}{}^I \;/\!/\!/\; x_{Fa}$$

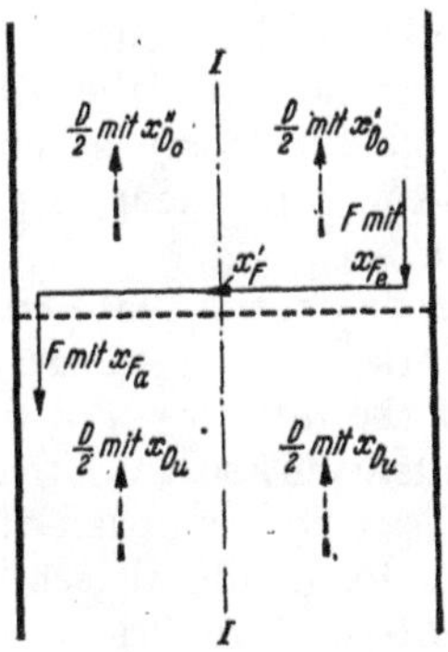

FIGURE 169

Bipartition scheme

Symbols indicate individual streams with corresponding concentrations

Let x_{Du} and x_{Fe} = concentrations of low boiler in the vapor and in the liquid reaching the plate. The theoretical enrichment in low boiler of the vapor then

$$a = x_{D_o}{}^I - x_{Du}$$

Actually, the enrichment in low boiler in the vapor only amounts to b mol per cent. The enrichment ratio (interchange ratio)

$$s = \frac{b}{a}$$

is of some help in determining the number of plates of a rectifier; (Sections D, IV, 1, 2, 3 and E, I). However, it will not permit the indication of the degree of perfection of a plate, or even of its efficiency; as a matter of fact, s may be much greater than unity. The extent to which it would increase in the case of perfect operation may be found from the maximum enrichment under ideal conditions. From the latter, one can also derive the conception of efficiency.

2. Graphical method of determining the theoretical enrichment, the theoretical enrichment ratio, and the efficiency, in the case of cross-flow of liquid and vapor, on a rectifying plate[74]

In a rectifying column, the vapor flows between the plates parallel to the axis of the column. In the downspouts, the liquid also moves parallel to the column axis; on the plate itself, however, it moves vertically to the axis; Figure 169. The entire mass of liquid moves crosswise on the plate, so that it encounters the part of the vapor near its entering point first, and that near its exit point last.

Assume that the vapor on the plate considered is of uniform composition at all points where it enters the plate. The molar heat of evaporation of the mixture is again assumed to be independent of the concentration, so that the masses of liquid and of vapor entering and leaving are constant.

Assume, further, a longitudinal section *I–I* through the column. Then, the interchange described between vapor and liquid will happen on the right-hand and left-hand sides of the section. Both the right-hand and the left-hand side of the plate receive a mass of vapor $D/2$ mols per hour, with the same concentration x_{Du} mol per cent. In the middle plane, the liquid concentration amounts to x_F'. In the left-hand section, the rising vapor has concentration x_{Do}'', in the right-hand one, x_{Do}'. With perfect mixing of liquid and perfect interchange in both sections, we have

$$x_{Do}'' \;///\; x_{Fa} \quad \text{and} \quad x_{Do}' \;///\; x_F' \tag{170}$$

For the right-hand side of the plate, F mols per hour being refluxed, we have the following quantity equation

$$F \times x_{Fe} + \frac{D}{2} \times x_{Du} = F \times x_F' + \frac{D}{2} \times x_{Do}'$$

from which

$$\frac{D}{2}\,(x_{Do}' - x_{Du}) = F(x_{Fe} - x_F') \tag{171}$$

Analogously, for the left-hand section

$$\frac{D}{2}(x_{Do}'' - x_{Du}) = F(x_F' - x_{Fa}) \tag{172}$$

If E mols of product per hour are withdrawn from the column

$$v = \frac{F}{E}$$

which is the reflux ratio, for which one has

$$D = E(v + 1)$$

This transforms equations (171) and (172) into

$$\frac{x_{Do}' - x_{Du}}{x_{Fe} - x_F'} = 2 \times \frac{v}{v+1} \tag{173}$$

$$\frac{x_{Do}'' - x_{Du}}{x_F' - x_{Fa}} = 2 \times \frac{v}{v+1} \tag{174}$$

The average concentration x_{Do}^{II} of the vapor rising from the plate follows from

$$\frac{D}{2} \times x_{Do}' + \frac{D}{2} \times x_{Do}'' = D \times x_{Do}^{II}$$

$$x_{Do}^{II} = \frac{x_{Do}' + x_{Do}''}{2} \tag{175}$$

This concentration may be found graphically on the basis of equations (173), (174) and (175); Figure 170. Next to the equilibrium curve, the enrichment line of a binary mixture is shown. Its slope α against the abscissa follows from equation (68)

$$\tan \alpha = \frac{v}{v+1}$$

The relation between x_{Du} and x_{Fa} is supplied by point A located on the enrichment line, while a parallel to the ordinate passing through abscissa value x_{Fa} will intersect the value for concentration x_{Do}'' at point B of the equilibrium curve. If a straight line having a slope of

$$\tan \alpha' = 2 \tan \alpha \tag{176}$$

is passed through point A, the parallel to the abscissa passing through B will supply point C with abscissa value x_F'. Equation (174) is then satisfied.

Since $x_{Do}' \,///\, x_F'$, point D will yield x_{Do}'. Point D, with

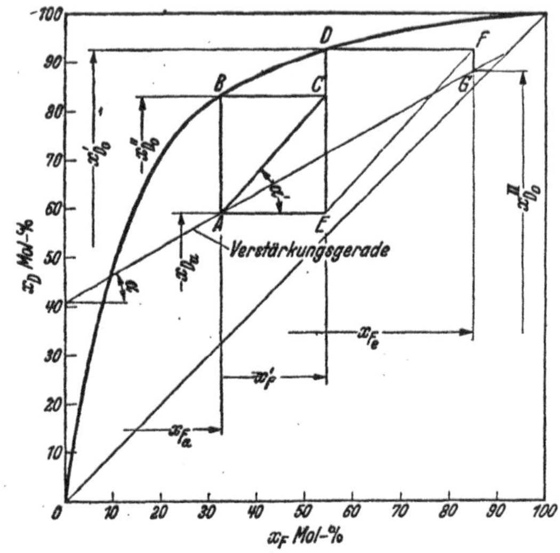

FIGURE 170

Determination of liquid and vapor concentrations for the bipartition scheme

Verstärkungsgerade—Enrichment line

abscissa value x_F', is located on the equilibrium curve. Point E is the intersection of the horizontal through A with the vertical through C. A straight line through E with slope α' intersects the parallel to the abscissa passing through D at point F having abscissa value x_{Fe}, which is in agreement with equation (172).

Abscissa value x_{Fe} will yield point G on the enrichment line, corresponds to G vapor concentration $x_{Do}{}^{II}$, which is the arithmetic mean between x_{Do}' and x_{Do}''.

While, in the case of perfect liquid mixing on a plate and of perfect material interchange, the vapor rising from a plate will have a concentration of $x_{Do}{}^{I}$, where $x_{Do}{}^{I} /// x_{Fa}$, splitting the column into two halves will yield an average concentration of $x_{Do}{}^{II}$ in the vapor rising upwards, for which

$$x_{Do}{}^{II} > x_{Do}{}^{I} \tag{177}$$

It is seen that cross-current is more efficient than liquid mixing.

Instead of dividing the column into two sections, one may also divide it into three, through each of which the same amount of vapor $D/3$ then passes. Assuming that in each of the sections liquid mixing and perfect interchange take place, the vapor rising from the plate in each part will be in equilibrium with the liquid leaving that part. Using the relations given in Figure 171, we have

$$x_{Do}' \,/\!/\!/\, x_F'; \quad x_{Do}'' \,/\!/\!/\, x_F''; \quad x_{Do}''' \,/\!/\!/\, x_{Fa} \tag{178}$$

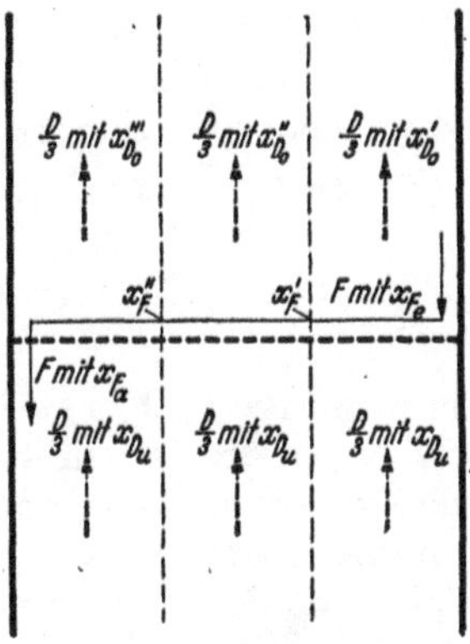

FIGURE 171

Tripartition scheme

Symbols refer to streams and respective concentrations.

For this tripartition, new equations corresponding to (173) and (174) appear

$$\frac{x_{Do}' - x_{Du}}{x_{Fe} - x_F'} = 3 \times \frac{v}{v+1} \tag{179}$$

$$\frac{x_{Do}'' - x_{Du}}{x_F' - x_F''} = 3 \times \frac{v}{v+1} \tag{180}$$

$$\frac{x_{Do}''' - x_{Du}}{x_F'' - x_{Fa}} = 3 \times \frac{v}{v+1} \tag{181}$$

The average concentration of the vapor rising from a plate becomes

$$x_{Do}^{III} = \frac{x_{Do}' + x_{Do}'' + x_{Do}'''}{3} \tag{182}$$

A graphical method, based on equations (179) to (182), may be applied analogously to the procedure explained for the case of the bipartition in Figure 170.

For the same final plate concentration (plate effluent concentration) x_{Fa}, and the same reflux ratio

$$x_{Do}{}^{III} > x_{Do}{}^{II} > x_{Do}{}^{I} \tag{183}$$

and

$$(x_{Do}{}^{III} - x_{Do}{}^{II}) < (x_{Do}{}^{II} - x_{Do}{}^{I}) \tag{184}$$

Splitting the column section into four equal parts would, consequently, yield an average concentration $x_{Do}{}^{IV}$, for which

$$x_{Do}{}^{IV} > x_{Do}{}^{III} > x_{Do}{}^{II} > x_{Do}{}^{I} \tag{183a}$$

and

$$(x_{Do}{}^{IV} - x_{Do}{}^{III}) < (x_{Do}{}^{III} - x_{Do}{}^{II}) < (x_{Do}{}^{II} - x_{Do}{}^{I}) \tag{184a}$$

Going farther, one might derive $x_{Do}{}^{V}$, $x_{Do}{}^{VI}$, $x_{Do}{}^{VII}$, etc. values. The larger the number of partitions, the smaller the differences between two subsequently obtained concentrations.

By means of this method, the average concentrations in the vapor rising from a plate could be determined for an alcohol-water mixture, when the reflux ratio v is 1, the distillate concentration x_E, is 75 mol per cent of alcohol, and the plate effluent concentration x_{Fa}, is 30 mol per cent of alcohol, for increasing numbers of partitions. The results are presented in Figure 172.

The abscissa indicates the number of partitions, and the ordinate the corresponding average concentrations in the vapor above the plate. For the effluent concentration given, $x_{Fa} =$ 30 mol per cent, the corresponding point of the enrichment line (Figure 170, point A) yields an alcohol concentration x_{Du} of 52.4 mol per cent in the vapor entering the plate. The parallel to the ordinate through abscissa value $x_{Fa} = 30$ mol per cent yields an intersection (B in Figure 170) on the equilibrium curve with the ordinate value $x_{Do}{}^{I} = 56.5$ mol per cent. In the case of liquid mixing across the entire plate, the column section being considered as a single part only, computation yields a theoretical enrichment

$$a = x_{Do}{}^{I} - x_{Du} = 56.5 - 52.4 = 4.1 \text{ mol per cent}$$

also shown in Figure 172.

A bipartition yields, according to Figure 172, an average concentration of $x_{Do}{}^{II} = 57.2$ in the vapor rising from the plate;

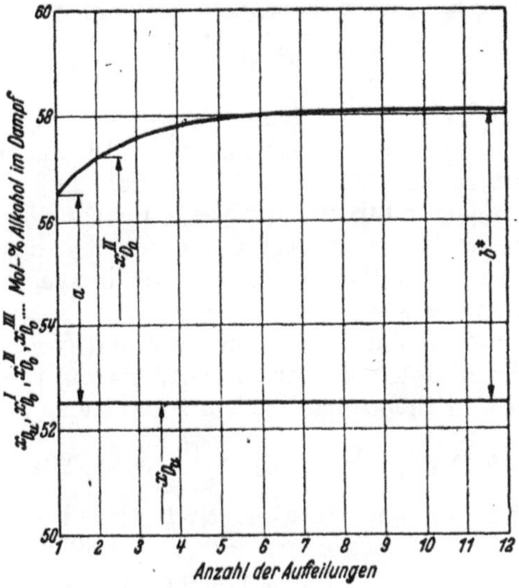

FIGURE 172

Average vapor concentration plotted against number of subdivisions

Anzahl der Aufteilungen—Number of subdivisions
Mol % Alkohol im Dampf—Mol per cent alcohol in the vapor phase

with a six-fold partition the average $x_{Do}^{VI} = 58$ mol per cent would result.

According to Figure 172, the curve indicating the interrelation of vapor concentrations and column partitions asymptotically approaches $x_{Do}^{\infty} = 58.1$ mol per cent which value could be obtained with an infinite number of partitions. It is seen that a satisfactory approximation to x_{Do}^{∞} is achieved by eight-fold partition.

With cross-current flow, and assuming the vapor formed is always in equilibrium with the liquid from which it was generated, the maximum possible enrichment in low boiler in the vapor therefore follows from

$$b^* = x_{Do}^{\infty} - x_{Du} \qquad (185)$$

Equal concentration of the vapor entering the plate at all points can always be obtained by mixing the vapor before it enters the plate. Using optimum enrichment as given by equation (185), one obtains a theoretical enrichment ratio of

$$s^* = \frac{b^*}{a} \tag{186}$$

Figure 173 shows s^* as a function of effluent concentration x_{Fa} for different reflux ratios, for an ethanol-water mixture, and for a product concentration x_E of 75 mol per cent of alcohol. In the case of vapor mixing, the value for the enrichment ratio s^* by far exceeds unity in certain ranges of concentration. It has a minimum for all reflux ratios, and increases for low and for high concentrations.

Efficiency η of a plate may now be found from the ratio between actual and optimum enrichment, or from the interrelation of the actual and theoretical enrichment ratios

$$\eta = \frac{b}{b^*} \times 100 = \frac{s}{s^*} \times 100 \text{ (per cent)} \tag{187}$$

Assume that for a mixture of ethyl alcohol and water, with $v = 4$, $x_E = 75$ mol per cent, and $x_{Fa} = 50$ mol per cent of alcohol, enrichment ratio s has a value of 0.9. The belief that the efficiency then equals 90 per cent is often encountered but is erroneous. Figure 173 reveals that, under the conditions

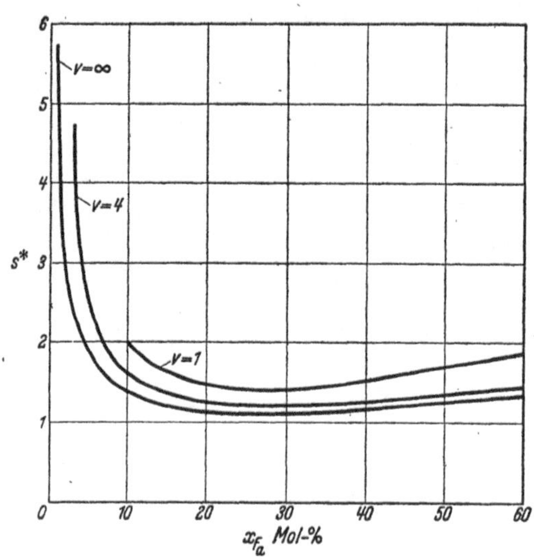

FIGURE 173

Theoretical enrichment ratio s^* plotted against final plate concentration x_{Fa}, Ethanol-water, x_E = 75 mol per cent

given, the theoretical enrichment ratio equals 1.32, so that the efficiency actually becomes

$$\eta = \frac{0.9}{1.32} \times 100 = 68 \text{ per cent}$$

The difference between the numerical values of η and $s \times 100$ depends largely on the final plate (effluent) concentration. In the case mentioned, the difference $100s - \eta$ may be much greater than the computed one which equals 90–68 = 22, if the final concentration x_{Fa} is low; see Figure 174. Therefore, the

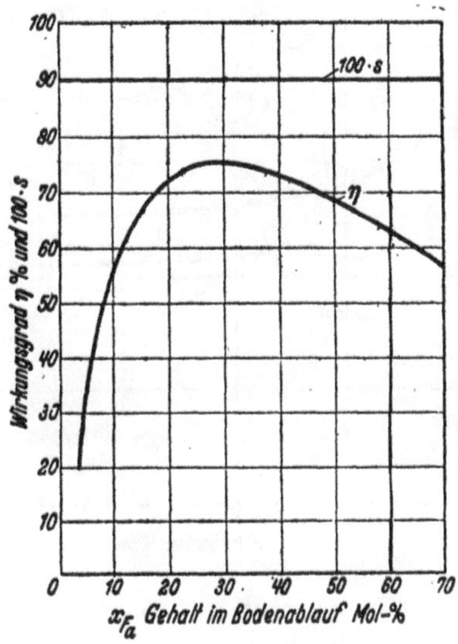

FIGURE 174

Discrepancy between 100·s and actual efficiencies

Ethanol-water; 75 mol per cent low boiler in distillate; reflux ratio $v = 4$.
Gehalt im Bodenablauf Mol % — Mol per cent in liquid leaving plate
Wirkungsgrad η % und 100 · s — Per cent efficiency, η and (100 · s)

belief that $s \times 100$ could be regarded as an efficiency figure, is definitely wrong.

On a rectifying plate, cross-current flow between liquid and vapor is achieved more perfectly the longer the path of the liquid, and better mixing of the liquid is thereby prevented. This is shown by comparative tests, the results of which are presented in Figure 175. The abscissa indicates the reflux ratio, and the ordinate the enrichment ratio found, as well as the final plate concentration.

These tests were run on a mixture of ethyl alcohol and water with a column 110 mm in diameter. A long liquid path was obtained by the use of an annular sieve-type plate. The efficiency of the latter was better than that of a sieve plate without flow control.

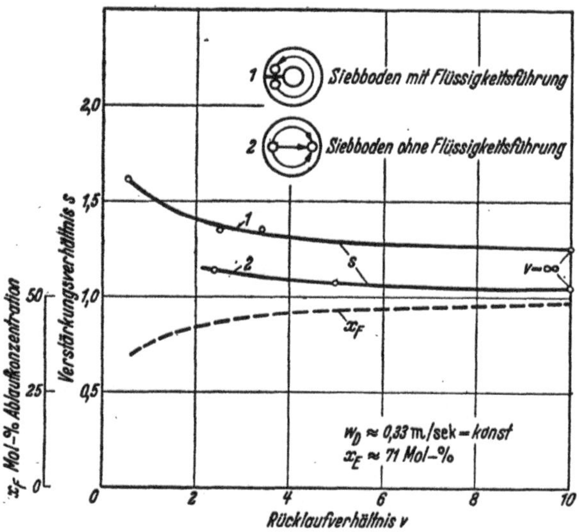

FIGURE 175

Terminal concentration and enrichment ratio plotted against reflux ratio, on sieve plates designed for controlled and uncontrolled liquid flow, respectively

Siebboden mit Flüssigkeitsführung—Sieve plate with controlled flow
Siebboden ohne Flüssigkeitsführung—Sieve plate with uncontrolled flow
Rücklaufverhältnis—Reflux ratio
Verstärkungsverhältnis—Enrichment ratio
Mol % Abflusskonzentration—Mol per cent terminal concentration

Figure 175 reveals the same interdependence of s and v shown in Figure 173. However, in the case of columns of large diameter, the unfavorable influence of an irregular liquid flow across the plate superposes itself on the favorable influence of the long liquid path. This situation can be improved by skilful design; see section H, I, e.

It will be assumed that the vapor is introduced to a rectifying plate at uniform concentration across this plate. This, however, will only apply to the bottom plate of a column which receives its vapors immediately from the still. The vapors will escape from the plates with a composition that varies along the path of the liquid. Other things being equal, this variation will depend on the direction in which the liquid flows on the individual plates.

As a rule, the reflux reaches the plate at a point near the circumference of the shell, where it is opposite to the next downpipe. The liquid then moves in a direction opposite to that of the liquid on the plates above and below. This counterflow condition is illustrated in Figure 176, which also indicates the concentrations which result by tripartition.

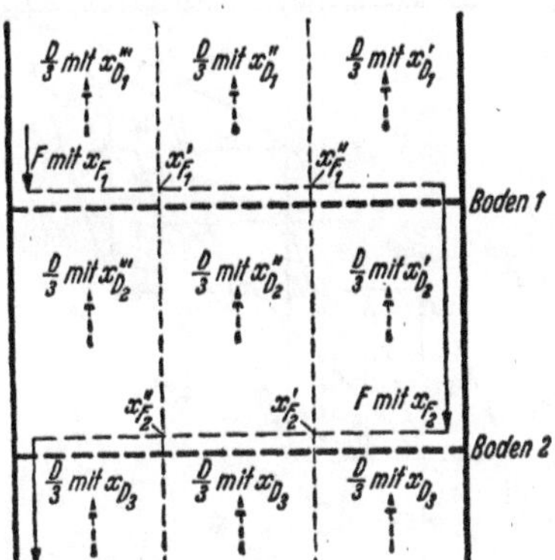

FIGURE 176

Alternating liquid flow scheme

(Reading up, mass of liquid F has a concentration of x_{F3}, etc., mass of vapor $D/3$ (one-third of total) a concentration of x_{D3}, etc.)
Boden—Plate

If it is assumed that the vapor rising from a plate in a given section is in phase equilibrium with the liquid leaving that section, then

$$x_{D1}''' \,///\, x_{F1}'; \quad x_{D1}'' \,///\, x_{F1}'; \quad x_{D1}' \,///\, x_{F2}, \quad \text{etc.} \qquad (188)$$

and, on the basis of equations (179) to (181)

$$\frac{x_{D1}' - x_{D2}'}{x_{F1}'' - x_{F2}} = 3 \times \frac{v}{v+1} \tag{189}$$

$$\frac{x_{D1}'' - x_{D2}''}{x_{F1}' - x_{F1}''} = 3 \times \frac{v}{v+1} \tag{190}$$

$$\frac{x_{D1}''' - x_{D2}'''}{x_{F1} - x_{F1}'} = 3 \times \frac{v}{v+1} \tag{191}$$

If the concentration x_{D3} of the vapor reaching all points of the bottom plate is known, the vapor concentrations above plates 2 and 1 (Figure 176) in the different sections, as well as the liquid concentrations at the points where the reflux leaves these sections can be found by the procedure underlying Figure 170.

Figure 177 shows an application of this method in a tripartition of the column section, and for a reflux ratio v of ∞, for which the enrichment line and the diagonal of the plot fuse.

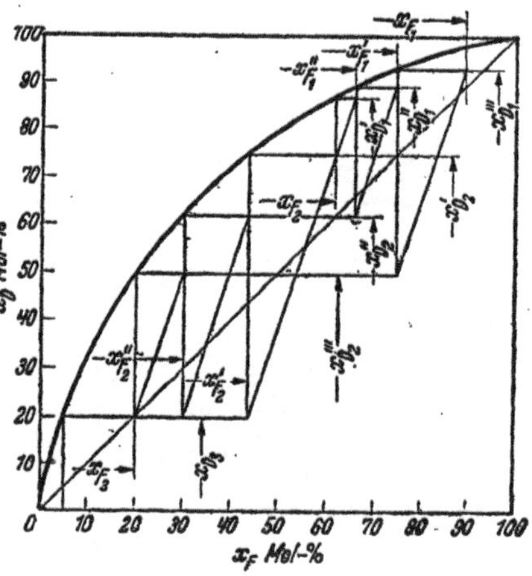

FIGURE 177

Case of tripartition and alternating liquid flow

Determination of liquid and vapor concentrations; infinite reflux ratio.

In Figure 177 the concentrations corresponding to Figure 176 have been indicated. As the number of partitions increases, the average concentrations of the low boiler in the vapor approach the value which would be reached by an infinite number of partitions.

It had been shown by Figure 172 that this final value holds

(with sufficient accuracy) even with a finite number of partitions which is not too large. The variations of liquid and vapor concentration across a plate are obtained by plotting the different liquid and vapor concentrations for the different sections against the path of the liquid.

The enrichment on a plate and the variations of liquid and vapor concentrations may be found for parallel flow of the liquid in the same way as for the counter-flow of the liquid. In this case, according to Figure 178, the liquid flows in the same direction on all of the plates. Figure 179 shows that this

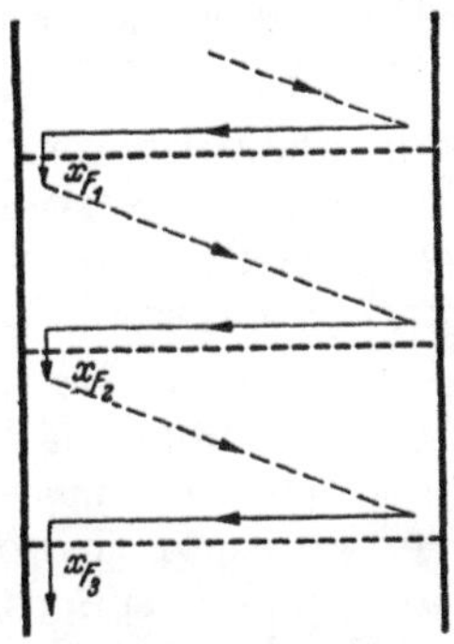

FIGURE 178

Diagrammatic representation of unidirectional liquid flow in a column

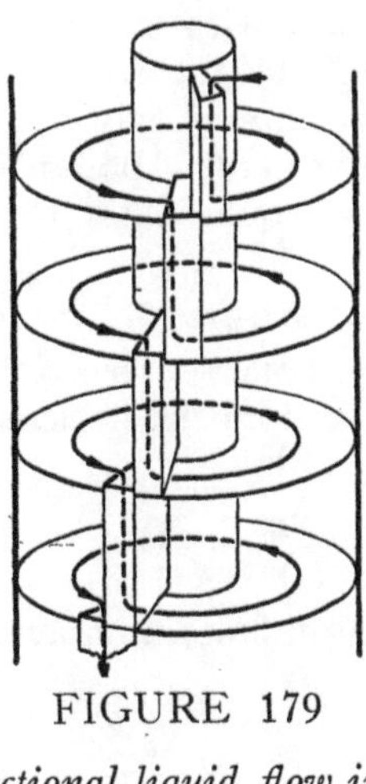

FIGURE 179

Controlled and unidirectional liquid flow in a sieve-plate column

type of flow may also be achieved by the use of ring-type plates which have inlet and exit close together with a wall between them.

For the cases of liquid path and vapor passage mentioned, the optimum enrichment in low boiler in the vapor and the variation of the concentrations above the plates may be found not only by graphical, but also by purely arithmetical methods.

While the former method served to find the theoretical enrichment ratio in the case of vapor mixing for an industrially important mixture (Figure 173), a mathematical treatment follows for different reflux ratios, depending on the final plate concentrations.

3. Computation of the Theoretical Enrichment and of the Variation in Concentration Across Exchanger Plates, as Well as the Efficiency in the Case of Cross-Current Flow Between Liquid and Vapor

The theoretical enrichment may be found by computation only, provided a number of assumptions is permitted.[75]

If the same order of ideas as outlined in the previous paragraph is applied, a certain simplification becomes feasible.[76] Again, the section of a column is supposed to be split up into n sections. Starting with the effluent point, one obtains a first, second, . . . kth and eventually nth part. F mols per hour of liquid, containing x_{F_e} mol per cent of low boiler are fed to the plate; product concentration is x_{F_a} mol per cent. The liquid reaches the kth part with x_{Fk} mol per cent and leaves it with x_{Fk-1} mol per cent. Of the entire mass of vapor (D mols per hour) rising in the column, only a fraction (D/n mols per hour) with x_{Du} mol per cent low boiler will enter the kth part under consideration; an equal mass of vapor leaves it with x_{Do} mol per cent.

a. Mathematical concept; the limiting case of infinite distribution

The total mass of low boiler entering the kth part equals the amount leaving it. Therefore the basic equation corresponding to relations (179) to (181) is

$$\frac{x_{Do} - x_{Du}}{x_{F(k)} - x_{F(k-1)}} = n\frac{F}{D} \tag{192}$$

To pass to the case of an infinite number of partitions, partition ratio

$$t_{(k)} = \frac{k}{n} \tag{193}$$

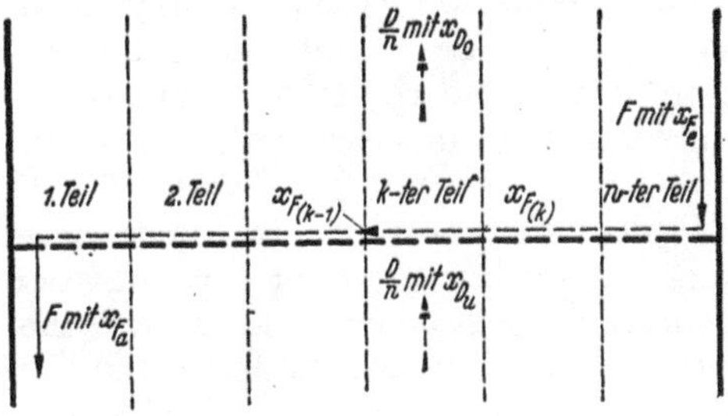

FIGURE 180

Partitioning of column cross-section

1. Teil—First subdivision
2. Teil—Second subdivision

k-ter Teil—*k*-th subdivision
n-ter Teil—*n*-th subdivision
F mit x_{Fa}—*F* with corresponding value x_{Fa}
etc. etc.

is introduced. It represents the ratio of the number of the part under consideration to the sum of these parts. Supposing that the path of the liquid is of equal length in every part of the section, and that its length amounts to l; the numerical value of t_k is also defined by the ratio of the length of the path in the k parts to the total length of the path in n parts

$$t_{(k)} = \frac{k \times l}{n \times l} \tag{193a}$$

The difference of the partition ratio of two adjacent parts is

$$t_{(k)} - t_{(k-1)} = \frac{1}{n} \tag{194}$$

If the n value from equation (194) is introduced into equation (192), one obtains the basic equation

$$\frac{x_F(_k) - x_F(_{k-1})}{t(_k) - t(_{k-1})} = \frac{D}{F}(x_{Do} - x_{Du}) \tag{195}$$

The left-hand side of this equation represents a difference quotient. It is converted into the differential quotient dx_F/dt if the number of partitions is made infinite. Thus

$$\frac{dx_F}{dt} = \frac{D}{F}(x_{Do} - x_{Du}) \tag{196}$$

In consideration of equation (193a), t may be defined as the ratio between the distance of the point under consideration from the bottom drain, and the total length of the liquid path. At the bottom drain, its numerical value is 0, and at the feed-point it equals unity.

Equation (196) is the most general mode of expressing the relation between the outline of vapor concentrations on a plate, and vapor concentrations below and above the same plate. x_F is the concentration of the low boiler in the liquid at a given point on the plate, and x_{Du} and x_{Do} are the concentrations in the vapor vertically below and above the same point.

The equation is employed to find the enrichment on a plate for the two cases of counter-flow and parallel flow on a plate. In the case of vapor mixing, however, if the vapor reaches all points of the plate with the same concentration, equation (196) may be integrated by separating the variable. Thus

$$\int_{x_{Fe}}^{x_{Fa}} \frac{dx_F}{x_{Do} - x_{Du}} = \frac{D}{F} t \tag{197}$$

The upper limit x_{Fe} of the integral on the left-hand side of equation (197) represents the concentration of low boiler in the liquid entering the plate; the lower limit x_{Fa} represents the concentration in the bottom effluent.

b. Evaluation of results

Vapor mixing. The use of equation (197) in finding the outlines of x_F and x_{Do} for a given x_{Fa} will be discussed for the case of an ethanol-water mixture, the equilibrium curve of which is presented in Figure 181. The concentration of low boiler in the product x_E is assumed to be 75 mol per cent of alcohol, and the molar reflux ratio v to be 4. These two quantities determine points A and B in Figure 181; connecting these forms the enrichment line.

The latter has a slope of α against the horizontal, for which

$$\frac{F}{D} = \tan \alpha \tag{198}$$

$\tan \alpha$ can be read off, and is a known constant. In Figure 181 $\tan \alpha = 4/5 = 0.8$. The final effluent concentration x_{Fa} is 24 mol per cent. Its abscissa value intersects the ordinate value x_{Du} at point C of the enrichment line. For all x_F values ex-

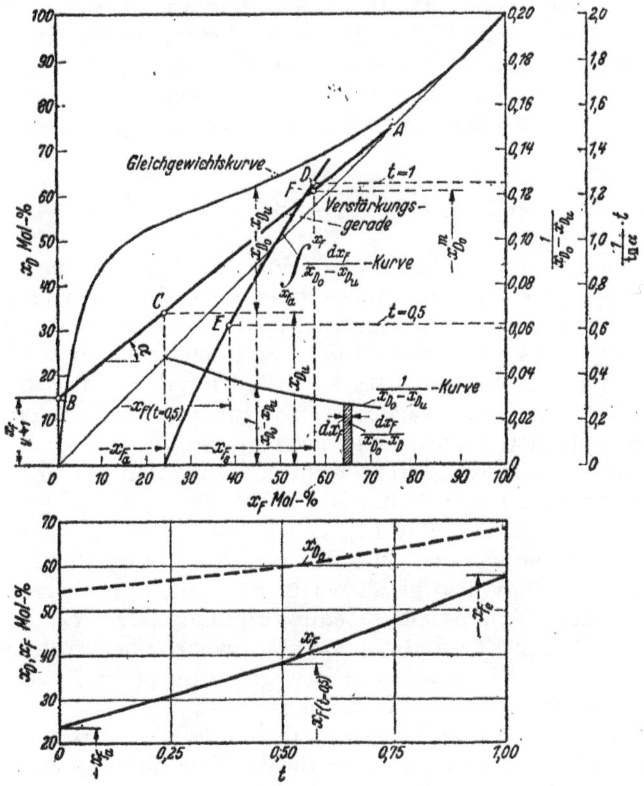

FIGURE 181

Case of thorough mixing of vapors. Plot for determining liquid concentrations

Kurve—Curve
Gleichgewichtskurve—Equilibrium curve
Verstärkungsgerade—Enrichment line

ceeding x_{Fa} the difference $x_{Do} - x_{Du}$ may be read as the ordinate section between the equilibrium curve and the horizontal at a distance x_{Du} from the ordinate origin.

Interchange is assumed to be perfect, x_{Do} being in phase equilibrium with x_F at any given point of the plate. One might introduce an individual efficiency figure by substituting for x_{Do} some value that is smaller than that assumed for the perfect interchange under phase equilibrium conditions.

The reciprocals of $x_{Do} - x_{Du}$ are plotted over the x_F values in question; this yields a $1/(x_{Do} - x_{Du})$ curve; see Figure 181. Below the curve, the abscissa element dx_F cuts off an area

$$\frac{dx_F}{x_{Do} - x_{Du}}$$

If one determines the areas below the $1/(x_{Do} - x_{Du})$ curve from x_{Fa} to x_F, and plots these values as ordinates over x_F, the integral curve shown in Figure 181 is obtained. In accordance with equations (197) and (198), the numerical values of these areas represent the quantity $t/\tan\alpha$. For $t = 1$ this becomes $1/0.8 = 1.25$, and is reached by the integral curve at point D, which supplies the plate feed-concentration x_{Fe}.

Analogously, any desired liquid concentration may be determined for, e.g., $t = 0.5$, that is, for half the length of the liquid path, by means of point E. If the x_F corresponding to each t is plotted, the x_F curve in the lower part of Figure 181 results. The x_{Do} line may also be shown in the same figure, since for a given t there will be phase equilibrium between x_{Do} and x_F. It is unnecessary to find the average vapor concentration x_{Dom} by measuring the areas, since it is given by point F where x_{Fe} intersects the enrichment line; Figure 181.

Parallel and counter-current liquid flow compared. Vapor mixing will be encountered on the bottom plate of any interchanging column. The plates higher up may be designed in such a way that the liquid on all plates moves in the same direction, or that the liquid on any given plate moves in a direction opposite to that of the liquid on the plates below and above the one considered.

Because x_{Du} is not constant, but will vary if t varies, equation (196) must be employed to find the distribution of the concentrations on a given plate for these two cases.

Figure 182, which holds for the same conditions as Figure 181, shows how equation (196) is applied in the case of parallel liquid flow. Liquid concentration x_{F1} is given as a function of t for the bottom plate for the case of vapor mixing. The dotted line for x_{Du1} represents the vapor concentration above the first (bottom) plate, x_{Du1} and x_{F1} representing concentrations in phase equilibrium at any time. With these data, the outline

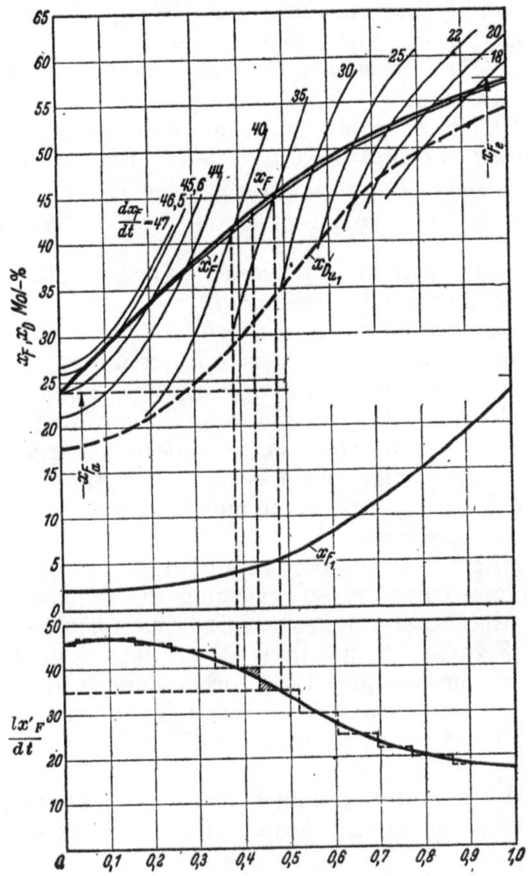

FIGURE 182

Determination of the trend of liquid concentration for the case of unidirectional flow

Mixture ethanol-water; reflux ratio $v = 4$; distillate concentration $x_E = 75$ mol per cent ethanol

of the concentrations in liquid and vapor on and above the second (penultimate) plate has to be found.

The inlet concentration of the first plate equals the outlet concentration x_{Fa} of the second one. This establishes the starting point of the line indicating the liquid concentrations on the second plate. The ordinate value is x_{Fa} with $t = 0$.

Equation (196) is solved by the procedure indicated by Runge, using isoclines.[77] These will represent in the t vs. x_F plot the locus of all those points which have the same dx_F/dt value. Thus, to draw the line for $dx_F/dt = 30$, read a x_{Du} of 35.05 for $t = 0.5$. Since $D/F = 1/\tan \alpha$ is given as 1.25

$$1.25(x_{Do} - 35.05) = 30$$

From this, $x_{Do} = 59.05$ mol per cent. Vapor of this concentration is in equilibrium with liquid of $x_F = 36.6$ mol per cent. This value is plotted against $t = 0.5$. Then, the x_F values corresponding to further t values are determined for the same dx_F/dt ratio. The connecting line of these points yields the isocline for $dx_F/dt = 30$. In the same fashion, further isoclines are plotted at certain intervals, as shown in Figure 182. At the point of intersection of an isocline with the x_F curve, the slope of the latter is given by the numerical value corresponding to the isocline considered.

Thus, beginning at the point defined by x_{Fa} at $t = 0$, one can draw a polygon curve by so arranging the individual straight pieces that they intersect in the middle between two isoclines. As seen in Figure 182, one thus obtains a x_F' curve, which is almost the x_F curve required. Another sufficiently satisfactory approximation is obtained by determining the interrelation of dx_F'/dt and t from the intersections of the x_F' curve with the isoclines.

Then, the curve thus obtained, shown in the lower part of Figure 182, is integrated graphically, by substitution of a sequence of steps. The excess areas should cancel each other, as indicated by shading in Figure 182. After assuming an integration plot, one obtains the x_F curve required, shown as a heavy line in the upper part of the figure. It is seen that the difference between the x_F' and the x_F curve is actually insignificant. Therefore, it is unnecessary to repeat the integration process for the purpose of increased accuracy. The course of the vapor concentration above the penultimate plate depends on the phase equilibrium of the liquid and vapor.

Conditions are simpler in the case of counter-current liquid flow. The tangent at the lowest point of the x_F curve is horizontal because, in equation (196), $x_{Do} - x_{Du}$ and consequently dx_F/dt become zero.

c. Concentration curve, and theoretical enrichment ratio in the separating column

Figure 183 shows the variations in liquid and vapor concentrations in a column for the different conditions possible, which are liquid mixing, vapor mixing, liquid counter-flow, and liquid parallel flow. The upper part of the figure indicates in a diagrammatic way how these different conditions can be achieved, while the lower part shows concentrations as a function of the liquid path.

In all cases, the liquid leaves the bottom plate with a concentration of 1.87 mol per cent of ethyl alcohol in the ethanol-water mixture. Distillate concentration x_E is 75 mol per cent alcohol, and the reflux ratio v is 4. Curve x_{F1} indicates the course of the liquid concentration on the bottom plate, x_{F2} is that above the penultimate plate, x_{D1} stands for the vapor concentration above the bottom plate, x_{D2} for that above the penultimate plate, etc. The average low-boiler concentrations in the case of vapor mixing are x_{D1}^m, x_{D2}^m, etc. Which of these four cases is the most favorable may be readily recognized from the location of point Z which indicates the low-boiler concentration in the liquid reaching the penultimate plate.

The comparison between the different types of flow becomes especially convincing if the theoretical enrichment in low boiler in the vapor for the different conditions is divided by the enrichments in case 1, liquid mixing. The theoretical enrichment ratios thus obtained, $s_F^\times$, $s_D^\times$, $s_{geg}^\times$, $s_{gl}^\times$ have been plotted against the final plate-effluent concentrations in Figure 184. For a certain concentration of liquid, these curves show a minimum. In the figure, curve 4 is the uppermost one. This means that parallel liquid flow will yield highest efficiency. Curve 1, liquid mixing for a theoretical enrichment ratio $s_F^\times$ of 1 illustrates the least favorable condition; the vapor rising from a plate is in phase equilibrium with the liquid leaving the same plate. The course of the concentrations on rectifying plates has been determined by tests, and the results are in good agreement with the theoretical lines in Figure 183.

The method indicated may be applied to the enriching column as well as to the stripping column. Figure 185 gives data applying to both types, the mixture handled being composed of benzol and toluol, and vapor mixing being present. The feed is supposed to be at boiling temperature. In both columns,

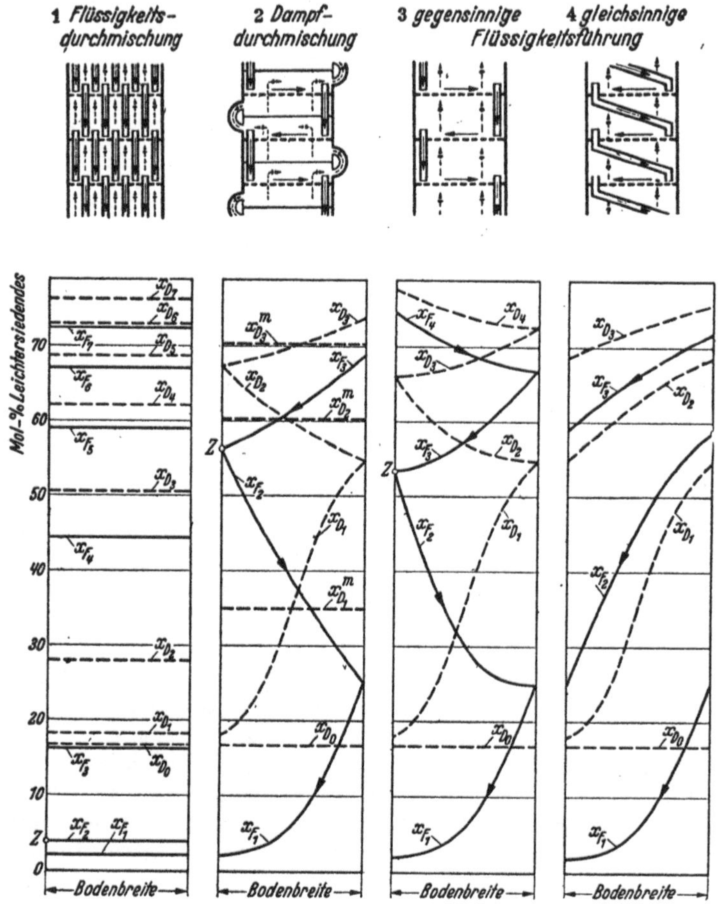

FIGURE 183

Possible conditions of flow in a rectifying column. Concentrations of low boiler in liquid and in vapor as functions of length of distance travelled by liquid

Concentrations of low boiler in liquid and in vapor as functions of length of distance travelled by liquid

1. Flüssigkeitsdurchmischung—Case of thorough mixing of liquid
2. Dampfdurchmischung—Case of thorough mixing of vapor
3. Gegensinnige Flüssigkeitsführung—Liquid flow in alternating directions
4. Gleichsinnige Flüssigkeitsführung—Unidirectional flow of liquid

Bodenbreite—Width of plate
Mol % Leichtersiedendes—Mol per cent of low boiler
Full lines ——— refer to low boiler in the liquid phase
Broken lines - - - - refer to low boiler in the vapor phase

---

the theoretical enrichment ratio decreases as the final plate effluent concentration increases.

d. Actual efficiency

On the basis of equation (187), the actual efficiency generally results as the ratio of the actual enrichment to the theoretical

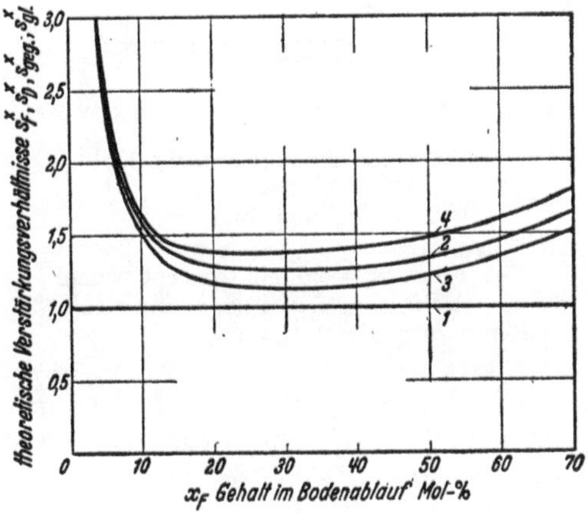

FIGURE 184

Enriching column. Theoretical enrichment ratios for different conditions of fluid flow:

Ethanol-water; reflux ratio $v = 4$; distillate $x_E = 75$ mol per cent ethanol.

1. Mixing of liquid
2. Mixing of vapor
3. Alternating liquid flow
4. Unidirectional liquid flow

Gehalt im Bodenablauf—Terminal plate concentration
Theoretische Verstärkungsverhältnisse—Theoretical enrichment ratios

enrichment, or actual to theoretical enrichment ratio. Since, other things being equal, the theoretical enrichment ratios resulting will be different for the four cases described, each of these four cases will yield a different actual efficiency for a

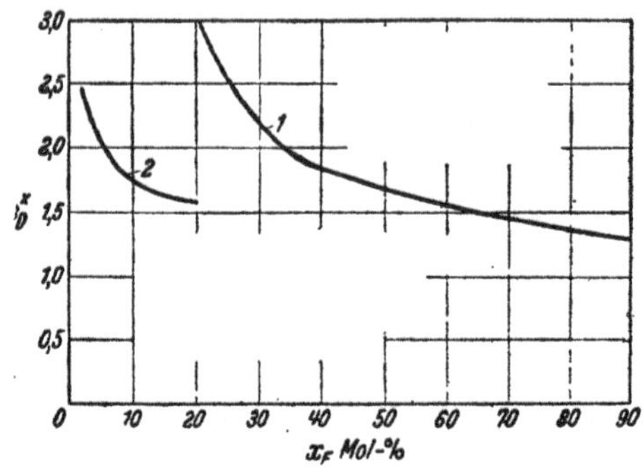

FIGURE 185

Theoretical enrichment ratios for the enriching and stripping columns

Case of vapor mixing; benzol-toluol
1. Enriching column
2. Stripping column
Distillate x_E = 95 mol per cent benzol
Feed x_M = 20 mol per cent benzol
Drain x_A = 2 mol per cent benzol
Reflux ratio v = 4

given constant actual enrichment ratio, which equals the quotient of the constant actual enrichment ratio and the varying theoretical enrichment ratios.

V. Rectifying plate design

As far as downcomers are concerned, there is no difference in design between sieve plates and bubble-cap plates. Structurally, sieve plates, plain sheets of metal with holes in them, are extremely simple. Bubble caps, composed of vapor nozzles and caps or tunnels, are available in a good many different types, dimensions, as is seen in Figures 134, 151 and 152. The influence of design upon efficiency has been shown in Figure 153. The large variety of types tends to demonstrate that no generally recognized rules for best design have been found.

The outer circumference of the caps may be smooth, serrated, or slotted, and slots may be open or closed. A smooth circumference offers the advantage that in the case of restricted vapor velocities and densities, the vapor is distributed more evenly. This point is especially important in large-diameter columns.

At higher loads, however, serrations or slots will cause different bubbles to have paths of varying length through the liquid layer. The author has run comparative tests with plates of the type shown in Figure 151 and plates of the same general design without serrations. In both cases, the vapor velocity was more than 0.3 m per second, and in both cases the efficiency was found to be the same. In these tests, a mixture of ethyl alcohol and water was separated under atmospheric pressure.

Figures 186 to 194, show a number of different cap designs.

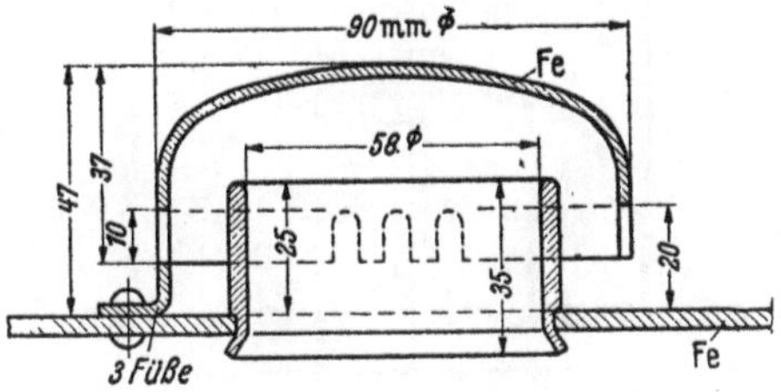

FIGURE 186

Crenelated bubble caps, squat design (steel)

Füsse—Lugs

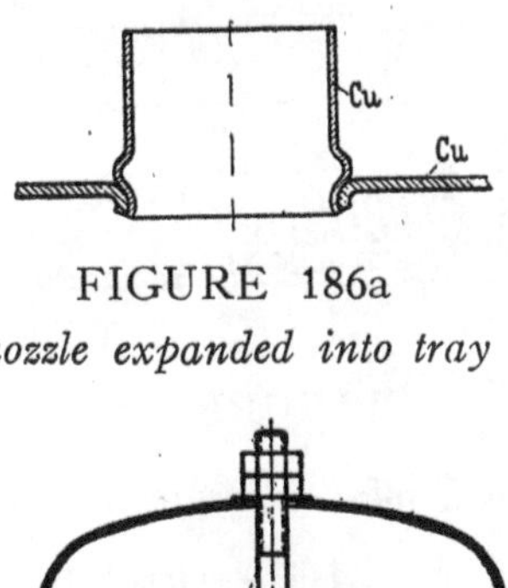

FIGURE 186a

Riser nozzle expanded into tray (copper)

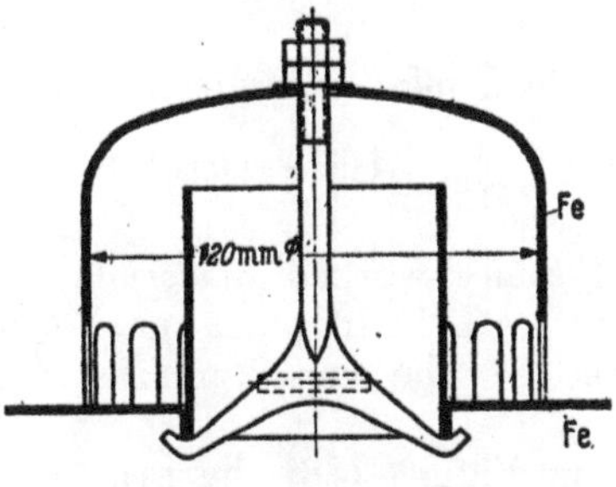

FIGURE 187

Slotted cap (steel)

The low cap, Figure 186, saves material as compared to higher caps, and permits the liquid to cover the space above it, which increases the area of interchange between phases. Sometimes, however, it is necessary to limit the number of caps to one or perhaps a few, to eliminate the risk of clogging. Large-size caps may not have lugs, but may simply be held in their place by their weight, the necessary clearance being maintained by inside ribs.

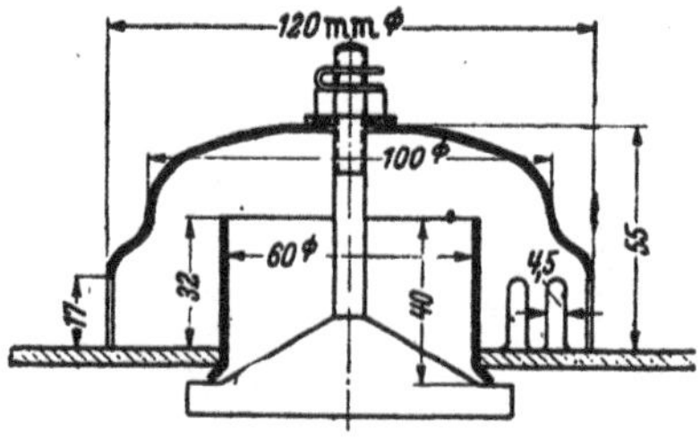

FIGURE 188

Slotted cap with flared lower portion (steel)

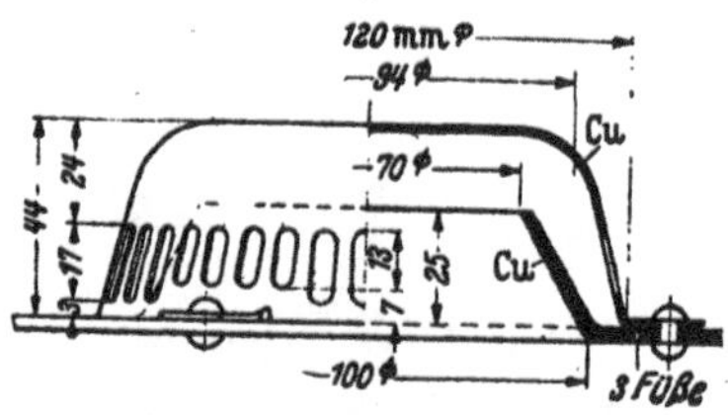

FIGURE 189

Slotted cap with solid circumference at base (copper)

Füsse—Lugs

In the case of easily worked materials such as copper, the vapor nozzle may be held in place as shown by Figure 186a. Again, in Figure 187 the vapor nozzle is supported by the through bolt.

In the design in Figure 188, the principle was, on the one hand, to provide a maximum circumference for the vapor to enter the liquid, and on the other, to allow as much space as

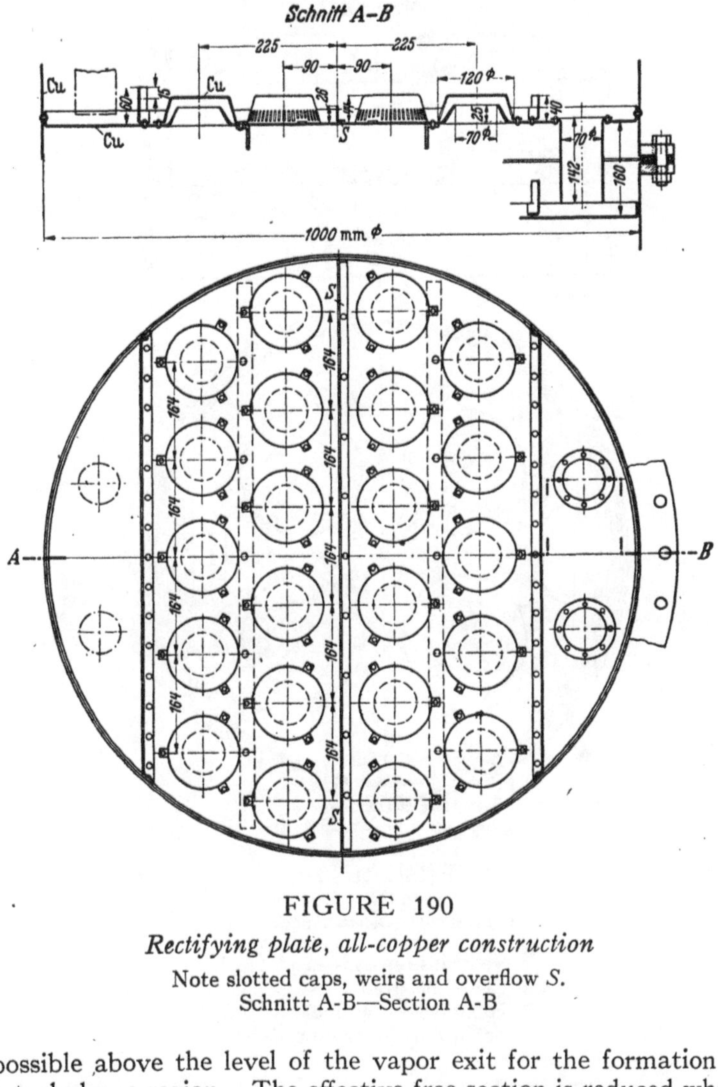

FIGURE 190

Rectifying plate, all-copper construction

Note slotted caps, weirs and overflow *S*.
Schnitt A-B—Section A-B

possible above the level of the vapor exit for the formation of a turbulence region. The effective free section is reduced where the slots are closed at the bottom, at low loads operation will be almost as in the case of a sieve plate, except for the non-draining properties of the bubble cap construction. Any reduction of the passage area will increase the pressure loss in

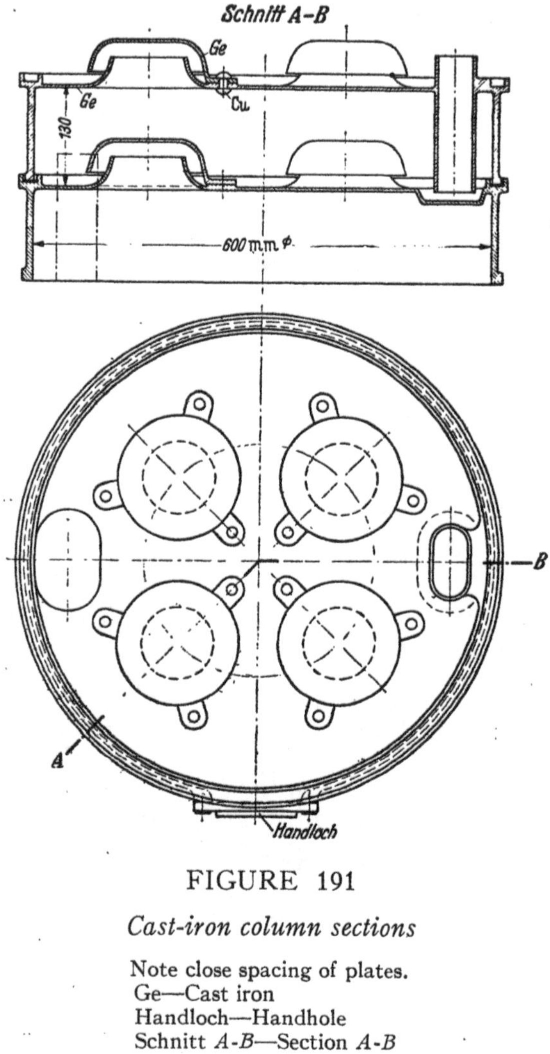

FIGURE 191

Cast-iron column sections

Note close spacing of plates.
Ge—Cast iron
Handloch—Handhole
Schnitt *A-B*—Section *A-B*

the vapor. This means an increased boiling temperature, which is undesirable, especially in vacuum units. On the other hand, it assures a more regular distribution of the vapor.

In most cases, caps made of rolled material are attached to the plates by three riveted lugs.

FIGURE 192

Rectifying plate equipped with slotted bubble cap

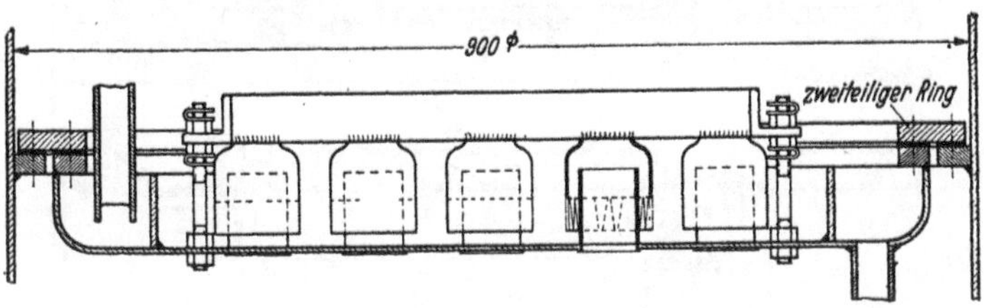

FIGURE 193

Bubble-cap plate with caps suspended from crossbars

Zweiteiliger Ring—Ring cut in two parts

Two principles are basic in the design of interchanger plates. One is to avoid liquid mixing, and the other to achieve a cross-current flow. However, despite the long liquid path inherent in cross-current flow, the liquid should be uniformly distributed on the plate, and the vapor should also be distributed evenly. In the case of long liquid paths caused by large column diameters the liquid will eventually form a higher level near the inlet than

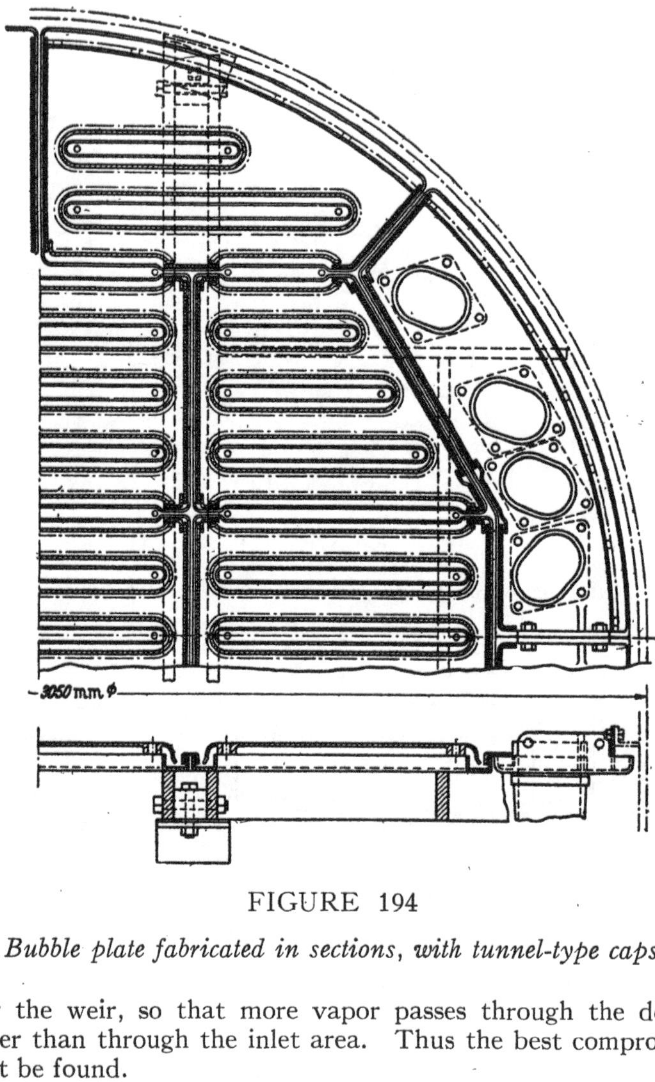

FIGURE 194

Bubble plate fabricated in sections, with tunnel-type caps

near the weir, so that more vapor passes through the downcomer than through the inlet area. Thus the best compromise must be found.

The first of these two principles is demonstrated by the use of long weirs and bays, as shown in Figures 134, 151, 152, 190, 192 and 193. In the design in Figure 193 a row of bubble caps is attached to a bar permitting their vertical adjustment. A relative arrangement of inlet and outlet points as in Figure

192 may also be provided. In the case of very large column diameters (2 to 3 m) the weir and bay constructions follow the outline of the shell; see Figure 194. The plates of similar, mostly open-air, columns, are made of several parts so dimensioned[78] that they can be introduced through manholes. In consideration of the principles explained, a liquid flow scheme as in Figure 195 must not be employed in columns exceeding 1 m in diameter.

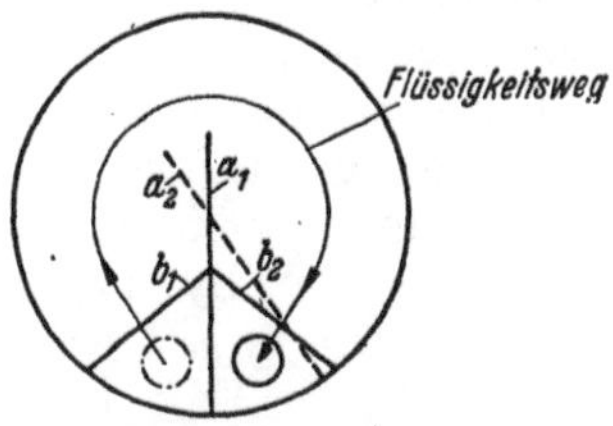

FIGURE 195

Diagrammatic representation of unidirectional liquid flow

Flüssigkeitsweg—Path of liquid

In the figure, a_1 is a partition between two plates, b_1 and b_2 are the bay and weir of a plate, and a_2 is the partition in the space below the plate under consideration. In this way, parallel liquid flow is achieved. As a rule, the counter-flow arrangement is selected, because it will yield the simplest arrangement and construction. To shorten the path of the liquid on the plates, the liquid may also be conducted radially, i.e., it may flow from the center to the periphery on one plate, and from the periphery to center on the next plate.

J. RECTIFICATION IN PACKED COLUMNS

I. General remarks

If an individual plate is considered, vapor and liquid will usually be found to move in cross-current flow. Nevertheless, if the entire column is viewed, the main direction of flow is downwards for the liquid, upwards for the vapor, i.e., the two phases move in counter-current. The only purpose of the plates is to provide an ample contact surface between phases, and other favorable conditions for an exchange of heat and of material.

However, both a counter-current flow between liquid and vapor and a contact surface between the two phases may be achieved by packing, supported on a grating in the column. The liquid is fed as uniformly as is feasible over the entire cross-section of the column, and trickles downwards, thus wetting the packing surfaces. The vapors (-r gases) flow in the opposite direction. Assuming satisfactory wetting of the packing surface, the latter gives a direct gage of the extent of the contact.

The path of the phases in the packing layer should be as long as is feasible, and flow conditions should favor optimum exchange of heat and material. It was Raschig who found that these conditions are best met by rings of some appropriate substance, having a diameter equal to their height. Such rings will, if present in bulk, arrange themselves so that the cylinder axis is in an inclined position. That compels both phases to follow a zigzag path through the packing layer. The main points for the selection of packings are as follows:

1. Useful area per unit volume
2. Material and cost
3. Flow of phases, and their distribution in the packing layer
4. Pressure loss in vapor

Their relative importance depends on individual conditions encountered. There are a good many types of packings proposed in an effort to comply as far as possible with one or several of these points.

The useful surface is usually given as m^2/m^3 of packing space, and is a function of the size and shape of the packing elements. Figure 196 shows three typical makes. The Prym ring differs

FIGURE 196

Typical tower packings

from the Raschig ring by its inner bridge. Saddle elements were designed to reduce pressure loss. Characteristic data of packing elements, as taken from catalogs, are:

Porcelain Raschig rings

Dimensions		Surface
15 × 15 mm,	2 mm wall thickness	330 m^2 per m^3
25 × 25	2 mm wall thickness	220 m^2 per m^3
35 × 35	3 mm wall thickness	160 m^2 per m^3
50 × 50	4 mm wall thickness	110 m^2 per m^3

For 25-mm rings of wrought iron with a wall thickness of 0.8 mm a solid volume equalling 8 per cent of the packing volume is stated.

Porcelain saddles

Size	Surface
6 mm	1150 m^2 per m^3
15 mm	450 m^2 per m^3
25 mm	260 m^2 per m^3
50 mm	120 m^2 per m^3

Metal Prym rings

Size	Surface
2 mm	2000 m^2 per m^3
6 mm	1000 m^2 per m^3
8 mm	700 m^2 per m^3
15 mm	400 m^2 per m^3
25 mm	255 m^2 per m^3
45 mm	140 m^2 per m^3

The greatest latitude in the selection of the packing material is probably allowed by the Raschig ring, which may be made of rolled or cast metals, porcelain, stoneware, glass, carbon, rubber, etc. The price depends largely on the choice of material.

Prym rings may be made of metal; saddles may be made of metal or porcelain, but not of stoneware. Besides the types mentioned above, there are ball packings, hollow balls, tri-cornered designs, wire spirals, lattice-work, coke or carborundum slabs, etc., all of which cannot be treated in detail.

The basic advantage of packing elements is the wide range of the materials of which they are fabricated, and their low pressure loss, which is especially important where distillation or rectification is effected at low absolute pressures.

The layout of distilling equipment featuring packed columns

is the same as that of continuous or discontinuous units with interchanger plates. The only difference is in the inner construction of the column, the packing elements supported by grating. Figure 197 shows the layout of a batch-type packed column for vacuum operation.

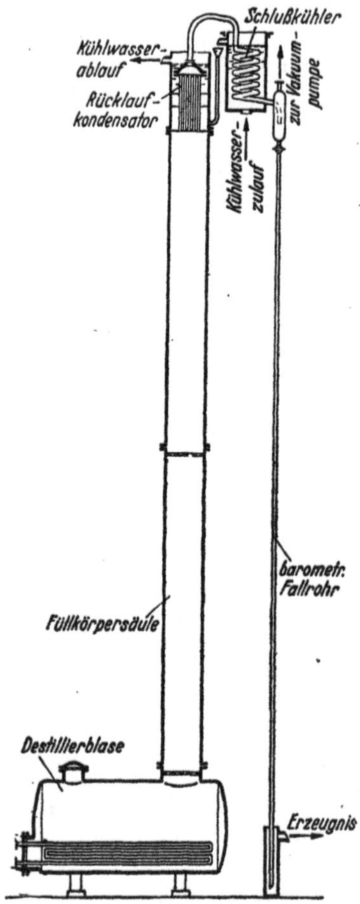

FIGURE 197

Vacuum rectifier with packed column, for batch operation

Destillierblase—Still kettle
Füllkörpersäule—Packed column

Rücklaufkondensator—Reflux condenser
Schlusskühler—Final cooler
Barometr. Fallrohr—Barometric leg
Zur Vakuumpumpe—To vacuum pump
Kühlwasserzulauf—Cooling water inlet
Kühlwasserablauf—Cooling water exit
Erzeugnis—Product

II. Determining the column height by means of the corresponding theoretical plate number; different types of packings

A basis for comparison of the efficiency of a packed column with that equipped with plates may be obtained by stating that number of plates which would yield, with liquid mixing and perfect interchange, the same effect as a layer of packing of a given height. Figure 198 shows how this comparative number of theoretical plates is found.

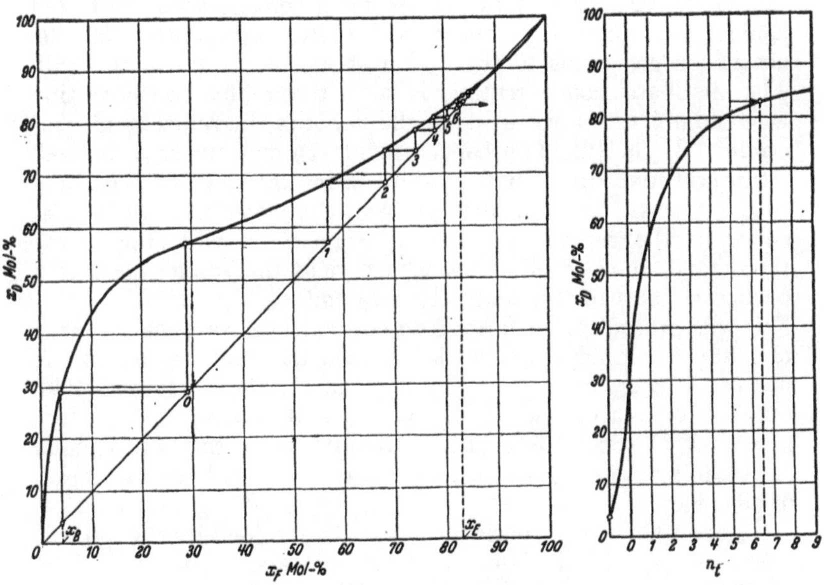

FIGURE 198

Determination of the theoretical number of plates. Ethanol-water; reflux ratio v = ∞; atmospheric operation

One assumes an enrichment ratio v of ∞, which makes the enrichment line coincide with the diagonal in the x_F *vs.* x_D plot. In the left-hand side of the latter, the equilibrium curve of an ethanol-water mixture has been shown. If concentration x_B mol per cent of low boiler in the still is given, the ordinate point 0 corresponds to the concentration of the vapor leaving the still, and its abscissa value indicates the concentration of the reflux from the lowest theoretical plate flowing into the still pot.

The ordinate value of point 1 supplies the vapor concentration above the first plate (bottom), the ordinate value of point 2 the alcohol concentration in the vapor above the second plate, etc. Points 0, 1, 2, 3, etc., are located on the enrichment line, and represent the corner points of the sequence of steps between the equilibrium curve and enrichment line, which generally serves to determine the theoretical number of plates; see sections D, IV and E, I.

In the right-hand part of Figure 198, the low-boiler concentration of the vapor above the plates under consideration is plotted against the number of theoretical plates, n_t; point 0 on the abscissa corresponds to the region above the liquid in the still. The distillate concentration is also the reflux concentration when complete condensation of the vapor in the reflux condenser occurs. If the distillate concentration x_E mol per cent, as well as the still concentration, is known, the intersection of the parallel to the abscissa at a distance x_E from the latter with the x_D, n_t curve in the right-hand part of Figure 198 will supply the theoretical number of plates which yield the same effect as a packed column for concentrations x_B and x_E.

If this procedure for finding an equivalent theoretical number of plates is employed, there is no need for rounding off to the nearest integer, as in the case of a bubble plate column. The number of plates (operating with liquid mixing and perfect interchange) required to yield a definite separating effect, such as obtaining x_E from x_B mol per cent, serves directly as a criterion of efficiency.

Figure 199 shows experimental results obtained by the author. The column, with a diameter of 400 mm and a packing height of 1000 mm, was superposed immediately above the still. A mixture of ethyl alcohol and water was treated; reflux was complete ($v = \infty$). The reflux was fed evenly by a Segner wheel distributor. The liquid contained about 10 per cent by weight of alcohol.

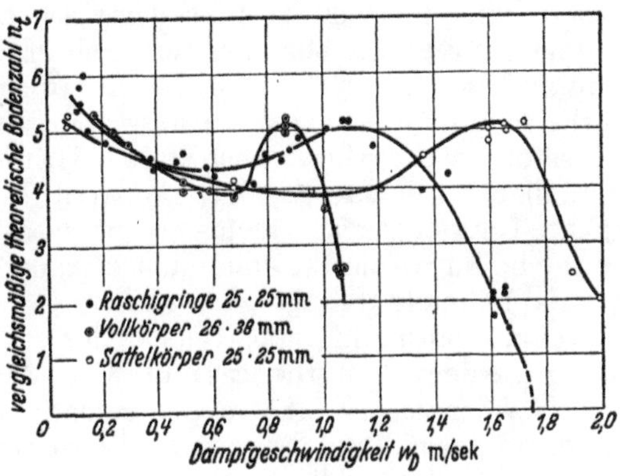

FIGURE 199

Equivalent theoretical number of plates plotted against load for three types of porcelain packing

Column diameter 400 mm, packing height 1 m; ethanol-water; reflux ratio $v = \infty$.
Dampfgeschwindigkeit—Vapor velocity
Vergleichsmässige theoretische Bodenzahl—Equivalent theoretical number of plates
Raschigringe—Raschig rings
Vollkörper—Solid elements according to Figure 200
Sattelkörper—Saddles

Results are shown in Figures 198, 199, 201 and 202. The vapor velocity w_D in the free column section has been plotted on the abscissa of Figure 199 and the equivalent number of theoretical plates, n_t, on the ordinate. The larger n_t becomes, the higher the efficiency of the packing. Each of the curves applies to a definite type and size of packing, made of unglazed porcelain. The solid element is shown in Figure 200.

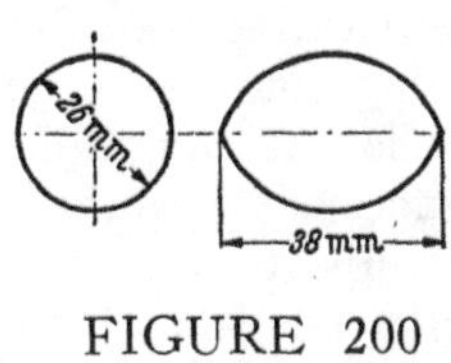

FIGURE 200

Tower packing according to German Patent 591,820

As load increases, efficiency at first decreases because the period of contact between the phases becomes short; Figure 199. After a minimum has been reached (in the case of solid packing and saddles the minimum occurs when n_t is approximately 3.9, and, in the case of Raschig rings, when n_t is 4.3), the improved interchange at higher velocities becomes apparent. Simultaneously, the phase-contact surface seems to increase due to the dispersion of the liquid within the column, according to observations which could be made through glass shells.

There is a definite optimum vapor velocity for each type and size of packing element. In the case of 26 × 38 mm solid elements, this is when $w_{D_0} = 0.85$ m per second; for Raschig rings it is 1.1 m per second and for saddles, 1.6 m per second.

Thus, for a given output saddles yield a smaller column section than do Raschig rings. However, one must also consider the investment, especially in view of the fact that Raschig rings can be made of stoneware which is inexpensive, while saddles cannot be made of stoneware.

In selecting the section of a column, one should thus assume, wherever the load fluctuates, a vapor velocity below the optimum, as efficiency drops rapidly after the optimum velocity has been exceeded.

According to Figure 199, the three types of packings, of equal sizes and at optimum load, give an equivalent number of theoretical plates of $n_{t0} = 5.1$. If one assumes an enrichment ratio s of 0.8, corresponding to a vapor velocity of 0.7 m per second, for a plate spacing of 200 mm in a bubble-cap column, the actual number of plates, 5.1/0.8, will correspond to the theoretical value 5.1. Thus, under the conditions prevailing during the tests, a bubble-cap-type column would have to be higher than the packed column, for equal efficiency.

It seems possible that the surface condition of the packing elements influences their separating effect, inasmuch as a rough, porous surface would be wetted more evenly by a liquid than a smooth surface. Comparative tests by the author, in which porous and glazed Raschig rings were employed (Figure 201), demonstrate, however, that as compared to other factors, the surface condition has practically no influence.

What is particularly important is the influence of the size of packing elements upon efficiency and permissible load, for any given type. The results in Figure 202 show that smaller packing elements yield a better separation, but that, when

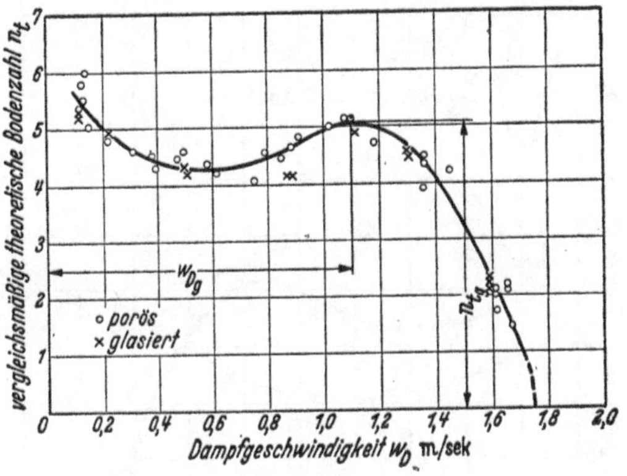

FIGURE 201

Relative effectiveness of unglazed and glazed porcelain Raschig rings of 25-mm size

Column diameter 400 mm, packing height 1 m; ethanol-water; reflux ratio $v = \infty$.
Dampfgesschwindigkeit—Vapor velocity
Vergleichsmässige theoretische Bodenzahl—Equivalent number of theoretical plates
Porös—Unglazed
Glasiert—Glazed

they are used, the optimum superficial vapor velocity is less than for larger elements.

The shape of the experimental curves in Figure 202 is due to the contact-time and the intensity of interchange which is also dependent on the velocity, and furthermore there is the phenomenon that at velocities exceeding the optimum one, the liquid becomes only partly mixed within the packing layer. At certain points it is entrained upwards, while at others it can trickle downwards. This fact could be observed during the tests. After the optimum velocity has been exceeded, the liquid in the lower central part of the column is carried up. In the upper part, it accumulates over the entire section, and flows downwards, mainly in the vicinity of the shell. Below the grating of the experimental unit another glass-shell section was provided, and it was seen that, at velocities above the optimum, the bulk of the liquid came down in large streams at the circumference.

The more important data of Figure 202 are presented in

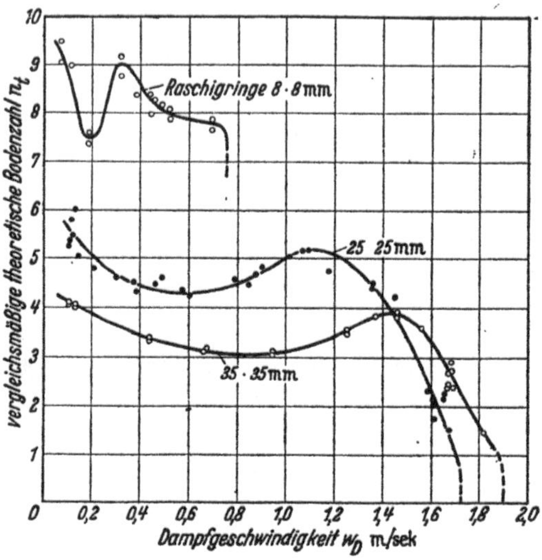

FIGURE 202

Effectiveness of packings of a given type, but different sizes, at various loads

Unglazed porcelain Raschig rings; column diameter 400 mm; packing height 1 m; ethanol-water; reflux ratio $v = \infty$.
Dampfgeschwindigkeit—Vapor velocity
Vergleichsmässige theoretische Bodenzahl—Equivalent number of theoretical plates
Raschigringe—Raschig rings

Table 4. The products ($n_{tg} \times w_{Dg}$) and ($O \times w_{Dg}$) are indicated. They permit approximate prediction of conditions for packing-element sizes other than those on which Table 4 is based. w_{Dg} is the optimum vapor velocity for which the equivalent theoretical number of plates reaches its maximum n_{tg}. For Raschig rings of 25-mm diameter, w_{Dg} becomes 1.1 m per second and n_{tg}, 5.1. Furthermore, O m²/m³ is the surface of the packing elements per unit of volume.

It is seen how the optimum vapor velocity increases and the optimum theoretical number of plates decreases as the packing-element size increases. Product $n_{tg} \times w_{Dg}$ is the same for rings of 25-mm and 35-mm size, while product $O \times w_{Dg}$ approaches 240 for all of the packing elements investigated.

TABLE 4

Optimum velocities w_{Dg}, equivalent theoretical number of plates n_{tg} and typical data for Raschig rings made of porcelain

Dimension mm	w_{Dg} m/sec	n_{tg}	$(n_{tg} \times w_{Dg})$	$(O \times w_{Dg})$
8	0.34	9	3	$600 \times 0.35 = 210$
25	1.1	5.1	5.6	$220 \times 1.1 = 242$
35	1.45	3.85	5.6	$160 \times 1.45 = 232$

This figure permits the estimation of the optimum theoretical number of plates from the surface O of a packing element, after the optimum velocity and product $(n_{tg} \times w_{Dg})$ have been estimated. In this way at least an approximation of the data required for computation and operation is obtained. They hold for the conditions of the experimental runs, in which a batch containing 10 per cent of ethyl alcohol was made to yield an average distillate concentration of 90 per cent by weight of alcohol. Thus, the average density of the vapor was 1.3 kg per cubic meter. It will be explained in section J, IV, 2 how the optimum vapor velocity through layers of packing is determined from the pressure loss in the vapor.

Given concentration $x_B^\times$ of a mixture to be separated, the procedure illustrated in Figure 198 will permit the determination of a theoretical number of plates $n_t^\times$ which is required to obtain the desired distillate concentration $x_E^\times$. If tests have established that the effect of a packing layer 1 m in height corresponds to a theoretical number of plates n_t (as shown in Figures 199, 201, 202), then a concentration increase from $x_B^\times$ to $x_E^\times$ will necessitate a packing height, measured in m, of

$$H_F = \frac{n_t^\times}{n_t} \tag{199}$$

This relation will supply accurate H_F values if n_t is the average for the concentration range of the mixture to be separated between $x_B^\times$ and $x_E^\times$. Otherwise, one must be satisfied with the approximation method for determining the number of plates; sections D, IV and E, I. Here, the theoretical number of plates had to be found; the mean enrichment ratio (section D, IV, 3 a) was assumed to be independent of the concentration and character of the materials involved. In other words, the theoretical number of plates n_t corresponding to the effect of a packing layer 1 m in height is also considered to be independent of the concentration and nature of the substances to be separated. However, tests corroborating the validity of such assumptions do not seem to be available at this time.

The results shown in Figures 199, 201 and 202 hold for a packing height of 1 m, and it may be assumed that the distribution of the phases in motion was reasonably uniform. With high packings, these assumptions will no longer hold; this means that the corresponding theoretical plate number is smaller; see also section J, III.

III. Determination of column height from the heat transmission coefficient between phases; liquid distribution within the column

1. Determining the packing height

For rectifiers equipped with packed columns, the same material and heat-balance equations apply as for columns equipped with plates. Therefore, equation (60) for the enrichment line also holds in the case of packed columns.

Then, in equation (60), x_D mol per cent is the average concentration of low boiler in the vapor, in any section of column, and x_F mol per cent is the average concentration in the liquid for the same horizontal section. The enrichment line is shown in Figure 203, together with the equilibrium curve of an ideal mixture, in an x_F vs. x_D plot with origin O_1. The enrichment line connects points A and B. Point C on the enrichment line gives the relation between concentrations x_D and x_F in a certain section of the column. These concentrations do not yield an equilibrium; the vapor temperature exceeds the liquid temperature. The difference between vapor temperature t_D and liquid temperature t_F may be found as follows, provided t_D and t_F represent condensation and boiling temperatures, which means that the vapor is not superheated and the condensate is not cooled below boiling temperature.

Boiling-point curve and dewpoint curve are also plotted in Figure 203, point O_2 being the origin. Curve CD will supply the condensation temperature t_D corresponding to x_D, and curve CEF yields boiling temperature t_F corresponding to x_F. Difference $\vartheta = t_D - t_F$ is also indicated in Figure 203. This difference

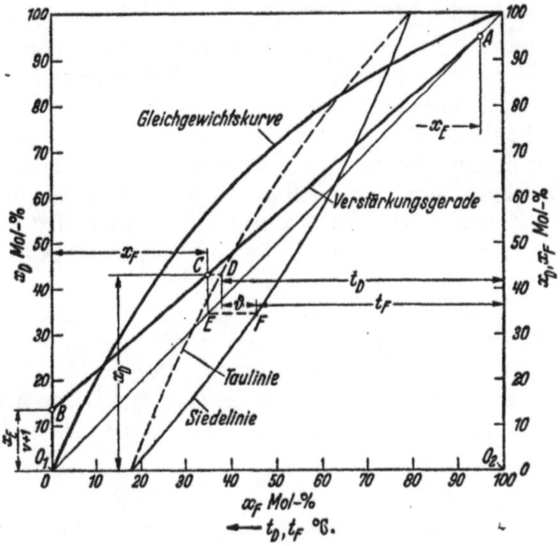

FIGURE 203

Determination of the temperature drop between vapor and reflux

Gleichgewichtskurve—Equilibrium curve
Verstärkungsgerade—Enrichment line
Taulinie—Dewpoint curve
Siedelinie—Boiling-point curve

causes heat to be transmitted from the vapor to the liquid. Therefore, some vapor condenses.

The heat of condensation penetrates into the liquid film, up to a point where the boiling temperature corresponding to the liquid concentration prevails. At this point, liquid is evaporated, and the newly generated vapor escapes from the liquid trickling down and mixes with the rest of the vapor mass. The liquid boundary layer, through which the newly formed vapor has to pass, has a higher temperature than the main bulk of the liquid stream. If the heat and material transmission figure between the vapor and its surface of contact with the liquid are much greater than the corresponding figures between this surface of contact and the liquid, which is assumed in this case, the temperature and concentration of the mass of vapor in a given section, up to the point of contact with the liquid, may be regarded as constant; in the liquid itself, beginning at the surface of contact, there will be a drop in temperature and a rise in concentration. At the surface of contact, liquid and vapor will be in phase equilibrium during condensation.

On the basis of this picture, one may derive the following method for finding the height of a column,[79] and establishing by test the heat-transmission coefficient between vapor and liquid. Results of other tests used to find the liquid distribution in packed layers provide check on the accuracy of data thus found.

Assume that the mass of vapor, D (mols per hour), passing through the column section along an infinitely small height of column dH, with concentration x_D yields the infinitely small mass of condensate dC with concentration $xF^{\times}$. This, then, will cause the formation of an infinitely small mass of vapor dD_v with a concentration $x_D^{\times}$. The equilibrium conditions in force are then

$$x_D^{\times} \,///\, x_F \quad \text{and} \quad x_F^{\times} \,///\, x_D$$

By this partial condensation, the remainder of the vapor $(D - dC)$ is then enriched up to $x_D + dx_{Dn}$, and according to equation (27)

$$dx_{Dn} = \frac{dC}{D}(x_D - x_F^{\times}) \tag{200}$$

Packing surface dF corresponds to the infinitely small height of column dH, with the relation

$$dF = K \times dH \tag{201}$$

K (square meters) is thus the surface of a packing volume with a height of the packing layer of 1 m. This surface depends on the column diameter d_s (meters) and surface O (square meters) per cubic meters of packing volume

$$K = O \times \frac{d_s^2\pi}{4} \tag{202}$$

For porcelain Raschig rings of 25-mm dimensions, $O = 220$ sq m per cu m. With this type of packing, and a column diameter d_s of 0.5 m we have

$$K = 220 \times \frac{0.5^2\pi}{4} = 43.2 \text{ m.}$$

The shell surface, along which some liquid might also flow downwards, is very small as compared with the surface of the packing elements. In the case of a column of 1-m diameter and Raschig ring packing of 25 mm size, the shell area amounts to only 1.8 per cent of the packing surface.

If the heat transmission coefficient between vapor and liquid contact surface (based on the unit packing area) is α (kcal/m^2 $\times$ hr $\times$ °C), and the molar heat of evaporation r (kcal per mol), the mass of condensate formed along dH will be, according to equation (201)

$$dC = \frac{K \times dH \times \vartheta \times \alpha}{r} \tag{203}$$

Because of the constancy of the molar heat of evaporation, the amount of condensate formed will equal that of the mass of newly-formed vapor

$$dC = dD_v \tag{204}$$

which means that D as well as the amount of reflux F (mols per hour) are constant. From the material-interchange process described, a mass equation for height dH is obtained

$$(D - dC)(x_D + dx_{Dn}) + (dD_v \times x_D^{\times}) = x_D + dx_D \tag{205}$$

where dx_D is the total enrichment in low boiler in the vapor, along height dH, caused by partial condensation and re-evaporation.

From this (by means of equations (200), (203) and (204)) the height required to obtain enrichment dx_D is derived

$$dH = \frac{D \times r}{K} \times \frac{1}{x_D^{\times} - x_F^{\times}} \times \frac{1}{\vartheta} \times \frac{1}{\alpha} \times dx_D \tag{206}$$

If, for instance, the still delivers vapor of concentration x_{DB} to a packed column which is supposed to yield a distillate concentration x_E, the height required is determined from

$$H = \frac{D \times r}{K} \int_{x_{DB}}^{x_E} \frac{1}{x_D^{\times} - x_F^{\times}} \times \frac{1}{\vartheta} \times \frac{1}{\alpha} \times dx_D \tag{207}$$

where r and K are constants, and constant D may be replaced with $[E \times (v + 1)]$, E mols per hour being the mass of distillate.

Figure 204 shows how to determine the concentration difference $(x_D^{\times} - x_F^{\times})$ corresponding to a given concentration x_D. The relation between vapor and liquid concentrations x_D and x_F is supplied by point C of the enrichment line, while x_D and $x_F^{\times}$, as well as x_F and $x_D^{\times}$ represent interrelated equilibrium concentrations.

If it is desired to use a batch still with a packed column to decompose a binary mixture having a batch concentration of

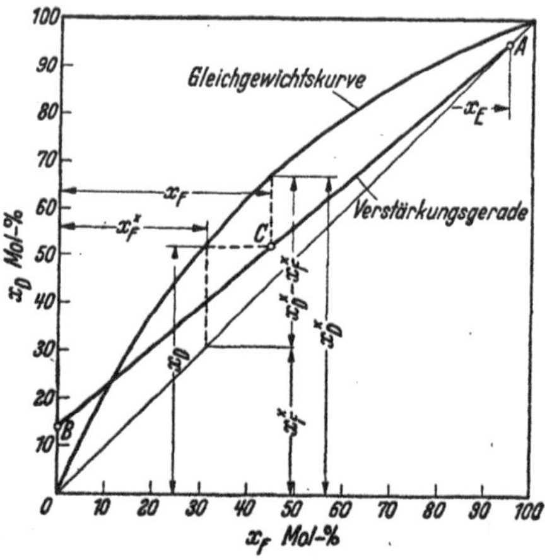

FIGURE 204

Determination of the $x_D{}^{\times}$ and $x_F{}^{\times}$ values for the total enrichment

Gleichgewichtskurve—Equilibrium curve
Verstärkungsgerade—Enrichment line

x_{FB}, so as to obtain a distillate concentration of x_E, values ϑ and $(x_D{}^{\times} - x_F{}^{\times})$ are determined for different x_D values by means of the equilibrium curve, boiling-point and dewpoint lines of the mixture and the enrichment line as shown by Figures 203 and 204 and the right-hand side of Figure 205. The reflux is supposed to have no enriching effect; in other words, the vapor leaving the packed column and the reflux leaving the condenser should be of identical composition.

The vapor of concentration x_{DB} leaving the still is in equilibrium with the still-liquid, which has a concentration of x_{FB}. Reflux ratio $v = F/E$ is found by the method used for bubble-plate designs; see section D, V.

In the left-hand part of Figure 205, the reciprocals of ϑ, of $(x_D{}^{\times} - x_F{}^{\times})$, of a mean film coefficient α assumed to be constant, and of $[(x_D{}^{\times} - x_F{}^{\times}) \times \vartheta \times \alpha]$ have been plotted against x_D. The shaded strip of the graph, of width dx_D, represents value

$$df = \frac{1}{(x_D{}^{\times} - x_F{}^{\times}) \times \vartheta \times \alpha} \times dx_D$$

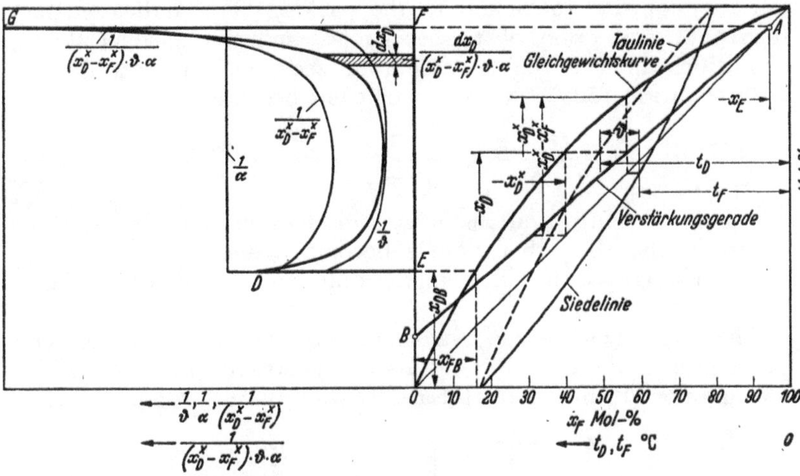

FIGURE 205

Plot for determining the height of a batch-type packed column

Taulinie—Dewpoint curve
Gleichgewichtskurve—Equilibrium curve
Verstärkungsgerade—Enrichment line
Siedelinie—Boiling-point curve

and area *DEFG* below the $\{1/[(x_D^{\times} - x_F^{\times}) \times \vartheta \times \alpha]\}$ curve represents the integral value required

$$f = \int_{x_{DB}}^{x_E} \frac{1}{(x_D^{\times} - x_F^{\times})\vartheta\alpha} \times dx_D \tag{208}$$

Thus the height of the packing layer becomes

$$H = \frac{D \times r}{K} \times f \tag{209}$$

This method permits the determination of the packing height of both enrichment and stripping columns. One must know only the value of the film coefficient from actual experience.

2. The film coefficient between the vapor and the contact surface of the liquid

The method outlined demands that the film coefficient between vapor and liquid should be known. Moreover, the variation of this coefficient in the packing layer indicates the degree of perfection of the exchange, and the ensuing flow-conditions in the different column sections. In accordance with equation (206), the film coefficient becomes

$$\alpha = \frac{D \times r}{K} \times \frac{1}{x_D^\times - x_F^\times} \times \frac{1}{\vartheta} \times \frac{dx_D}{dH} \qquad (210)$$

If vapor concentration x_D is measured experimentally at different levels, the other values become available by plotting x_D against H, which will yield the differential quotient dx_D/dH in equation (210).

The ϑ and $(x_D^\times - x_F^\times)$ values are obtained by following the procedure in Figures 203 and 204. The results obtained with a column of 110-mm diameter are shown in Figure 206 and 207.[80]

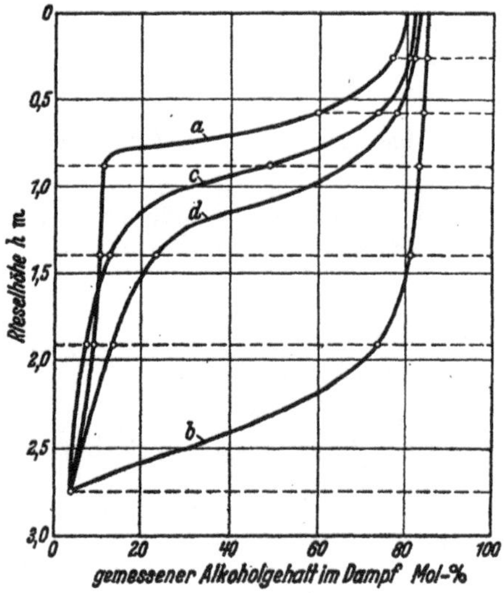

FIGURE 206

Alcohol concentration in vapor as a function of trickling height

Gemessener Alkoholgehalt im Dampf, Mol%—Alcohol concentration measured in vapor phase, mol per cent

Rieselhöhe—Trickling height

Curve	*a*	*b*	*c*	*d*
Reflux ratio	5.53	15.65	4.65	7.67
Vapor velocity, $_Dw$, meters per second	0.163	0.116	0.192	0.256

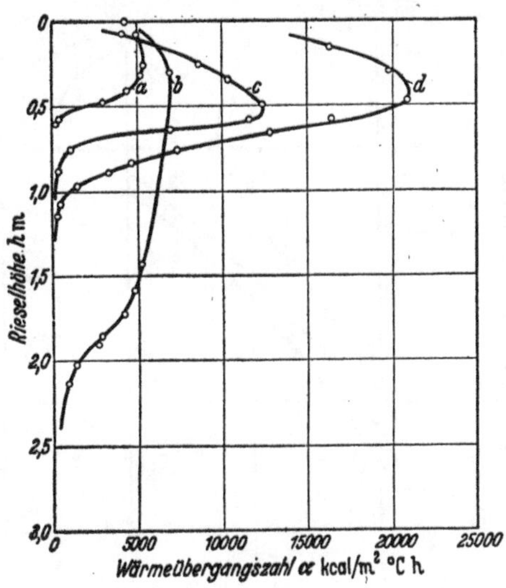

FIGURE 207

Surface coefficient of heat transfer as a function of trickling height (6 mm Prym rings)

Wärmeübergangszahl—Surface coefficient of heat transfer
Rieselhöhe—Trickling height

Curve	*a*	*b*	*c*	*d*
Reflux ratio	5.53	15.65	4.65	7.67
Vapor velocity, w_D, meters per second	0.163	0.116	0.192	0.256

The reflux of the ethanol-water mixture investigated was purposely fed by a pipe to the packing in the central portion of the column. In this case, the top packing layers are not wetted uniformly over the entire column section, and thus a very low film coefficient will result for the top section of the column and the operating variables concerned.

As the trickling height, which is the distance between the liquid feed point and a section under consideration increases, α will at first increase rapidly. It then reaches a maximum at a given point and decreases quickly.

This shape of the α curve is explained by the distribution of the liquid in the packing. Since there is good agreement between experimental results and theoretical figures concerning

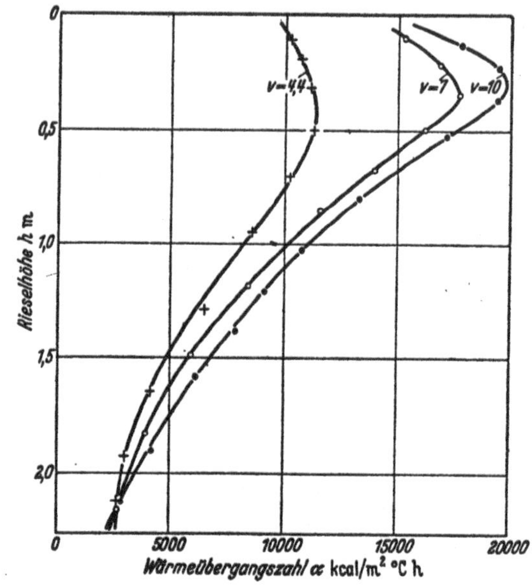

FIGURE 207a

Surface coefficient of heat transfer α as a function of trickling height

Column diameter 300 mm, packing 8 mm size porcelain Raschig rings. Distillate contains 93 per cent by weight of alcohol in ethanol-water mixture. Vapor velocity 0.27 m per second.
Wärmeübergangszahl—Surface coefficient of heat transfer
Rieselhöhe—Trickling height

liquid distribution and film coefficient, equation (206) is probably correct.

In Figure 208, the flow density (mass of liquid flowing per unit of time through 1 sq m of column section) has been plotted for different points of the packing, and therefore as a function of the distance of that point from the column centerline and of the trickling height. The tests were performed on water with a 110-mm column.

Similar curves, in Figures 206, 207 and 208, were obtained with columns 300 mm in diameter.[81] Figure 208 shows that the liquid fed to the center of the column by means of a pipe moves towards the shell of the column as the trickling height

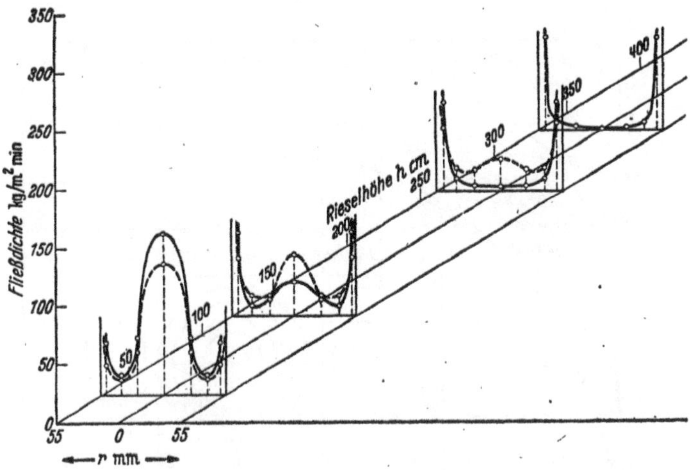

FIGURE 208

Flow densities as functions of trickling height and distance from column centerline

Fliessdichte—Flow density
Rieselhöhe—Trickling height
——— Porcelain Raschig rings 15 × 15 mm
- - - - Porcelain Raschig rings 8 × 8 mm

increases and that large rings will conduct the liquid outward more readily than will small ones.

Analogous results for different conditions are contained in Figure 209, where the trickling height is the abscissa, and the distance from the center of gravity the ordinate. According to Figure 209, the center of gravity is obtained by considering the masses of liquid flowing through certain annual sections of the column as forces P_1, P_2, P_3, etc., attacking at the center of the stretched annular area. The resultant force then acts in the center of gravity which is separated from the column center by distance x_s. In the figure, d_s is the column diameter, and d is the size of the porcelain Raschig rings. A comparison of the curves shows that large rings will tend to distribute the liquid outwards and that, beginning at a certain trickling height, the distance remains constant.

A value for x_s of $0.67 \times d_s/2$ corresponds to a uniform liquid

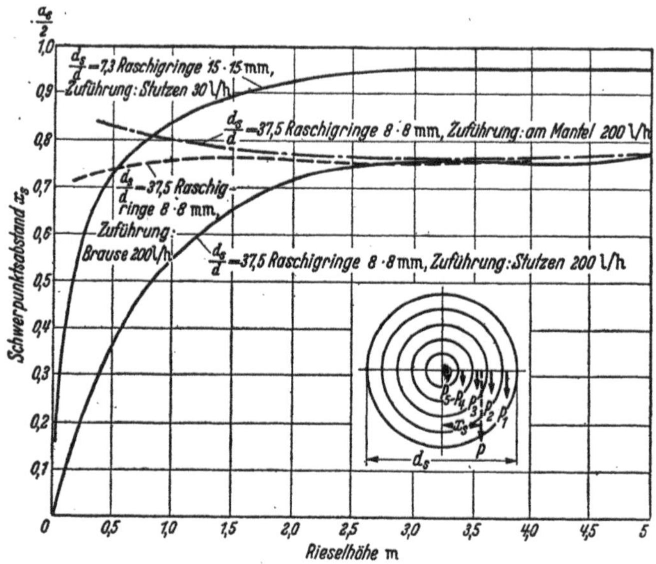

FIGURE 209

Dislocation of the center of gravity of a liquid body through packing layers

Rieselhöhe—Trickling height
Schwerpunktsabstand—Distance from center of gravity
Raschigringe—Raschig rings
Zuführung: Stutzen 30 l/h—Feed through connection at rate of 30 l per hour
Zuführung am Mantel 200 l/h—Feed through shell at rate of 200 l per hour
Zuführung: Brause 200 l/h—Feed through showerhead at rate of 200 l per hour
Zuführung: Stutzen 200 l/h—Feed through connection at rate of 200 l per hour

distribution. The test was run with water. The displacement of the center of gravity was found to be independent of the amount of liquid introduced.

One reason for the particular movement of liquid observed is that the packing elements will arrange themselves at the shell in a certain order, so that larger passages may result. Figure 210 shows that rings, for instance, will usually arrange themselves in such a way that either their outer wall or their faces lie against the shell. There is nothing to divert the path of the liquid away from the shell. Within the bulk of the packing, the arrangement of the elements is more regular in

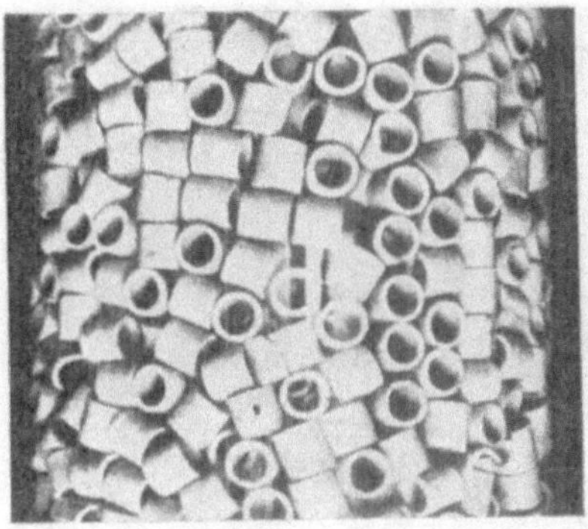

FIGURE 210

Typical distribution of packing rings against the inside of a column shell

the outer zones than in the central one; this condition increases the outward flow of the liquid as a whole.

Considering these facts, the ratio of column diameter to packing-element size should not be less than 10. Experiments concerning the loss of pressure in packed columns (Section J, IV) have corroborated this. It is also stated that, in the case of very large column diameters, the liquid will not move towards the shell to any considerable extent.[82]

In any case, there is one undesirable phenomenon; this is the formation of rivulets due to the surface tension of the liquid. It causes the latter to accumulate in little streams instead of uniformly wetting the packing; this condition becomes worse as the trickling height increases.

Figure 211 shows water at 20°C leaving the grating of a column 1.5 m in height; this liquid was introduced by a shower nozzle at a rate of 200 l per hour, and the column, 300 mm in diameter, was filled with 8-mm porcelain Raschig rings. If air is drawn through the layer at a rate of 0.5 m per second, the number of liquid streams leaving the layer decreases from about 60 to 30, and the latter arrange themselves mainly near the circumference. Possibly this formation of rivulets depends upon size and shape of packing elements, as well as upon the nature and mass of the liquid.

FIGURE 211

Typical distribution of the liquid passing through the grate supporting the packing in a column

If some device which prevents rivulet formation is installed in the column, efficiency increases. In rectification, this means a higher film coefficient of heat transmission. In Figure 212, curve a shows the variation of the film coefficient. If no distributor device were present, the curve would follow outline a_o. The device itself is shown in Figure 213[83]; its use increases α from 5000 to about 13,000 kcal/sq m $\times$ h $\times$ °C. The area between the ordinate for the trickling height and the a_o curve on the one hand, and the b curve on the other, offers a yardstick for the saving in column height made possible by the device. Comparable results were obtained with packed columns with diameters greater than 110 mm; Figure 212 was based on a column of this size.

It is seen that efficient operation of a packed column will demand not only uniform distribution of the liquid by shower nozzles, spray plates, spray wheels, etc., but also an even distribution of the liquid within the packing, and suppression of rivulets.

IV. Pressure loss in packed columns

Packed columns serve, if wetted by a liquid, to rectify materials and to promote other material-interchange processes. When dry, they will separate suspended matter from gases or vapors. In both cases, it is essential to know what the pressure loss of the vapor or gas when passing through the packing will be. Under the first condition, this knowledge will be required for

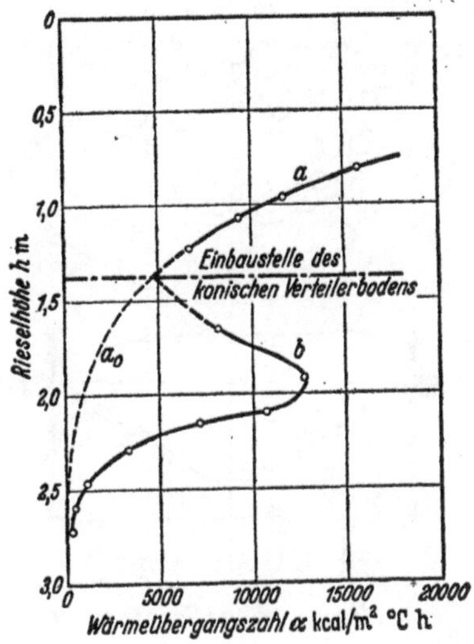

FIGURE 212

Increase in the surface coefficient α due to the use of a distributor cone

a = α-values measured above distributor
b = α-values measured below distributor
a_0 = Curve resulting when distributor is omitted
Einbaustelle des konischen Verteilerbodens—Level at which distributor was mounted in column
Wärmeübergangszahl—Surface coefficient of heat transfer
Rieselhöhe—Trickling height
Porcelain Raschig rings 8 × 8 mm; reflux ratio v = 8; vapor velocity 0.22 m per second

determining the boiling temperature in the still of a packed-column unit; under the second condition, it will be necessary in order to find the power for operating the blower which forces the gas through the packing.

1. Dry packings

The most important variable influencing the pressure loss is the gas or vapor velocity w (meters per second). For the purpose of computation, it is based upon the free column section. Tests run by the author at Karlsruhe Polytechnic Institute

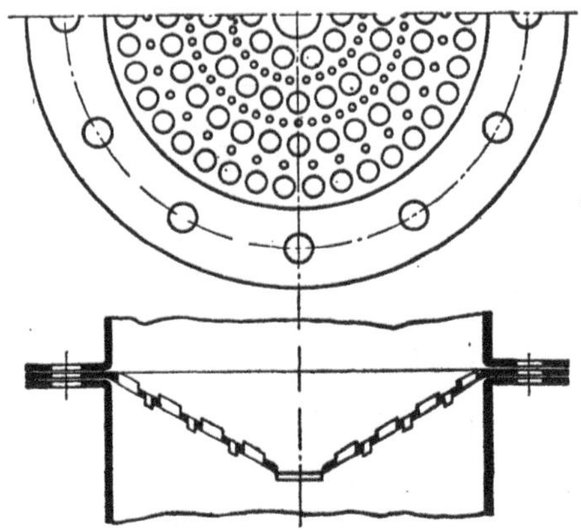

FIGURE 213

Prym liquid re-distributor cone

have yielded the data plotted in Figures 214 and 215.[84] Plotting the pressure loss against the velocity in a plot using a logarithmic scale yields a straight-line relationship.

These tests apply to a column with a diameter d_s of 300 mm, and atmospheric air having a density of 1.2 kg per cubic meter. For porcelain rings of 8- to 25-mm size, the pressure loss increases on an average as $w^{1.85}$, for saddles of the same size as $w^{1.82}$.

As a general thing, the velocity exponent decreases as the packing size decreases; at the same time the pressure loss increases. This is shown in Figure 216, where the pressure loss has been plotted against the reciprocal value of the packing size d (meters). Generally speaking, the pressure loss increases as the packing size decreases, other things being equal. This will also be true of wetted packings.

The influence of gas viscosity on the pressure loss is negligible. However, the density (kilograms per cubic meter) of the vapor or gas, plays a distinct part, as seen from Figure 217.

If H = height of the packing layer (meters),

ΔP = pressure loss, (millimeters H_2O),

then, for dry Raschig-type porcelain rings of 8- to 25-mm size

$$\Delta P = 0.27 \times H \times \frac{w^{1.85} \times \gamma^{0.83}}{d^{1.27}} \text{ (millimeters } H_2O) \quad (211)$$

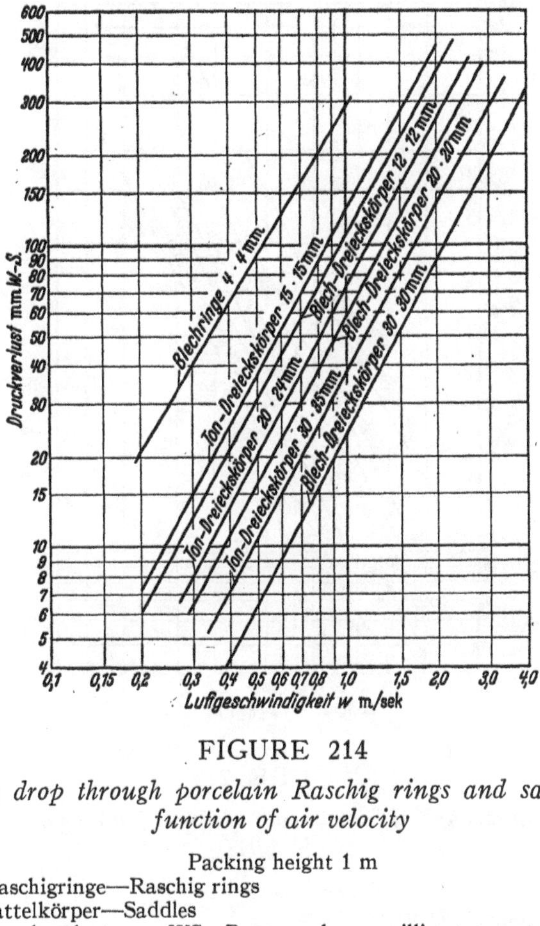

FIGURE 214

Pressure drop through porcelain Raschig rings and saddles as a function of air velocity

Packing height 1 m

Raschigringe—Raschig rings
Sattelkörper—Saddles
Druckverlust, mm WS—Pressure drops, millimeters water
Luftgeschwindigkeit, m /sek—Air velocity, meters per second

and for porcelain saddles of the same sizes

$$\Delta P = 0.13 \times H \times \frac{w^{1.82} \times \gamma^{0.83}}{d^{1.4}} \text{ (millimeters } H_2O) \qquad (212)$$

Thus, using 25-mm Raschig rings (d = 0.025 m) there is a pressure loss of ΔP = 33 mm H_2O per meter of packing layer (H = 1 m) for atmospheric air (γ = 1.2 kg per cubic meter).

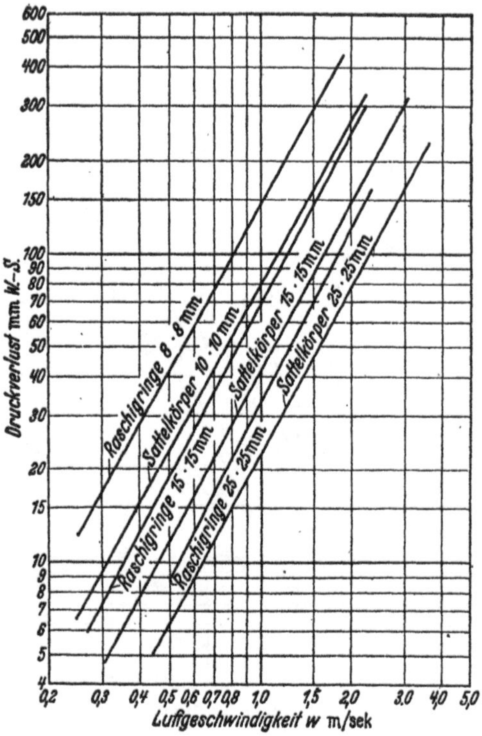

FIGURE 215

Pressure drop through different types of Prym packing as a function of air velocity

Packing height 1 m
Blechringe—Sheet metal rings
Ton-Dreieckskörper—Triangular clay prisms
Blech-Dreieckskörper—Triangular sheet metal prisms
Druckverlust, mm WS—Pressure drop, millimeters water
Luftgeschwindigkeit, m /sek—Air velocity, meters per second

The pressure loss of a well-designed grate amounts to about 20 per cent of that of 1 m of packing. This value will also hold for wetted packings. Fouling and a certain percentage of fracturated packing elements may increase the pressure loss by 20 to 40 per cent or more. Thus, 2 per cent of broken

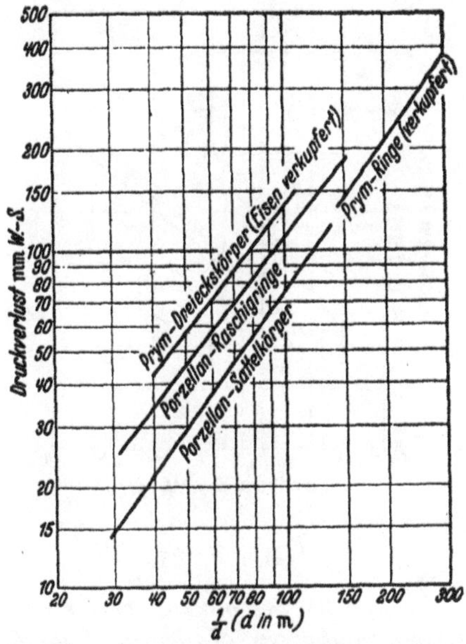

FIGURE 216

Pressure drop through packing as a function of the reciprocal value of packing element size d

Air velocity 1 m per second, packing height 1 m.
Prym-Dreieckskörper (Eisen verkupfert) —Prym prisms (copper-plated steel)
Porzellan-Raschigringe—Porcelain Raschig rings
Porzellan-Sattelkörper—Porcelain saddles
Prym-Ringe (verkupfert)—Copper-plated Prym rings
Druckverlust, mm WS—Pressure drop, millimeters water

packing elements were found to increase the pressure loss by 20 per cent.

The smaller the ratio z of column diameter d_s to packing size d becomes, the larger will be the clearance which will form near the column wall, and the more air will flow along the wall. This blowby will cause the pressure loss to decrease appreciably for z values below 10. For this reason, a larger ratio should be adopted in designs if possible.

A determination of the pressure loss in packed columns is simplified by the use of Table V in the Appendix, giving graphs based on the equations above.

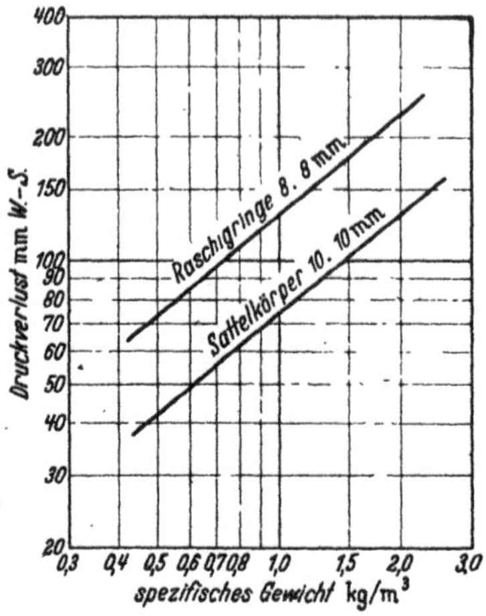

FIGURE 217

Pressure drops through 8 × 8 mm Raschig rings and through 10 × 10 mm saddles, as a function of gas density

Gas velocity 1 m per second, packing height 1 m.
Raschingringe—Raschig rings
Sattelkörper—Saddles
Druckverlust, mm WS—Pressure drop, millimeters water
Spezifisches Gewicht, kg/m^3—Density, kilograms per cubic meter

2. Wet packings

Wet packing causes a greater pressure loss than dry packing. The loss increases if the amount of liquid increases. Still, the vapor or gas velocity is the main factor. Figure 218 shows that, to begin with, the pressure loss increases with increasing w up to a certain point where the trend of the curves changes. The first sharp bend indicates the lower limiting velocity w_u, beginning at which the liquid begins to accumulate within the packing. A large variety of experiments involving different packings, gases and liquids shows that this velocity always corresponds to a pressure loss of 50 to 70 mm H_2O per meter of H_2O; packing height.

Other observations indicate that rectifying columns will operate at best efficiency with a pressure loss of 50 to 70 mm H_2O; therefore the load should be adjusted correspondingly.

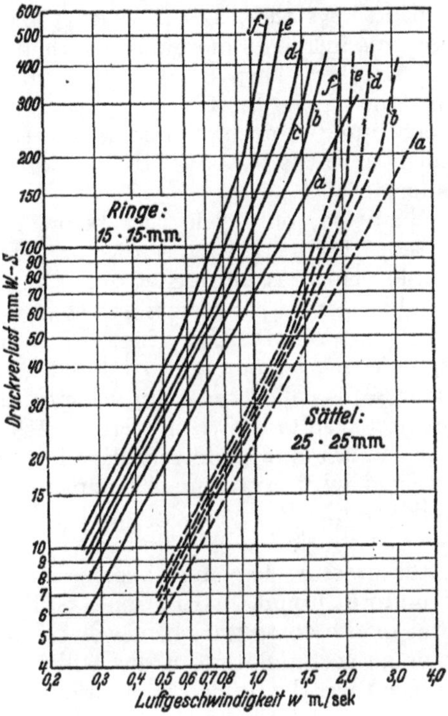

FIGURE 218

Pressure drop through irrigated porcelain Raschig rings and saddles as a function of air velocity

Packing height 1 m; liquid used: water

——— Rings, 15 × 15 mm
- - - - Saddles, 25 × 25 mm

a. Dry packing
b. Water rate of 1 cu m per square m per hour
c. Water rate of 2 cu m per square m per hour
d. Water rate of 4 cu m per square m per hour
e. Water rate of 8 cu m per square m per hour
f. Water rate of 12 cu m per square m per hour

Ringe—rings
Sättel—Saddles
Druckverlust, mm WS—Pressure drop, millimeters water
Luftgeschwindigkeit, m/sek—Air velocity, meters per second

Figure 218 also shows that, as the velocity increases beyond w_u, the pressure loss increases more quickly than with velocities below w_u until another sharp bend in the curves is reached, which indicates the upper limit velocity w_o. At loads above this range, much liquid is accumulated within the packing and partly entrained upwards. Above the packing, a turbulent liquid layer is formed.

For actual operation, only velocities below the lower limit w_u are considered. Figure 219 presents, for different types of packings, suitable velocities plotted against the mass of liquid (expressed as cubic meters per square meter of column section per hour). For porcelain Raschig rings, of 25-mm size, in a flow of atmospheric air, w_u amounts to about 1 m per second. The range of $w < w_u$ is called the main velocity range.

The experimental results shown in Figures 218 and 219 hold in the case that the liquid is uniformly distributed in the layer. They were obtained with a column 300 mm in diameter, over which the liquid was distributed by a wheel.

If the liquid distribution is non-uniform, particularly when rivulets form, the pressure loss may become smaller than it is in the case of uniform liquid distribution. Still, this influence can be neglected in the computation of wet columns because a closer packing or an accumulation of dirt will cause a more appreciable difference than the liquid distribution.

Apart from the mass of the liquid, the viscosity of a liquid will also influence the pressure loss. For the main velocity range, we can write

$$\Delta P = k_b \times H \times w^n \text{(millimeters } H_2O) \qquad (213)$$

where k_b = resistance coefficient for wetted packings. This is the pressure loss for a layer 1-m thick and a gas velocity of 1 m per second. It is derived from the empirical equation

$$k_b = k_i + 0.005\, k_i \times B \qquad (214)$$

B is the mass of liquid (cubic meters per square meter per hour), and k_i is the resistance coefficient for wetted (not actually trickled-over) packings, hence, for $B = 0$. It depends upon the pressure loss of the dry packing, the density γ (kilograms per cubic meter) of the gas or vapor, and upon the viscosity of the liquid

$$k_i = k_t \times \gamma^{0.83}(1 + 0.124E) \qquad (215)$$

E is the viscosity of the liquid in degrees Engler, and k_t is the resistance coefficient of the dry packing. The latter figure is the pressure loss in millimeters of H_2O for a layer of dry packing 1 m high, for a velocity w of 1 m per second, and for a specific weight of the gas γ of 1 kg per cubic meter. For rings and saddles, it may be computed from equations (211) and (212).

If the viscosity is stated in poises ($\eta g \times cm^{-1} \times sec^{-1}$), then

$$k_i = k_t \times \gamma^{0.83}(1 + 1.25\eta) \tag{215a}$$

The velocity exponent n in equation (213) has an average value of 1.9. If the liquid trickling down is water, one may write

$$k_i = 1.4k_t \tag{215b}$$

In computing the pressure loss in irrigated packing, one starts by determining the resistance coefficient k_t for dry packing from equations (211) and (212). Then, k_i follows from equations (215), (215a) or (215b), and k_b from equation (214). This, in turn, yields the pressure loss from equation (213).

The tests and computations described have been condensed into Table V which comprises 4 sections. The pressure loss for a given set of conditions may be taken from the table. The graph holds for the main velocity range only ($w < w_u$), that is, for those cases where the pressure loss per meter of packing height does not exceed 50 to 70 mm H_2O, in other words, for the range which is important in practice.

Numerical example. 2300 cu m per hour of a gas having a density γ of 0.8 kg per cubic meter are passed through a column of 1000-mm diameter filled with Raschig rings 25 mm in size. 16 cu m per hour of liquid with a viscosity of 6° Engler are introduced. The velocity w thus becomes 0.82 m per second, and the flow of liquid B 20.5 cu m per square meter per hour.

The dotted line in the graph (Table V) shows a pressure loss of 53 mm per meter of packing height. On the basis of the packing size given, section 1 gives the specific weight of the gas, section 2 considers the viscosity of the liquid, section 3 the mass of liquid, and section 4 the gas velocity. The ordinate of section 4 yields the pressure loss in question.

In agreement with the results in Figure 219, the lower limiting velocity will also increase with increasing packing-element size. For best rectification efficiency, the column section thus becomes smaller for large-size packing than for smaller-size packing. The variation of the packing height required, however, goes in the reverse direction, in accordance with Figure 202. One will

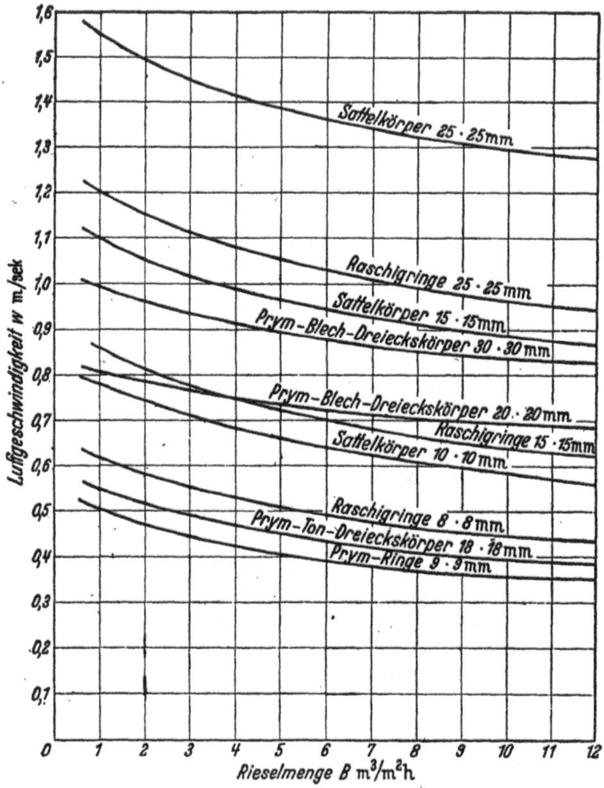

FIGURE 219

Lower limiting velocities at given water rates, for selected types of packing

Rieselmenge, m^3/m^2 h—Water rate, cubic meters per square meter per hour
Luftgeschwindigkeit w m /sek—Air velocity w, meters per second
Sattelkörper—Saddles
Raschigringe—Raschig rings
Prym Blech-Dreickskörper—Prym triangular sheet metal prisms
Prym Ton-Dreieckskörper—Prym triangular clay prisms
Prym-Ringe—Prym rings

probably select large size packing elements where the throughput is large and where there is a risk of clogging or of contamination.

K. DETAILS

I. Heat exchangers

Besides the column, heat exchangers are the main item in a rectifying unit. As in the case of the simple batch still, vapor must also be generated from liquid in the case of continuous equipment. Figures 43, 46, 47, 48 and 54 show typical heaters of batch-type units. In the case of continuous outfits involving a rectifying and a stripping column, steam is the usual heating medium in those cases where the effluent liquid consists of water. This heating steam is introduced by a sparger pipe below the bottom plate of the column.

Where it is desirable to prevent a contact between the effluent and the heating steam, a coil (Figure 220) will serve for small

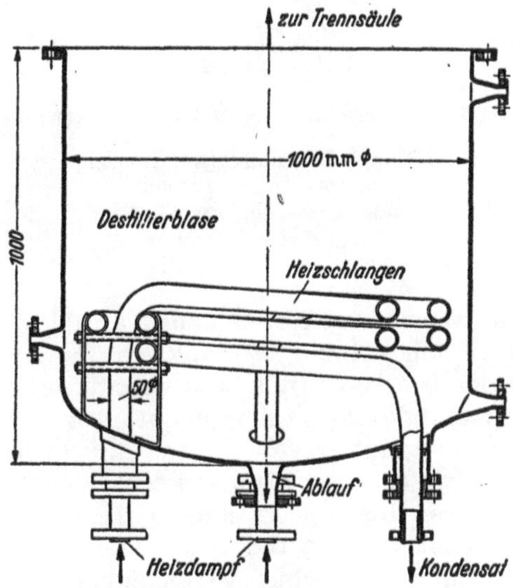

FIGURE 220

Base heater (coil-type) of a continuous rectifier

Destillierblase—Still pot
Heizschlangen—Heating coils
Heizdampf—Heating steam
Ablauf—Drain
Kondensat—Condensate
Zur Trennsäule—To separating column

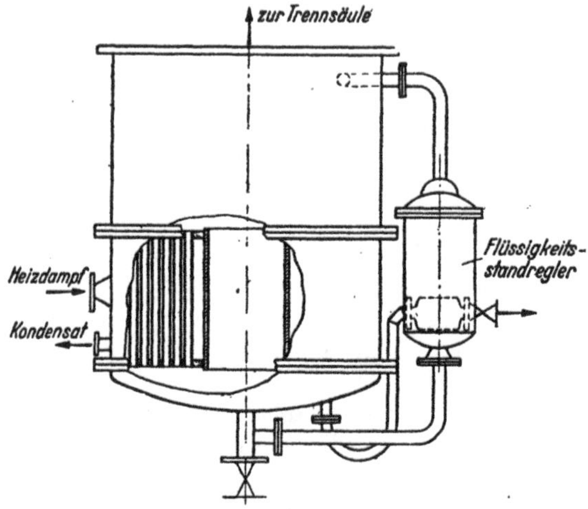

FIGURE 221

Standard vertical-tube heater section of a continuous rectifier

Flüssigkeitsstandregler—Leveler
Heizdampf—Heating steam
Kondensat—Condensate
Zur Trennsäule—To column

outputs, and a tubular heater or calandria arranged immediately below the rectifying column is employed for larger outputs. Liquid circulation is assured by a large central downpipe; Figure 221. The apparent liquid level is maintained by a controller; a float actuates a regulating valve in the effluent line.

Figure 222 shows a heater arranged separately and equipped with vertical tubes; this type is used in the case of very large outputs. A control valve in the return line permits the maintenance of that liquid level which yields highest efficiency. The apparent level is obtained in a gage pipe connected to the spaces below the lower and above the upper tube plate.[85]

The vapors from the column are condensed, wholly or partially, in the reflux condenser, which may be either directly superposed on the column, or arranged laterally. Of the former type, Figures 223 and 224 show designs featuring vertical or horizontal tubes. Under special conditions, as in the alcohol industry, these tubes are made of copper and expanded into cast-iron walls.

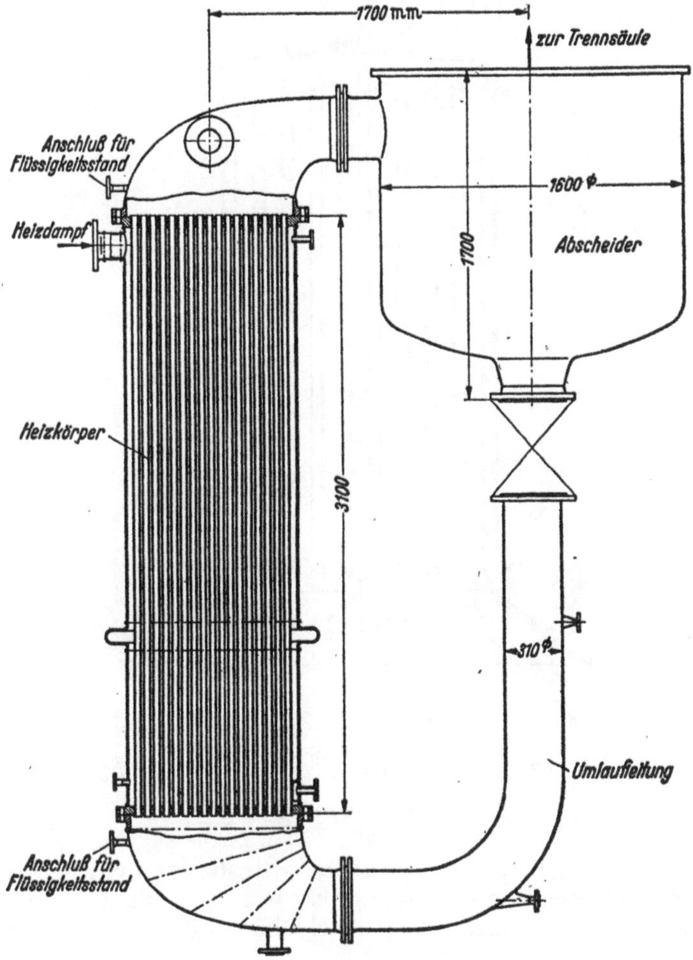

FIGURE 222

Vertical-tube natural-circulation heater of a continuous rectifier, arranged parallel to the column

Heizkörper—Heater unit
Umlaufleitung—Downtake
Abscheider—Separator
Zur Trennsäule—To separating column
Anschluss für Flüssigkeitsstand—Level gage connection
Heizdampf—Heating steam

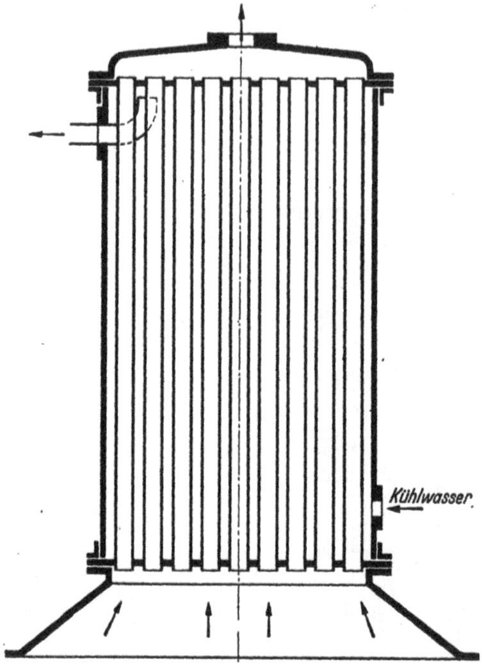

FIGURE 223

Reflux condenser with vertical tubes, for mounting on top of a column

Kühlwasser—Cooling water

The same general arrangements will also apply where the reflux condenser is erected separately from the column. Figure 54 shows, for instance, a separately installed reflux condenser with straight tubes.

Final coolers for medium and large outputs are usually straight-tube heat exchangers (Figure 54), or, for smaller outputs, coils (Figure 197). The heat-exchange surface may be found by the usual procedure.

II. Control equipment

The most important controller is that which maintains the output at a desired level. The higher the vapor velocity, and

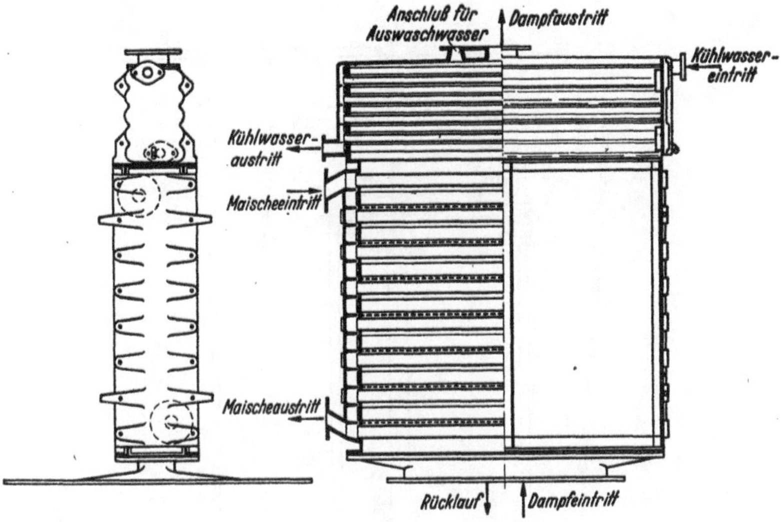

FIGURE 224

Reflux condenser with horizontal tubes, for mounting on top of a column

Dampfeintritt—Vapor inlet
Rücklauf—Reflux
Maischeeintritt—Feed inlet (Mash)
Maischeaustritt—Feed outlet (Mash)
Kühlwassereintritt—Cooling water inlet
Kühlwasseraustritt—Cooling water outlet
Dampfaustritt—Vapor outlet
Anschluss für Auswaschwasser—Backwash connection (for cleaning)

consequently the output, the higher becomes the pressure loss in the vapor passing through the rectifying column. The pressure difference between the vapor space in the still and that in the condenser is therefore some measure of the load condition. If, for instance, the final cooler of a bubble-plate column is open to the atmosphere, output control means keeping the excess pressure in the vapor space of the still kettle above the atmospheric pressure. In many cases the effluent consists of water. Then, the float controller shown in Figure 225 will yield satisfactory results.

The excess pressure in the still acts upon the liquid in the

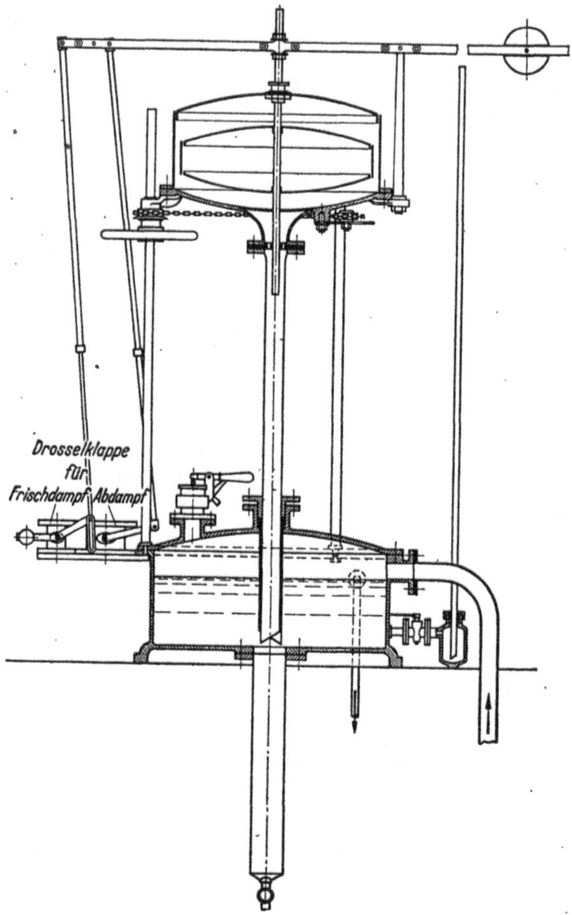

FIGURE 225

Automatic steam-pressure controller

Drosselklappe—Throttle valve
für Frischdampf—for live steam
für Abdampf—for exhaust steam

lower space; it is indicated by a gage pipe and also by the difference in levels between the upper and the lower space. Both are connected by an equalizer pipe. If the pressure in the still tends to rise, the level in the upper space will rise similarly. The float in the latter then actuates a lever which handles a regulating valve in the steam line, and throttles the flow of steam. The reverse conditions in the still will cause an opposite effect. The upper space can be moved up or down to adjust the controller.

As in the case of packed columns there is a definite relation between effect and pressure loss, the pressure loss should be kept at that level which yields highest efficiency; see Section J, II, IV.

L. MOLECULAR DISTILLATION

Molecular distillation is different from ordinary distillation since the evaporation of the liquid is effected in such a way that no boiling occurs.[86] The vapor is liberated from the liquid surface in the same way as a liquid being evaporated by a gas stream. The maximum evaporation velocity V (grams per second per square centimeter) at which boiling does not take place, depends on the pressure p (millimeters of Hg), the molecular weight μ of the material, and its absolute temperature T(°K)[87]

$$V = 5.83 \times 10^{-2} \times p \times \sqrt{\frac{\mu}{T}} \qquad (216)$$

In the case of triolein, which at 250°K has a vapor pressure of 0.0043 mm Hg, V becomes 0.0326×10^{-2} gram per second per square centimeter, so that, theoretically, it is possible to evaporate 2 g per minute on a surface of 100 sq cm.

However, the ideal evaporation velocity, according to equation (216) exists with one proviso, which is, that none of the vapor molecules which leave the surface of the liquid return to it. This is only possible if these molecules are intercepted on their way, that is, condensed by a surface. This means that the velocity of molecular distillation depends on the velocity of the condensation.

Therefore, the relative location of the condensing and evaporating surfaces must be such that the conditions under which equation (216) holds are fulfilled as far as possible. This is obtained by arranging the condensing and evaporating surfaces in parallel, and by making the distance between them somewhat smaller than the free length of the path of the molecules. The average free length of the path of molecules depends on the pressure and molecular weight of the material. As pressure decreases, the free length of the path increases. In actual practice, one maintains a pressure of about 10^{-4} mm Hg in the space between evaporation and condensation surfaces. Thus, the distance between these surfaces becomes 2 to 3 cm.

The pressure mentioned can only be maintained over the evaporating surface if the path of the vapors between the point of their generation up to the point where they condense is 2 to 3 cm long in this case. The quantity of gases liberated from the liquid must be correspondingly small, which demands previous degassing. But even for this small mass of gas, a short distance must be available between liquid surface and the connection of the vacuum pump. The material to be distilled may contain water or volatiles, which must be stripped under vacuum in a separate procedure, prior to distillation. The temperature difference between evaporating and cooling surface must also suffice to condense all of the vapor present. As a rule, a temperature difference of 100°C is adequate. If one considers that the molecules leave the liquid surface, which is the same as the evaporating surface, at different angles, one will realize that when molecular distillation is carried out, 90 per cent of the ideal evaporation velocity can be achieved at best.

Although some molecules between the evaporating and the cooling surface may hit each other, the bulk of them will move in one direction. Therefore, there is, properly speaking, neither a stationary pressure of the molecules distilling nor an equilibrium between phases. Furthermore, under the conditions of molecular distillation, there is no definite distillation temperature which can be compared to a boiling point coordinated with a definite pressure. Distillation becomes possible at any temperature, and the components evaporate independently of each other. If the temperature of the liquid distilling is increased, the only consequence is an increased distillation velocity.

By means of molecular distillation it is possible to distill liquids which would decompose under the temperatures and conditions usual in distillation or rectification, or which would not distill at all. This is due not only to the low absolute pressure, but also to the fact that between the evaporation and the cooling surface the molecules cannot collide to any appreciable extent, as they move essentially in one direction only. By molecular distillation, it is, for instance, possible to prepare special oils and fats from mineral-oil residues. It has further been of use in separating bituminous materials and coal-tar products, for treating all triglyceride oils, and in the preparation of vitamins.

Molecular distillation may be operated either as a batch or as a continuous process; the latter is the preferable one. Figure 226 shows the design of a distillation tube of a continuous unit

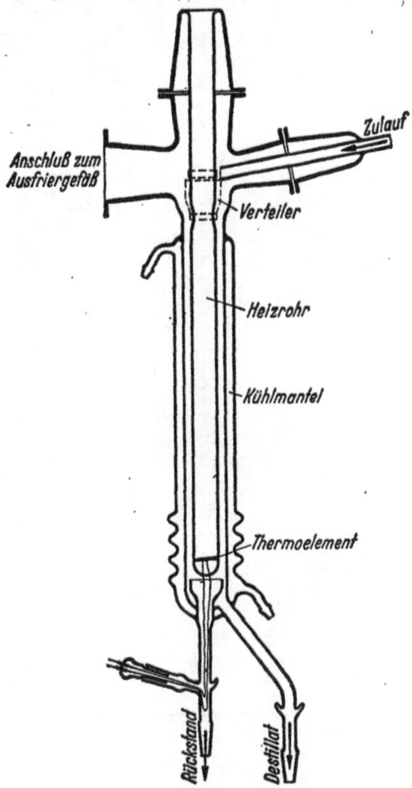

FIGURE 226

Molecular distillation tube for continuous operation

Zulauf—Feed connection
Anschluss zum Ausfriergefäss—Connection to freezing vessel
Verteiler—Distributor
Heizrohr—Heater tube
Kühlmantel—Cooling jacket
Rückstand—Residue
Destillat—Distillate

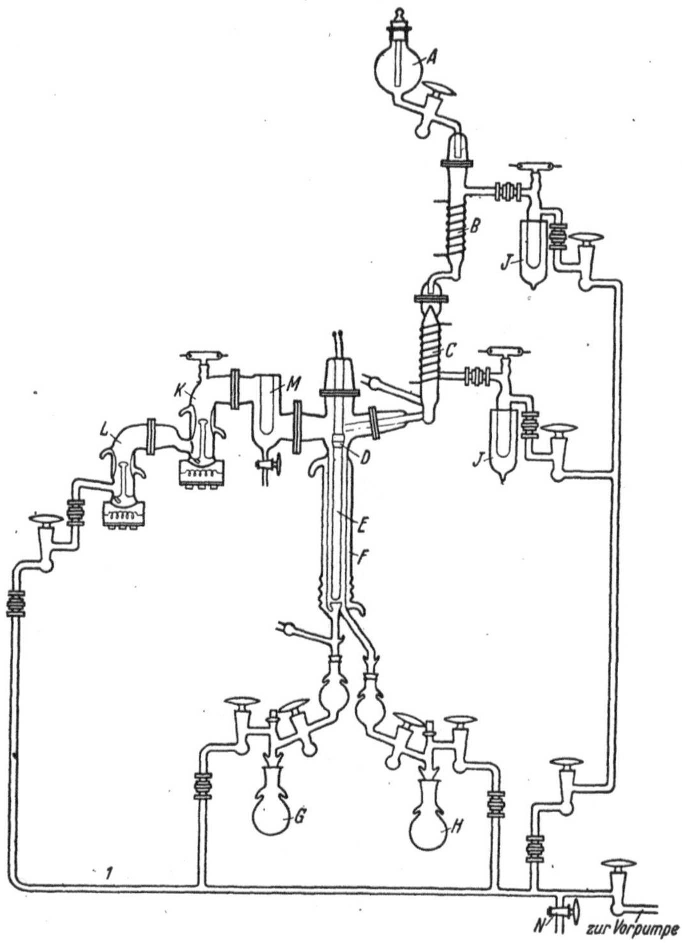

FIGURE 227

Layout of a continuous molecular distillation unit with one distillation tube

Zur Vorpumpe—To forepump (booster)

The liquid to be decomposed leaves the supply flask A and passes first into two electrically heated predegassers B and C, the former of which is filled with Raschig rings made of glass. The liquid to the latter passes over a filter cone of glass. Distributor D spreads the liquid out to form a film over the outer surface of heating tube E which is 400 mm long and surrounded by cooling jacket *F*. The residue accumulates in vessel *G*, the distillate in vessel *H*.

The primary degassers are connected to freezing pockets *J* which dip into Dewar flasks containing liquid air, and to the line towards to the primary vacuum pump. Oil-diffusion pumps *K* and *L*, which together provide the high degree of vacuum required, are connected to freezing jacket *M* and to the distilling tube.

The main piping system *1* is usually assembled in situ from Duran glass tube of 15–16 mm outer diameter. At the lowest point of the main piping, cock *N* is provided. Oil eventually accumulating in the system can be drained through *N*.

---

made of glass. The degassed liquid is fed to the outer side of an electrically heated tube, spreads to form a film and flows downwards. The residue leaves the lower end of the tube, the temperature of which is measured by a thermoelement. The distillate is condensed on the cylindrical surface arranged concentrically with the heating tube and cooled by water. It is collected at the lower end of the cooling jacket.

The entire layout of a continuous unit for molecular distillation is presented in Figure 227. The unit consists of glass parts. The jackets of primary degassers *B* and *C* are equipped with a heating element, as are distillation tube *E* and diffusion pumps *K* and *L*. The distillation tube and diffusion pumps require cooling water. Ahead of the diffusion pumps, and of the forepump, which is oil-filled and of the rotary type, freezing vessels filled with liquid air are arranged which prevent vapors from entering the pumps.

A unit of the type shown in Figure 227 serves to decompose a liquid into two components. Where splitting up into more than two constituents is required, several distillation tubes are arranged in series. The residue from one distillation tube forms the feed to the next tube. As a rule continuous molecular distillation units made of glass have outputs of 50–250 g per hour.*

*Data from Jenaer Glaswerk, Schott and Gen., Jena.

M.* APPENDIX

I. Equilibrium data for binary mixtures; interrelation between vapor composition and boiling point at constant pressure p

The lower boiler is always named first.

The number of the binary mixture in the following tabulation corresponds to the number of the equilibrium curve in Graphs I, II, III or IV.

GRAPH I

1. Carbon tetrachloride—Ethyl acetate[88]
$p = 745$ mm Hg

Per cent by weight carbon tetrachloride in liquid phase	Per cent by weight carbon tetrachloride in vapor phase	Boiling Point °C
0	0	76.50
10	10.7	76.06
20	21.1	75.56
30	31.2	75.10
40	41.2	74.74
50	51.1	74.35
60	60.4	74.07
70	70.4	74.10
80	80.4	74.34
90	90.2	74.89
100	100	75.92

2. Ethyl bromide—Benzol[88]

p = 760 mm Hg

Per cent by weight ethyl bromide in liquid phase	Per cent by weight ethyl bromide in vapor phase	Boiling Point °C
0	0	80.25
10	26.6	74.86
20	46.8	70.00
30	61.4	65.39
40	72.0	60.84
50	80.4	56.52
60	86.4	52.59
70	91.0	48.78
80	94.6	45.10
90	97.6	41.49
100	100	38.38

3. Chloroform—Methanol[88]

p = 757 mm Hg; azeotropic point at 87.4 per cent by weight chloroform, 53.43°C

Per cent by weight chloroform in liquid phase	Per cent by weight chloroform in vapor phase	Boiling Point °C
0	0	64.86
10	25.2	64.05
20	41.7	62.38
30	54.1	60.68
40	64.0	59.07
50	72.3	57.52
60	78.0	55.94
70	81.5	54.52
80	84.2	53.66
90	88.7	53.65
100	100	61.37

4. Ethanol—Trichlorethylene[89]

p = 760 mm Hg; azeotropic point at 28 per cent by weight ethanol, 70.8°C

Per cent by weight ethanol in liquid phase	Per cent by weight ethanol in vapor phase	Boiling Point °C
0.5	3.7	83.4
0.8	5.6	81.6
1.0	9.7	78.7
0.9	13.5	76.6
2.0	15.9	74.2
2.5	17.1	73.0
4.5	19.7	71.6
11.6	24.7	71.2
26.7	27.3	70.8
26.9	27.9	70.8
29.2	28.1	70.8
37.3	29.7	70.9
45.6	31.9	71.0
53.5	35.3	71.3
58.6	37.0	71.4
60.8	39.2	71.7
65.4	41.5	72.2
69.4	44.7	72.7
72.7	50.2	73.0
76.5	53.8	73.4
81.3	60.5	74.2
85.3	65.3	75.0
90.0	73.4	75.6
93.4	79.9	76.2

5. Methanol—Trichlorethylene[89]

p = 760 mm Hg; azeotropic point at 34 per cent by weight methanol, 59.4°C

Per cent by weight methanol in liquid phase	Per cent by weight methanol in vapor phase	Boiling Point °C
0.1	1.1	83.5
0.7	5.3	78.2
0.8	7.7	75.3
0.8	10.3	72.6
0.8	12.6	71.1
1.0	11.6	70.1
1.1	16.4	67.3
1.3	17.9	66.1
1.3	21.1	64.1
2.5	26.6	60.9
7.0	31.2	60.4
9.4	31.2	60.1
11.2	30.8	60.1
24.7	32.3	59.6
31.0	33.3	59.4
38.7	34.8	59.4
51.6	38.9	59.5
59.3	42.7	59.8
64.4	45.1	60.1
69.6	48.4	60.4
74.2	52.4	60.7
80.3	57.2	61.0
82.5	61.5	61.4
86.5	69.0	61.8
91.8	76.1	62.5
96.7	86.1	63.3

6. Ethanol—Benzol[89]

p = 760 mm Hg; azeotropic point at 32.5 per cent by weight ethanol, 67.9°C

Per cent by weight ethanol in liquid phase	Per cent by weight ethanol in vapor phase	Boiling Point °C
1.8	3.6	78.0
1.6	7.0	77.1
2.2	12.8	75.0
3.1	12.8	74.6
4.4	16.6	72.8
4.6	20.6	71.2
6.4	23.6	70.3
7.6	26.1	69.0
9.0	29.6	68.5
15.8	30.1	68.3
22.2	30.7	68.1
22.6	31.4	68.0
26.6	31.6	67.9
30.7	32.7	67.9
35.3	33.6	67.9
39.2	34.1	68.0
43.6	35.2	68.0
48.2	36.1	68.1
52.9	37.3	68.2
55.9	38.9	68.3
60.3	40.0	68.6
64.2	41.4	68.8
70.1	43.4	69.1
72.7	45.9	69.6
76.1	49.4	70.1
81.0	53.6	71.2
84.2	56.3	71.8
86.3	60.8	73.1
93.4	73.0	74.7
94.8	77.7	75.3
96.4	78.7	75.6
98.9	91.3	76.8

7. Methanol—Benzol[89]

p = 760 mm Hg; azeotropic point at 39 per cent by weight methanol, 57.9°C

Per cent by weight methanol in liquid phase	Per cent by weight methanol in vapor phase	Boiling Point °C
2.2	14.5	70.9
1.1	16.5	69.4
2.2	20.7	66.8
2.7	23.0	65.7
2.7	24.8	64.2
4.1	28.5	61.4
6.7	34.9	59.2
6.3	35.2	59.0
14.4	36.2	58.0
25.5	37.2	57.8
37.3	39.0	57.7
50.1	40.9	57.6
63.3	45.4	58.1
74.8	51.7	58.9
77.9	54.2	59.6
85.0	62.1	60.4
87.4	65.3	61.2
93.1	77.8	62.4
96.7	87.3	63.4

8a. *n*-Pentane—*n*-Heptane[90]

p = 10.3 atmospheres absolute

Per cent by weight *n*-pentane in liquid phase	Per cent by weight *n*-pentane in vapor phase	Boiling Point °C
0.0	0.0	202.7
7.4	18.2	190.8
15.3	33.3	180.2
23.5	46.4	170.7
32.5	58.0	162.2
42.0	67.8	154.6
52.1	76.4	147.7
62.8	83.3	141.6
74.3	89.4	136.0
86.5	94.8	130.9
100.0	100.0	126.1

8b. *n*-Pentane—*n*-Heptane[90]

p = 20.6 atmospheres absolute

Per cent by weight *n*-pentane in liquid phase	Per cent by weight *n*-pentane in vapor phase	Boiling Point °C
0.0	0.0	246.5
7.4	12.8	235.9
15.3	25.3	226.1
23.5	37.7	216.8
32.5	49.6	208.1
42.0	60.6	199.7
52.1	70.2	191.9
62.8	78.4	184.6
74.3	86.1	177.7
86.5	93.0	171.0
100.0	100.0	164.6

8c. *n*-Pentane—*n*-Heptane[90]

p = 31.1 atmospheres absolute

Per cent by weight *n*-pentane in liquid phase	Per cent by weight *n*-pentane in vapor phase	Boiling Point °C
19.8	19.8	253.5*
23.5	26.1	248.8
32.5	37.9	239.5
42.0	49.2	230.7
52.1	60.1	222.3
62.8	70.7	214.1
74.3	80.6	206.3
86.5	90.9	198.6
100.0	100.0	191.3

*Critical point

9. Phenol—*o*-Cresol[91]

$p = 760$ mm Hg

Per cent by weight phenol in liquid phase	Per cent by weight phenol in vapor phase	Boiling Point °C
0.0	0.0	191.8
18.1	22.7	
33.6	40.0	
34.75	40.4	
58.9	64.8	
75.5	80.0	
81.25	85.92	
93.6	95.0	
100.0	100.0	182.0

10. Phenol—*m*-Cresol[91]

$p = 760$ mm Hg

Per cent by weight phenol in liquid phase	Per cent by weight phenol in vapor phase	Boiling Point °C
0.0	0.0	202.1
16.0	26.2	
27.3	41.3	
51.2	64.2	
72.0	83.1	
91.25	94.75	
97.75	98.6	
100.0	100.0	182.0

11. Phenol—*p*-Cresol[91]

$p = 760$ mm Hg

Per cent by weight phenol in liquid phase	Per cent by weight phenol in vapor phase	Boiling Point °C
0.0	0.0	202.5
20.0	30.0	
21.6	32.0	
32.7	44.4	
56.9	69.8	
74.5	84.0	
91.5	94.5	
100.0	100.0	182.0

Phenol forms ideal mixtures with the cresols. The mixtures phenol—*m*-cresol and phenol—*p*-cresol have practically the same equilibrium curve.

12a. Propane—Isobutylene[92]

$p = 14.1$ atmospheres absolute

Per cent by weight propane in liquid phase	Per cent by weight propane in vapor phase	Boiling Point °C
0.0	0.0	85.5
8.0	16.7	77.1
16.3	31.7	70.5
25.1	43.6	65.5
34.3	54.0	61.0
43.9	63.3	57.1
54.1	72.3	53.2
64.5	80.4	50.3
75.7	87.1	46.6
87.5	93.3	42.7
100.0	100.0	38.3

12b. Propane—Isobutylene[92]

p = 21.1 atmospheres absolute

Per cent by weight propane in liquid phase	Per cent by weight propane in vapor phase	Boiling Point °C
0.0	0.0	105.8
8.0	14.9	98.6
16.3	28.5	92.6
25.1	40.2	87.1
34.3	50.6	82.7
43.9	60.2	78.2
54.1	69.2	73.8
64.5	77.6	69.8
75.7	85.4	66.0
87.5	92.5	62.7
100.0	100.0	58.2

12c. Propane—Isobutylene[92]

p = 28.2 atmospheres absolute

Per cent by weight propane in liquid phase	Per cent by weight propane in vapor phase	Boiling Point °C
0.0	0.0	122.0
8.0	13.1	116.0
16.3	25.4	109.8
25.1	36.8	103.7
34.3	47.1	98.2
43.9	56.9	93.8
54.1	66.2	89.4
64.5	75.1	85.5
75.7	83.6	81.6
87.5	91.7	77.1
100.0	100.0	72.6

12d. Propane—Isobutylene[92]

p = 35.2 atmospheres absolute

Per cent by weight propane in liquid phase	Per cent by weight propane in vapor phase	Boiling Point °C
0.0	0.0	137.2
8.0	11.7	129.0
16.3	22.7	123.0
25.1	33.1	116.4
34.3	43.4	111.3
43.9	53.5	106.4
54.1	63.1	102.2
64.5	72.5	97.6
75.7	81.6	93.8
87.5	90.5	90.5
100.0	100.0	87.1

12e. Propane—Isobutylene[92]

p = 42.2 atmospheres absolute

Per cent by weight propane in liquid phase	Per cent by weight propane in vapor phase	Boiling Point °C
10.1	10.1	139.0*
16.3	17.8	134.0
25.1	28.5	127.5
34.3	39.8	122.5
43.9	50.4	117.5
54.1	60.1	113.1
64.5	69.9	108.2
75.7	79.5	104.3
87.5	89.3	100.0
100.0	100.0	93.3

*Critical point

13a. Ethanol—Water[93]

p = 760 mm Hg; azeotropic point at 95.57 per cent by weight ethanol, 78.15°C

Per cent by weight ethanol in liquid phase	Per cent by weight ethanol in vapor phase	Boiling Point °C
0	0.0	100
1	9.5	98.75
2	19.3	97.65
3	27.1	96.65
4	33.4	95.8
5	38.0	94.95
7.5	46.35	93.05
10	52.0	91.45
15	59.5	88.95
20	64.8	87.15
25	68.6	85.75
30	71.4	84.65
35	73.3	83.75
40	74.7	83.1
45	75.9	82.45
50	77.1	81.9
55	78.2	81.45
60	79.4	81.0
65	80.7	80.6
70	82.2	80.2
75	83.9	79.8
80	85.9	79.35
85	88.3	78.95
90	91.3	78.5
91	91.95	78.42
92	92.65	78.33
93	93.43	78.27
94	94.2	78.2
95	95.04	78.15
100	100.0	78.3

13b. Ethanol—Water[93]

p = 50 mm Hg; no azeotropic point

Per cent by weight ethanol in liquid phase	Per cent by weight ethanol in vapor phase	Boiling Point °C
0	0	38.05
5	24.5	35.15
10	44.0	32.8
15	55.5	30.9
20	63.2	29.4
25	68.7	28.15
30	72.5	27.1
35	75.0	26.25
40	76.8	25.6
45	78.1	25.05
50	79.25	24.6
55	80.35	24.3
60	81.5	23.9
65	82.65	23.6
70	84.0	23.25
75	85.45	22.9
80	87.3	22.6
85	89.45	22.25
90	92.1	21.95
95	95.5	21.7
100	100.0	21.55

Graph II

14. Water—*n*-Butanol[94]

p = 760 mm Hg; azeotropic point at 42.5 per cent by weight water, 92.7°C

Per cent by weight water in liquid phase	Per cent by weight water in vapor phase	Boiling Point °C
0.0	0.0	117.4
1.1	8.1	111.5
1.3	9.3	110.6
1.5	10.4	109.6
1.8	11.7	108.8
7.7	29.2	97.9
8.4	30.2	97.2
9.1	31.5	96.7
9.7	32.2	96.3
19.3	40.5	93.5
20.0	40.9	93.4
23.2	42.4	92.9
23.9	42.5	92.9
24.5	42.1	92.9
24.9	42.4	92.8
90.5	42.6	92.7
92.1	43.1	93.0
93.0	43.3	92.8
94.0	45.7	93.4
94.4	46.8	93.4
95.4	50.5	93.7
96.5	52.5	95.4
97.7	65.0	96.8
98.6	76.1	98.3
99.3	83.5	98.4
99.7	92.6	99.4
100.0	100.0	100.0

15. Water—Isobutanol[94]

p = 760 mm Hg; azeotropic point at 33 per cent by weight water, 89.2°C

Per cent by weight water in liquid phase	Per cent by weight water in vapor phase	Boiling Point °C
0.0	0.0	107.1
3.7	14.0	97.7
4.1	15.0	97.1
4.4	15.9	96.6
4.8	16.4	96.0
13.8	29.0	90.3
14.2	29.4	90.2
14.6	29.6	90.2
14.5	28.9	90.1
23.9	32.0	89.5
24.3	32.1	89.5
24.6	32.1	89.5
24.9	32.1	89.5
26.6	32.7	89.5
26.9	32.7	89.5
27.2	32.7	89.4
27.4	32.5	89.4
29.6	32.9	89.4
29.7	33.2	89.4
29.9	33.1	89.4
33.0	33.0	89.2
33.0	32.9	89.2
33.0	32.8	89.2
32.9	32.6	89.2
83.5	32.8	89.5
85.3	33.0	89.5
88.0	32.9	89.6
88.2	33.4	89.5
90.3	33.2	89.5
91.5	33.3	90.1
92.2	33.9	89.9
94.2	37.9	91.5
95.2	39.8	91.9
96.3	46.6	93.4
97.5	55.2	95.1
97.9	58.4	95.9

Per cent by weight water in liquid phase	Per cent by weight water in vapor phase	Boiling Point °C
98.5	66.1	96.8
98.6	68.4	97.1
99.0	76.4	98.1
99.4	84.8	98.9
100.0	100.0	100.0

16. Bromine—Water[95]

p = 750 mm Hg

Per cent by weight bromine in liquid phase	Per cent by weight bromine in vapor phase
0.20	39.48
0.40	59.7
0.695	72.1
1.02	81.5
1.38	88.35
1.59	92.1
1.94	91.1
2.10	93.2
2.16	92.85
2.29	96.0
2.85	97.6
3.14	97.7

17. Hydrochloric acid—Water[96]

p = 751.3 mm Hg; azeotropic point at 20.4 per cent by weight hydrochloric acid

Per cent by weight HCl in liquid phase	Per cent by weight HCl in vapor phase	Boiling Point °C
8.67	0.21	101.5
12.45	0.75	103.1
15.97	2.80	105.5
18.09	7.55	107.2
20.09	18.49	107.8
20.79	22.97	107.6
21.36	27.54	107.5
21.79	31.00	107.3

18. Acetone—Methanol[96]

p = 755 mm Hg

Per cent by weight acetone in liquid phase	Per cent by weight acetone in vapor phase	Boiling Point °C
9.6	23.4	61.9
31.8	51.8	58.3
45.2	62.0	56.8
58.8	68.3	55.6
67.0	73.5	55.4
81.4	83.3	54.9
89.9	88.7	54.7
92.9	92.1	55.0
95.8	96.0	55.2

19. Ethyl acetate—Ethanol[97]

Per cent by weight ethyl acetate in liquid phase	Per cent by weight ethyl acetate in vapor phase	Boiling Point °C
0.0	0.0	78.3
0.85	1.9	77.7
3.1	6.7	76.4
3.7	7.4	76.1
5.0	10.1	75.6
6.85	12.7	75.0
7.1	13.3	74.9
9.6	15.9	74.2
12.55	20.1	73.6
13.15	20.8	73.5
15.9	23.7	73.1
17.6	25.9	72.8
21.7	29.05	72.4
28.8	33.7	72.0
48.1	44.15	71.8
62.4	55.7	72.6
77.9	67.75	74.4
82.5	72.8	75.0
85.7	78.4	75.5
95.3	87.4	76.7
100.0	100.0	77.1

20. Ethyl acetate—Acetic acid[98]

p = 760 mm Hg

Per cent by weight ethyl acetate in liquid phase	Per cent by weight ethyl acetate in vapor phase	Boiling Point °C
0.0	0.0	118.2
3.7	6.3	
4.0	6.9	
5.0	10.2	
5.1	14.0	
5.9	16.3	
6.4	14.3	
9.6	24.1	
10.0	25.4	
10.65	31.6	
12.0	30.4	
14.95	41.3	
17.5	46.9	
20.3	49.55	
20.9	53.8	
22.4	53.4	
29.05	65.0	
29.85	64.7	
38.85	73.05	
39.75	73.05	
49.0	82.7	
61.6	91.45	
62.3	90.5	
75.4	97.2	
81.5	98.4	
92.35	99.6	
94.1	99.4	
100.0	100.0	77.1

21. Acetone—*n*-Butanol[98]

$p = 760$ mm Hg

Per cent by weight acetone in liquid phase	Per cent by weight acetone in vapor phase	Boiling Point °C
0	0	117.4
2	15.5	
5	34.3	
10	54.9	
15	68.5	
20	76.6	
25	81.9	
30	85.7	
35	88.8	
40	91.4	
45	93.2	
50	94.5	
55	95.6	
60	96.2	
65	96.9	
70	97.4	
75	97.9	
80	98.4	
85	98.8	
90	99.2	
95	99.6	
100	100.0	56.3

22. *n*-Butanol—*n*-Butyl acetate[98]

p = 760 mm Hg; azeotropic point at 63.2 per cent by weight butanol, 116.2°C

Per cent by weight butanol in liquid phase	Per cent by weight butanol in vapor phase	Boiling Point °C
7.2	15.0	121.75
14.3	24.1	120.1
21.1	30.9	119.1
26.5	35.7	118.4
32.7	40.6	117.8
34.0	41.8	117.5
40.0	45.8	117.3
43.2	48.9	117.1
44.2	49.5	116.8
46.3	50.8	116.6
49.7	53.3	116.4
53.7	56.3	116.3
57.4	58.9	116.2
60.9	61.6	116.2
62.8	63.9	116.6
63.2	63.2	116.2
63.4	63.7	116.55
66.4	65.7	116.55
75.4	73.5	116.3
80.3	78.3	116.8
86.9	84.4	117.0
93.8	91.1	117.0
96.8	94.4	117.0
99.0	98.1	117.0

23a. Ammonia—Water[99]
p = 1 atmosphere absolute

Per cent by weight ammonia in liquid phase	Per cent by weight ammonia in vapor phase	Boiling Point °C
2	22.5	91.5
4	41.0	84.8
6	54.5	79.0
8	64.8	73.5
10	72.2	68.4
15	84.4	57.1
20	91.6	46.2
25	95.5	36.6
30	97.8	27.4
35	99.0	18.2
40	99.6	10.0

23b. Ammonia—Water[99]

p = 10 atmospheres absolute

Per cent by weight ammonia in liquid phase	Per cent by weight ammonia in vapor phase	Boiling Point °C
2	13.6	172.8
4	25.4	166.5
6	35.4	160.5
8	44.7	154.7
10	53.0	149.3
15	68.8	135.8
20	79.8	123.7
25	86.9	112.0
30	91.7	100.9
35	94.8	90.3
40	96.6	80.5
50	98.7	62.1
60	99.5	48.0
70	99.8	39.0
80	99.9	32.4

24a. Nitrogen—Oxygen[100]

p = 1 atmosphere absolute

Per cent by weight nitrogen in liquid phase	Per cent by weight nitrogen in vapor phase	Boiling Point °K
3.45	12.5	88.9
7.1	23.65	87.9
11.05	33.6	87.0
15.25	42.7	86.1
20.0	50.9	85.2
25.15	58.4	84.2
30.9	65.0	83.3
37.3	71.1	82.3
44.45	76.6	81.4
53.3	82.4	80.4
63.6	87.4	79.4
76.05	92.6	78.4
90.7	97.4	77.6

24b. Nitrogen—Oxygen[100]

p = 5 atmospheres absolute

Per cent by weight nitrogen in liquid phase	Per cent by weight nitrogen in vapor phase	Boiling Point °K
3.6	8.0	107.3
10.7	24.5	105.5
19.75	38.75	103.5
30.1	52.7	101.5
42.65	66.1	99.5
57.85	78.3	97.6
66.1	83.75	96.7
76.7	89.75	95.6
88.65	95.5	94.6

25. Acetaldehyde—Water[101]

p = 760 mm Hg

Per cent by weight acetaldehyde in liquid phase	Per cent by weight acetaldehyde in vapor phase	Boiling Point °C
0.0	0.0	100.0
1.1	44.8	93.5
2.3	70.9	82.0
9.3	88.0	63.0
21.35	95.2	43.0
37.9	97.0	33.0
51.1	—	29.0
70.9	—	25.3
78.5	—	24.6
88.0	—	23.8

26. Benzol—Toluol

p = 760 mm; ideal mixture

Per cent by weight benzol in liquid phase	Per cent by weight benzol in vapor phase	Boiling Point °C
0.0	0.0	110.6
8.65	18.5	106.1
17.5	33.8	102.2
26.6	46.5	98.6
36.1	57.5	95.2
45.8	67.3	92.1
55.9	75.7	89.4
66.5	83.0	86.8
77.3	89.5	84.4
88.3	95.0	82.3
94.0	97.5	81.2
100.0	100.0	80.2

27. *n*-Hexane—Benzol[102]

$p = 760$ mm Hg

Per cent by weight hexane in liquid phase	Per cent by weight hexane in vapor phase	Boiling Point °C
0.5	1.0	79.95
2.7	6.5	79.15
4.7	9.0	78.5
8.1	16.5	77.5
23.1	35.6	74.35
45.8	57.0	71.3
73.1	77.0	69.2
78.0	80.5	69.0
89.8	90.2	68.75
96.9	97.0	68.7
99.1	99.1	68.7

28. 2,2,4-Trimethyl pentane—*n*-Octane[103]

$p = 760$ mm Hg

Per cent by weight 2,2,4-trimethyl pentane in liquid phase	Per cent by weight 2,2,4-trimethyl pentane in vapor phase	Boiling Point °C
0.0	0.0	125.4
5.1	10.5	124.05
10.7	19.7	122.55
16.0	29.0	121.15
20.8	35.6	119.9
24.8	40.9	118.85
28.1	45.0	118.0
32.1	50.0	116.95
36.0	53.5	115.95
39.2	57.5	115.1
42.4	61.0	114.25
46.4	65.0	113.2
48.8	67.5	112.6

Per cent by weight 2,2,4-trimethyl pentane in liquid phase	Per cent by weight 2,2,4-trimethyl pentane in vapor phase	Boiling Point °C
56.0	73.5	110.7
58.8	75.5	110.0
65.8	81.0	108.1
69.8	83.5	107.1
73.9	85.4	106.0
78.8	89.0	104.75
83.5	91.5	103.5
89.2	94.5	102.0
93.3	96.9	100.95
100.0	100.0	99.2

The boiling-point curve for this mixture is a straight line.

GRAPH III

29. Carbon tetrachloride—Benzol[104]

$p = 760$ mm Hg

Per cent by weight carbon tetrachloride in liquid phase	Per cent by weight carbon tetrachloride in vapor phase	Boiling Point °C
0.0	0.0	80.0
23.9	27.0	79.3
35.2	38.7	78.8
40.8	44.5	78.6
45.3	48.4	78.5
53.0	56.0	78.2
57.4	60.3	78.0
68.7	70.5	77.6
76.2	77.7	77.4
83.7	84.4	77.1

30. Carbon tetrachloride—Toluol[104]

p = 762 mm Hg

Per cent by weight carbon tetrachloride in liquid phase	Per cent by weight carbon tetrachloride in vapor phase	Boiling Point °C
0.0	0.0	110.4
9.3	19.5	
24.5	41.8	
40.4	62.0	
55.7	75.1	
68.0	83.7	
75.0	87.8	
85.7	93.7	
96.7	98.4	75.9

31. Carbon disulfide—Carbon tetrachloride[104]

p = 760 mm Hg

Per cent by weight carbon disulfide in liquid phase	Per cent by weight carbon disulfide in vapor phase	Boiling Point °C
0.0	0.0	76.7
1.5	4.3	74.9
3.1	8.3	73.1
5.8	15.1	70.3
7.7	19.8	68.6
14.6	32.6	63.8
24.1	46.0	59.3
35.9	59.5	55.3
49.2	70.6	52.3
60.9	78.0	50.4
75.3	87.1	48.5
100.0	100.0	46.3

32. Carbon disulfide—Acetone[104]

p = 760 mm Hg; azeotropic point at 72.3 per cent by weight carbon disulfide

Per cent by weight carbon disulfide in liquid phase	Per cent by weight carbon disulfide in vapor phase	Boiling Point °C
0.0	0.0	56.2
2.5	10.8	54.0
6.2	23.0	51.4
16.9	41.6	46.6
23.1	51.1	44.0
35.1	59.5	41.4
44.6	63.95	40.3
51.7	66.1	39.8
60.3	68.9	39.3
71.2	72.0	39.1
83.2	76.0	39.3
84.3	77.5	39.6
90.7	80.7	40.5
97.7	91.1	43.5
100.0	100.0	46.3

33. Benzol—Ethylene chloride[104]

p = 760 mm Hg

Per cent by weight benzol in liquid phase	Per cent by weight benzol in vapor phase	Boiling Point °C
0.0	0.0	83.7
10.5	11.1	83.0
16.5	17.6	82.7
27.0	28.5	82.3
31.0	32.1	82.2
41.4	42.4	81.9
46.7	48.1	81.6
57.7	58.9	81.4
73.1	74.2	80.9
85.2	86.3	80.6
100.0	100.0	80.2

34. Water—Furfurol[104]

p = 760 mm Hg; azeotropic point at 64.8 per cent by weight water, 97.9°C

Per cent by weight water in liquid phase	Per cent by weight water in vapor phase	Boiling Point °C
0.0	0.0	161.7
0.4	2.0	158.8
0.8	4.2	154.8
1.2	9.5	146.0
1.6	28.5	122.5
2.0	44.1	109.5
4.4	60.0	100.6
7.4	64.0	98.7
15.7	64.8	97.9
64.8	64.8	97.9
81.7	64.8	97.9
90.2	68.1	98.07
94.9	76.1	98.56
100.0	100.0	100.0

35. Isopropanol—Water[104]

p = 760 mm Hg; azeotropic point at 88.6 per cent by weight isopropanol, 80.5°C

Per cent by weight isopropanol in liquid phase	Per cent by weight isopropanol in vapor phase	Boiling Point °C
0.0	0.0	100.0
3.3	43.8	95.0
6.3	63.1	90.0
9.3	71.6	86.7
17.5	77.4	83.5
37.1	81.0	81.5
58.8	82.2	81.0
77.1	85.1	80.7
88.6	88.6	80.5
93.1	91.8	81.0
96.8	94.2	82.3

36. Nitric acid—Water[104]

p = 760 mm Hg; azeotropic point at 68.5 per cent by weight nitric acid, 121.9°C

Per cent by weight nitric acid in liquid phase	Per cent by weight nitric acid in vapor phase	Boiling Point °C
24.2	2.2	106.5
32.9	5.9	112.0
50.0	19.9	118.5
61.0	41.1	121.6
68.5	68.5	121.9
70.2	84.1	121.0
75.3	91.75	118.0
79.9	96.7	112.0
84.9	97.7	99.0

37. Toluol—*n*-Octane[105]

p = 760 mm Hg

Per cent by weight toluol in liquid phase	Per cent by weight toluol in vapor phase	Boiling Point °C
8.0	13.7	123.15
17.0	25.4	120.85
25.7	35.8	119.0
29.8	40.6	118.2
35.6	47.2	117.2
42.5	53.3	116.0
47.3	57.2	115.3
51.8	61.3	114.6
56.3	65.0	113.95
61.4	68.6	113.35
67.7	74.0	112.7
71.6	76.7	112.3
75.8	80.0	111.95
79.0	82.6	111.65
82.8	85.8	111.35
88.6	90.5	110.95
93.1	94.3	110.7

38. Acetone—Chloroform[106]

p = 750 mm Hg; azeotropic point at 22 per cent by weight acetone, 63.93°C

Per cent by weight acetone in liquid phase	Per cent by weight acetone in vapor phase	Boiling Point °C
0	0.0	61.25
10	5.3	63.02
20	18.5	63.84
30	34.1	63.41
40	47.7	62.19
50	60.2	61.03
60	70.6	59.91
70	79.7	58.83
80	87.65	57.79
90	94.7	57.00
100	100.0	55.97

39. Carbon tetrachloride—Ethanol[106]

p = 745 mm Hg; azeotropic point at 84.2 per cent by weight carbon tetrachloride, 63.85°C

Per cent by weight carbon tetrachloride in liquid phase	Per cent by weight carbon tetrachloride in vapor phase	Boiling Point °C
0	0.0	77.91
10	40.0	74.82
20	54.6	72.44
30	64.7	70.25
40	72.0	68.35
50	76.8	66.64
60	79.4	65.32
70	81.5	64.42
80	83.2	63.88
90	87.1	64.30
100	100.0	75.92

40. Ethyl ether—Carbon tetrachloride[106]

p = 760 mm Hg

Per cent by weight ether in liquid phase	Per cent by weight ether in vapor phase	Boiling Point °C
0	0.0	77.75
10	32.2	73.70
20	52.8	69.73
30	65.4	65.54
40	74.6	61.44
50	81.5	57.16
60	87.3	53.09
70	92.0	48.98
80	95.72	44.84
90	98.28	41.35
100	100.00	34.75

41. Chloroform—Benzol[106]

p = 760 mm Hg

Per cent by weight chloroform in liquid phase	Per cent by weight chloroform in vapor phase	Boiling Point °C
0	0.0	80.65
10	13.6	79.86
20	27.2	79.03
30	40.6	78.13
40	53.0	77.15
50	65.0	75.95
60	75.0	74.60
70	83.0	72.84
80	90.0	70.48
90	96.1	67.00
100	100.0	61.45

GRAPH IV

42a. Water—Phenol[107]

p = 760 mm Hg; azeotropic point at 90.6 per cent by weight water

Per cent by weight water in liquid phase	Per cent by weight water in vapor phase
2.49	49.86
3.31	65.44
4.40	74.06
9.42	84.21
10.60	84.55
10.80	84.8
26.12	87.88
28.86	87.26
50.10	88.18
60.30	88.29
73.45	88.86
83.39	89.42
89.43	90.49
89.88	90.51
93.11	91.87
95.71	94.13
97.71	96.13

42b. Water—Phenol[107]

p = 260 mm Hg; azeotropic point at 95.6 per cent by weight water

Per cent by weight water in liquid phase	Per cent by weight water in vapor phase
1.54	41.10
4.95	79.72
6.87	82.79
7.73	84.45
19.63	89.91
28.44	91.05
39.73	91.15
82.99	91.86
89.95	92.77
93.38	94.19
95.74	95.64

42c. Water—Phenol

p = 40 mm Hg; no clearly defined azeotropic point

Per cent by weight water in liquid phase	Per cent by weight water in vapor phase
2.09	69.05
4.58	82.69
11.20	91.99
24.45	95.53
67.65	95.70
96.26	96.46
96.98	97.37
98.06	98.08

43. Methanol—Water[108]

p = 730 to 750 mm Hg

Per cent by weight methanol in liquid phase	Per cent by weight methanol in vapor phase
0	0.0
1	7.3
4	23.5
6	31.5
10	43.4
20	61.0
30	70.5
40	76.7
50	81.2
60	84.8
70	88.3
80	92.1
90	96.0
100	100.0

44. Water—Acetic acid[108]

p = 730 to 750 mm Hg

Per cent by weight water in liquid phase	Per cent by weight water in vapor phase
0	0.0
4	6.8
10	16.1
20	29.6
30	41.8
40	52.8
50	62.0
60	70.0
70	77.6
80	85.0
90	92.5
100	100.0

For data on equilibrium curves of water-acetic acid mixtures at low pressures, see also Keyes, *Ind. Eng. Chem.* **25,** 569 (1933).

45. Acetic acid—Acetic anhydride[109]

p = 750 mm Hg

Per cent by weight acetic acid in liquid phase	Per cent by weight acetic acid in vapor phase
10	23.2
20	42.3
30	56.5
40	65.2
50	72.4
60	78.6
70	84.1
80	89.7
90	95.0

46. Water—Formic acid[109]

p = 750 mm Hg; azeotropic point at 23.5 per cent by weight water

Per cent by weight water in liquid phase	Per cent by weight water in vapor phase
10	6.0
20	18.2
23.5	23.5
30	34.8
40	50.2
50	64.4
60	75.8
70	83.6
80	90.2
90	95.8

47. Benzol—Acetic acid[109]

p = 758 mm Hg; azeotropic point at 98.5 per cent by weight benzol

Per cent by weight benzol in liquid phase	Per cent by weight benzol in vapor phase	Boiling Point °C
0.0	0.0	118.7
1.2	7.7	115.1
3.6	29.2	109.0
9.4	45.9	100.0
21.4	65.4	91.5
40.2	75.5	85.8
59.7	83.3	82.8
90.7	94.6	80.0
96.1	97.0	79.8

48. Water—Sulfuric acid[111]

p = 760 mm Hg; azeotropic point at 1.5 per cent by weight water, 335°C

Per cent by weight water in liquid phase	Per cent by weight water in vapor phase	Boiling Point °C
1.5	1.5	335
2.0	8.6	331
4.0	34.8	311
6.0	54.0	293
8.0	70.0	278
10.0	82.3	262
12.0	91.1	249
14.0	94.3	236
16.0	95.9	224
18.0	97.0	213
20.0	97.8	203
25.0	99.0	182
30.0	99.7	165
35.0	100.0	151

49. *n*-Heptane—Toluol[112]

p = 760 mm Hg

Per cent by weight heptane in liquid phase	Per cent by weight heptane in vapor phase	Boiling Point °C
0.0	0.0	110.4
8.6	14.3	108.75
10.1	16.3	108.45
21.2	30.5	106.35
30.7	40.4	104.75
37.1	46.0	103.8
43.0	51.4	102.95
52.2	59.9	101.8
59.3	65.8	101.0
69.7	74.2	100.15
86.8	88.2	98.95
93.5	94.2	98.6
100.0	100.0	98.4

50. *n*-Heptane—Methyl cyclohexane[112]

p = 760 mm Hg

Per cent by weight heptane in liquid phase	Per cent by weight heptane in vapor phase	Boiling Point °C
0.0	0.0	100.8
3.15	3.6	100.7
5.9	6.3	100.62
9.7	10.5	100.52
13.6	14.6	100.40
18.3	19.5	100.30
22.0	23.4	100.20
27.6	29.4	100.04
32.2	33.8	99.93
36.8	38.7	99.80
40.7	42.5	99.70
46.1	48.0	99.50
50.7	52.7	99.30
56.5	58.3	99.10
60.4	62.3	99.00
65.2	67.1	98.90
71.4	73.3	98.80
76.1	77.6	98.72
80.0	81.4	98.70
84.6	85.7	98.60
88.2	89.3	98.57
90.8	91.5	98.52
93.2	94.1	98.50
95.5	96.35	98.46
98.0	98.6	98.42
100.0	100.0	98.40

51. Acetone—Water[113]

p = 760 mm Hg

Per cent by weight acetone in liquid phase	Per cent by weight acetone in vapor phase	Boiling Point °C
0	0.0	100.0
5	62.6	88.2
10	78.4	80.0
15	84.2	75.2
20	87.6	71.9
25	89.8	69.4
30	91.0	67.6
35	91.8	66.2
40	92.5	65.2
45	93.0	64.3
50	93.3	63.4
55	93.6	62.7
60	94.0	62.0
65	94.3	61.4
70	94.6	60.8
75	94.9	60.3
80	95.3	59.7
85	95.7	59.0
90	96.3	58.2
95	97.5	57.3
100	100.0	56.4

52. Methanol—Ethanol[113]

p = 760 mm Hg

Per cent by weight methanol in liquid phase	Per cent by weight methanol in vapor phase
0	0.0
5	7.5
10	14.2
15	20.9
20	27.4
25	34.0
30	40.4
35	46.7
40	52.8
45	58.6
50	63.6
55	67.9
60	71.9
65	75.6
70	79.3
75	82.8
80	86.3
85	89.9
90	93.4
95	96.8
100	100.0

II. Heat of vaporization ρ of the substances listed in the tables of equilibrium conditions for binary mixtures, at a pressure p of 760 mm Hg, as far as known

Substance	Heat of Vaporization kcal/kg	Substance	Heat of Vaporization kcal/kg
Acetaldehyde	136.2	Furfurol	107.5
Acetic acid	96.75	*n*-Heptane	76.3
Acetic anhydride	66.2	*n*-Hexane	79.3
Acetone	125.0	Hydrochloric acid	106.0
Ammonia	327.0	Isobutanol	138.1
Benzol	94.5	Isopropanol	159.35
Bromine	48.0	Methanol	262.8
n-Butanol	141.3	Methylcyclohexane	76.9
Butyl acetate	73.8	Nitric acid	115.0
Carbon disulfide	84.1	Nitrogen	47.6
Carbon tetrachloride	46.35	*n*-Octane	70.95
Chloroform	59.0	Oxygen	51.0
m-Cresol	100.6	*n*-Pentane	85.0
Ethanol	206.4	Phenol	122.0
Ethyl acetate	88.0	Propane	107.0
Ethyl bromide	59.9	Sulfuric acid	122.1
Ethyl ether	86.0	Toluol	86.8
Ethylene chloride	77.3	Trichlorethylene	57.24
Formic acid	118.0	Water	540.0

III. Specific heat and density of substances listed in the tables of equilibrium conditions for binary mixtures in liquid state

Substance	Temperature °C	Specific heat kcal/kg×°C	Temperature °C	Density kg/l
Acetaldehyde	—	—	20	0.783
Acetic acid	0	0.467	15	1.0545
	20	0.488	—	—
	40	0.505	—	—
	80	0.545	—	—
Acetic anhydride	—	—	15	1.0797
Acetone	0	0.506	20	0.791
	3–22.6	0.514	—	—
	24.2–49.4	0.538	—	—
Ammonia	−60	1.05	15	0.618
	0	1.10	—	—
	40	1.16	—	—
	80	1.29	—	—
Benzol	20	0.415	20	0.8791
	50	0.430	—	—
Bromine	13–45	0.107	25	3.102
n-Butanol	21–115	0.687	20	0.810
Butyl acetate	19.4	0.4985	—	—
Carbon disulfide	0	0.235	20	1.2634
	20	0.238	—	—
	30	0.242	—	—
Carbon tetrachloride	0	0.198	20	1.595
	20	0.201	—	—
	30	0.203	—	—
Chloroform	0	0.232	20	1.489
	30	0.234	—	—
m-Cresol	0–20	0.479	15.5	1.0387
o-Cresol	0–20	0.499	15.5	1.0516
p-Cresol	20–100	0.487	15.5	1.0388
Ethanol	0	0.545	15	0.7936
	25	0.595	—	—
	50	0.665	—	—
	100	0.82	—	—
Ethyl acetate	0	0.46	20	0.900
	20	0.475	—	—

Substance	Temperature °C	Specific heat kcal/kg×°C	Temperature °C	Density kg/l
Ethyl bromide	−20	0.206	20	1.4555
	5–10	0.216	—	—
	10–15	0.213	—	—
	15–20	0.214	—	—
Ethyl ether	0	0.521	15	0.72
	30	0.545	—	—
	80	0.687	—	—
	120	0.800	—	—
Ethylene chloride	20	0.299	0	1.28
	30	0.304	—	—
	50	0.313	—	—
	60	0.318	—	—
Formic acid	0	0.436	15	1.2273
	15.5	0.509	—	—
	20–100	0.524	—	—
Furfurol	0	0.367	—	1.159
	20–100	0.416	—	—
n-Heptane	0–50	0.507	20	0.684
	20	0.490	—	—
	30	0.518	—	—
n-Hexane	0–50	0.527	20	0.66
	20–100	0.600	—	—
	20	0.45	—	—
Hydrochloric acid (17%)	20	0.74	0	0.908
Isobutanol	0	0.505	20	0.802
	20	0.55	—	—
Isopropanol	0	0.555	15	0.7896
	20	0.630	—	—
Methanol	5	0.576	15.5	0.796
	20	0.60	—	—
	40	0.617	—	—
Methylcyclohexane	—	—	20	0.7693
Nitric acid (100%)	20	0.41	15	1.5217
Nitrogen	64–76° Kelvin	0.475	77° Kelvin	0.811
n-Octane	0–50	0.505	20	0.7026
	20–123	0.578	—	—

Substance	Temperature °C	Specific heat kcal/kg×°C	Temperature °C	Density kg/l
Oxygen	57–73° Kelvin	0.398	90.2° Kelvin	1.142
n-Pentane	0	0.51	—	0.626
	20	0.52	—	—
Phenol	14–26	0.561	25	1.071
Propane	0	0.576	15	0.541
Sulfuric acid	0	0.335	15	1.8357
	20	0.345	—	—
	40	0.355	—	—
Toluol	0	0.39	20	0.866
	20	0.40	—	—
	50	0.43	—	—
	100	0.47	—	—
Trichlorethylene	20	0.223	20	1.464
Water	15	1.00	15	0.999

IV. Molecular weights of the substances listed in the tables of equilibrium conditions for binary mixtures

Substance	Molecular Weight	Substance	Molecular Weight
Acetaldehyde	44.05	Furfurol	96.03
Acetic acid	60.03	Heptane	100.19
Acetic anhydride	102.05	Hexane	86.17
Acetone	58.08	Hydrochloric acid	36.46
Ammonia	17.03	Methanol	32.04
Benzol	78.11	Methylcyclohexane	98.09
Bromine	159.83	Nitric acid	63.01
Butanol	74.12	Nitrogen	28.02
Butyl acetate	116.09	Octane	114.14
Butylene	56.10	Oxygen	32.00
Carbon disulfide	76.13	Pentane	72.14
Carbon tetrachloride	153.84	Phenol	94.05
Chloroform	119.39	Propane	44.06
Cresol	108.06	Propanol	60.09
Ethanol	46.05	Sulfuric acid	98.08
Ethyl acetate	88.10	Toluol	92.13
Ethyl bromide	108.98	Trichlorethylene	131.40
Ethyl ether	74.12	Trimethylpentane	114.14
Ethylene chloride	98.95	Water	18.02
Formic acid	46.03		

V. Conversion Tables

In this translation of Kirschbaum's work, the metric system has been retained. Thus, the original simplicity and lucidity of the numerical relations were preserved, as well as economy and logic of figures, and no introduction of confusing conversion factors became necessary.

However, these tables are presented for the convenience of readers who wish to convert figures into the British system. The derived unit names are self-explanatory:

T	Tera	$= 10^{12}$	h	Hecto	$= 10^{2}$	m	Milli	$= 10^{-3}$
G	Giga	$= 10^{9}$	D	Deca	$= 10^{1}$	μ	Micro	$= 10^{-6}$
M	Mega	$= 10^{6}$	d	Deci	$= 10^{-1}$	n	Nano	$= 10^{-9}$
k	Kilo	$= 10^{3}$	c	Centi	$= 10^{-2}$	p	Pico	$= 10^{-12}$

Thus,
1 megawatt $= 10^6$ watts $= 1000$ kilowatts
1 kilopound $= 10^3$ pounds $= 1000$ pounds
1 hectoliter $= 10^2$ liters $= 100$ liters
1 decigram $= 10^{-1}$ gram $= 0.1$ gram

Temperatures:

$$°C/10 = (°K - 273)/10 = (°F - 32)/18 = (°R - 460)/18$$

$0°C = 32°F = 273°K = 492°R$
$100°C = 212°F = 373°K = 672°R$
$0°F = -17.8°C$

Weights:

1 kg = 2.2 lb = 0.001 metric ton
0.373 kg = 1 lb = 0.00373 metric ton
1000 kg = 2205 lb = 1 metric ton

For water at 4°C (greatest density):
1 liter = 1000 cc = 1.000 kg = 2.205 lb
1 cu m = 1 metric ton

Densities:

$1\ g/ml = 1\ g/cm^3 = 1\ kg/l = 1\ ton/m^3 = 62.4\ lb/cu\ ft = 8.35\ lb/gal$

Miscellaneous hydrometer scales:
If N = hydrometer reading, actual densities are, for

Balling, heavier than water: $200/(200 - N)$
lighter than water: $200/(200 + N)$

Bates, 60°F: $(1000 + 2.78\ N)/1000$

Baumé, 12.5°C, heavier than water: $145.88/(145.88 - N)$
lighter than water: $145.88/(135.88 + N)$

Baumé, 15°C, heavier than water: $146.3/(146.3 - N)$
lighter than water: $146.3/(136.3 + N)$

Baumé Rational scale, 15°C, heavier than water: 144.3/(144.3 − N)

Baumé, 17.5°C, heavier than water: 146.78/(146.78 − N)
lighter than water: 146.78/(136.78 + N)

Beck, heavier than water: 170/(170 − N)
lighter than water: 170/(170 + N)

Brix, heavier than water: 400/(400 − N)
lighter than water: 400/(400 + N)

Cartier, heavier than water: 136.8/(126.1 − N)
lighter than water: 136.8/(126.1 + N)

Gay-Lussac, heavier than water: 100/(100 − N)
lighter than water: 100/(100 + N)

Twaddell: (0.5 N + 100)/100

Volumes:

l		hl		cm³		gal		cu in.		cu ft
1 l	=	0.01 hl	=	1000 cm³	=	0.264 gal	=	61 cu in.	=	0.035 cu ft
3.785 l	=	0.03785 hl	=	3785 cm³	=	1 gal	=	231 cu in	=	0.1337 cu ft
0.016 l	=	0.00016 hl	=	16.39 cm³	=	0.0043 gal	=	1 cu in.	=	0.00058 cu ft
28.32 l	=	0.283 hl	=	28320 cm³	=	7.48 gal	=	1728 cu in.	=	1 cu ft

Lengths:

m		mm		in.		ft		yd
1 m	=	1000 mm	=	39.37 in.	=	3.28 ft	=	1.093 yd
0.0254 m	=	25.4 mm	=	1 in.	=	0.0835 ft	=	0.028 yd
0.305 m	=	305 mm	=	12 in.	=	1 ft	=	0.333 yd
0.9144 m	=	914.4 mm	=	36 in.	=	3 ft	=	1 yd

Areas:

sq m		sq cm		sq in.		sq ft		sq yd
1 sq m	=	1000 sq cm	=	1550 sq in.	=	10.76 sq ft	=	1.196 sq yd
0.00065 sq m	=	6.45 sq cm	=	1 sq in.	=	0.007 sq ft	=	0.00077 sq yd
0.093 sq m	=	929 sq cm	=	144 sq in.	=	1 sq ft	=	0.111 sq yd
0.836 sq m	=	8361 sq cm	=	1296 sq in.	=	9 sq ft	=	1 sq yd

Velocities:

m/sec		m/min		ft/sec		ft/min
1 m/sec	=	60 m/min	=	3.28 ft/sec	=	196.85 ft/min
0.0167 m/sec	=	1 m/min	=	0.055 ft/sec	=	3.281 ft/min
0.305 m/sec	=	18.3 m/min	=	1 ft/sec	=	60 ft/min
0.005 m/sec	=	0.305 m/min	=	0.0167 ft/sec	=	1 ft/min

Pressures:

1 kg/sq cm = 14.22 lb/sq in. = 735.6 mm Hg = 28.96 in. Hg = 10 m H_2O = 32.84 ft H_2O

1.033 kg/sq cm = 14.7 lb/sq in. = 760 mm Hg = 29.92 in. Hg = 10.33 m H_2O = 33.92 ft H_2O

0.07 kg/sq cm = 1 lb/sq in. = 51.71 mm Hg = 2.036 in. Hg = 0.7 m H_2O = 2.31 ft H_2O

0.0345 kg/sq cm = 0.491 lb/sq in. = 25.4 mm Hg = 1 in. Hg = 0.345 m H_2O = 1.13 ft H_2O

0.0305 kg/sq cm = 0.4335 lb/sq in. = 22.4 mm Hg = 0.88 in. Hg = 0.022 m H_2O = 1 ft H_2O

1.3596 kg/sq cm = 19.34 lb/sq in. = 1000 mm Hg = 39.37 in. Hg = 13.6 m H_2O = 44.64 ft H_2O

In engineering usage, 1 atmosphere (at) = 1 kg/sq cm. The use of the term *atmosphere* with reference to the standard pressure of 760 mm Hg is no longer customary in European engineering literature.

1 oz/sq in. = 44 mm water
1 lb/sq ft = 4.88 kg/sq m
1 torr = 1 mm Hg at 0°C
Atmospheres "gage" = Absolute atmospheres (ata) + 1; (atü in German literature = at Ueberdruck)
Low absolute pressures may be measured in microns Hg; 1 micron (μ) = 0.001 mm
Pressure gradients: 1 mm Hg/m = 0.012 in. Hg/ft
83.3 mm Hg/m = 1 in. Hg/ft

Heat and energy:

1 kcal/kg × °C = 1 pcu/lb × °C = 1 Btu/lb × °F. Specific heats are numerically equal in the British and the Standard system.
1 pcu/lb = 1 kcal/kg = 1 cal/g = 1.8 Btu/lb

1 pcu/hr × sq ft × °C = 1 Btu/hr × sq ft × °F = 4.88 kcal/hr × sq m × °C
1 kcal/sec × sq cm = 3600 kcal/hr × sq cm = 13,272,000 Btu/hr × sq ft
1 kcal/sec × sq cm × cm × °C = 3600 kcal/hr × sq cm × cm × °C = 2,903,000 Btu/hr × sq ft × in. × F°
1 Btu/hr × sq ft × ft × °F = 1 pcu/hr × sq ft × ft × °C = 12 Btu/hr × sq ft × in. × °F = 0.00413 cal/sec × sq cm × cm × °C
1 kcal/m × sec × °C = 2,420 Btu/ft × hr × °F
1 kcal/cu m = 0.112 Btu/cu ft
1 Btu/lb = 0.555 kcal/kg
1 horsepower-hour = 2546 Btu = 0.746 kw-hr
1 kw-hr = 3415 Btu
1 kcal = 427 mkg

1 kcal/cu m × °C = 0.0624 Btu/cu ft × °F
16 kcal/cu m × °C = 1 Btu/cu ft × °F

1 kcal/sq m × hr × (°C)^4 = 0.0351 Btu/sq ft × hr × (°F)^4
28.5 kcal/sq m × hr × (°C)^4 = 1 Btu/sq ft × hr × (°F)^4

1 kcal/sq m × hr × °C = 0.205 Btu/sq ft × hr × °F
4.88 kcal/sq m × hr × °C = 1 Btu/sq ft × hr × °F

1 kcal/sq m × hr × cm = 0.936 Btu/sq ft × hr × in.
1.07 kcal/sq m × hr × cm = 1 Btu/sq ft × hr × in.

1 kcal/m × hr × °C = 0.672 Btu/ft × hr × °F.
2.71 kcal/m × hr × °C = 1 Btu/ft × hr × °F

1 °C/m = 0.549°F/ft
1.82 °C/m = 1°F/ft

Viscosities:

1 poise = 100 centipoises = 1 g/cm × sec = 0.0000672 lb/ft × sec
1 centipoise = 2.42 lb/ft × hr

BIBLIOGRAPHY

[1] Planck, *Vorlesungen ueber Thermodynamik. 8th ed.*, pp. 113, 176, Berlin & Leipzig, De Gruyter 1927.

[2] Lewis and Randall, *Thermodynamics.*

[3] Duhem, *Trav. et Mêm. des Facultés de Lille*, **13**, 75 (1894)

[4] Margules, *Wien. Ber.*, **104**, 1243 (1895)

[5] Hausen, *Rektifikation von Dreistoffgemischen, insbesondere von Sauerstoff-Stickstoff-Argongemischen.* Forschungsheft 1935, Ed. A, no. 1, p. 10. (Equations (7) are employed to determine the equilibrium conditions of the ternary mixture oxygen-nitrogen-argon. In these equations, let $C = 1$, and multiply the right-hand side of an *e*-function as the presence of argon interferes with the ideal behavior of the mixture.)

[6] Hausbrand, "Destillation mit eingeblasenem Wasserdampf," *Apparatebau* (1921)

[7] Stockhardt and Hull, "Vapor-liquid equilibria and boiling point composition relations for the systems *n*-butanol-water and isobutanol-water," *Ind. Eng. Chem.*, **23**, 1438 (1931)

[8] C. R. Berthelot, *Acad. Sci. Paris*, **126**, 1703 (1898)

[9] Galitzine, "Ueber Dalton's Gesetz," *Wied. Ann.*, **41**, 770 (1890)

[10] Perry, *Chem. Eng. Handbook*, and *International Critical Tables.*

[11] Von Rechenberg, *Einfache und fraktionierte Destillation in Theorie und Praxis*, p. 558, Miltitz, Schimmel & Co. 1923.

[12] Young-Prahl, *Theorie und Praxis der Destillation*, Berlin, Springer, 1932.

[13] Kirschbaum & Gerstner, "Gleichgewichtskurven, Siede-und Taulinien von Aethylalkohol-Wassergemischen bei Unterdruecken," *Verfahrenstechnik* no. 1, p. 10 (1939)

[14] Schreinemakers, "Tensions de vapeur de mélanges ternaires eau-acétone-phénol," *Arch. Neerlandaises* II, **8** (1903).

[15] Barbaudy, *Mémorial des sciences physiques*, 1928.

[16] Young and Fortrey, "Propriétés des mélanges d'alcools gras inférieurs et de benzène ainsi que des mélanges d'alcool, de benzène et d'eau," *J. Chem. Soc.*, **81**, 739 (1902).

[17] Merriman, *J. Chem. Soc.*, p. 628, 1913.

[18] Grumbt, "Siede- und Tauisobaren von Aethylalkohol-Wassergemischen fuer Drucke bis 15 at.," *Tech. Mech. Thermodynamik*, 1, nos. 9 and 10 (1930)

[19] Dodge and Dunbar, *J. Am. Chem. Soc.*, **44**, 608 (1927)

[20] Wucherer, "Messung von Druck, Temperatur und Zusammensetzung der fluessigen und dampffoermigen Phase von Ammoniak-Wassergemischen im Saettigungszustand," *Z. ges. Kaelteind.*, **39**, 97, 136 (1932)

[21] Cummings, "High-pressure rectification. Vapor-liquid equilibrium relations at high pressures," *Ind. Eng. Chem.*, **23**, 900 (1931)

[22] Cummings, Stones, Volante, "High-pressure rectification. *n*-pentane-*n*-heptane system," *Ind. Eng. Chem.*, **25**, 728 (1933)

[23] Von Keussler, *Die technische Erzeugung von absolutem Alkohol durch Druckdestillation des Gemisches Alkohol-Wasser-Benzol zur Herstellung eines billigen Kraftstoffes*, Thesis, Darmstadt, 1926.

[24] Kirschbaum, *Die Verstaerkung durch teilweise Kondensation binaerer Gemische und ihre Beruecksichtigung bei der Berechnung von Rektifizierapparaten*, Dechema-Moographie, Berlin, Verlag Chemie, 1930.

[25] Kirschbaum, "Stoff-und Waermeaustausch bei Zweistoffgemischen," *Ann. Physik*, **20** (1934)

[26] Fischer, "Die teilweise Kondensation von Dampfgemischen in der Destillationstechnik," *Arch. Waermewirt.*, **8**, 217 (1933).

[27] Merkel, "Die Rektifikation," *Arch. Waermewirt.*, 1, 13 (1929)

[28] Tyrer, *J. Chem. Soc.*, p. 1633, Sept. 1911; p. 81, Jan. 1912; p. 1104, June 1912.

[29] Eucken, *Lehrbuch der chemischen Physik*, p. 218, Leipzig, Akad. Verlagsges. 1930.

[30] Wilson & Bahlke, "The physical properties of the paraffin hydrocarbons," *Ind. Eng. Chem.*, **16**, 115 (1924)

[31] Thormann, *Destillieren und Rektifizieren*, p. 17, Leipzig, Spamer. 1928.

[32] Landolt-Boernstein, *Phys.-chem. Tabellen I*, p. 449.

[33] *International Critical Tables*, Vol. 3, pp. 115–129.

Perry, *Chem. Eng. Handbook*, p. 381 and ff., New York, McGraw-Hill, 1934.

[34] McCabe and Thiele, "Graphical design of fractionating columns," *Ind. Eng. Chem.*, **17**, 605 (1925)

[35] Murphree, "Graphical rectifying column calculations," *Ind. Eng. Chem.*, **17**, 960 (1925)

[36] Peters, "The efficiency and capacity of fractionating columns," *Ind. Eng. Chem.*, **14**, 476 (1922)

[37] Clark, "The plate efficiency of a continuous alcohol still," *Ind. Eng. Chem.*, **14**, 480 (1922)

[38] Kirschbaum, "Wirkung, Verkrustung und Druckverlust von Rektifizierboeden," *Verfahrenstechnik*, **5**, 139 (1937)

[39] Kirschbaum, "Berechnung der Bodenzahl von Rektifiziersaeulen," *Z. Ver. deut. Ing.*, **80**, 633 (1936)

[40] Linde, "Ueber Modellversuche an Rektifizierboeden," *Verfahrenstechnik*, 2 (1939)

[41] Kirschbaum, "Der Rektifiziervorgang auf Verstaerkungsboeden," *Z. Ver. deut. Ing.*, 401, **77** (1933)

[42] Kirschbaum, "Ueber den Wirkungsgrad von Rektifizierboeden," *Chem. Fabrik*, **6**, 431 (1933)

[43] Bosnjakovic and Grumbt, "Waermeinhalt fluessiger Aethylalkohol-Wassergemische," *Forsch. Gebiete Ingenieurw.*, **2** (Dec. 1931)

[44] Weir and Eaton, "Heat content of petroleum oil fractions at elevated temperatures," *Ind. Eng. Chem.*, **24**, no. 2 (1932)

[45] Watson and Nelson, "Improved methods for approximating critical and thermal properties of petroleum fractions," *Ind. Eng. Chem.*, **25**, no. 8 (1932)

[46] *Z. ges. Kaelteind.* **39**, no. 12 (1932)

Landolt-Boernstein, *Phys.-chem. Tabellen, I,* pp. 449, 457; *II*, pp. 1266, 1480.

[47] Schimunek, Schwab, and Tapp, "Der Einfluss der Zulauftemperatur bei der Verarbeitung von niedrigprozentigen Maischen auf Sprit," *Chem. Fabrik,* **10**, nos. 27–28 (1937)

[48] Othmer, "Vapor re-use process," *Ind. Eng. Chem.*, **23**, no. 12 (1936).

[49] Kirschbaum, Die Waermeaustauschapparate. Goeschen vol. 1067. Berlin & Leipzig, De Gruyter, 1933.

[50] Ger. Patent 656,828.

[51] Eucken & Jakob, Der Chemie-Ingenieur, Vol. I, part 3, section by Hausen. Leipzig, Akad. Verlagsgesellschaft, 1933.

[52] Bosnjakovic, "Technische Thermodynamik, Part 2," Dresden, Steinkopff 1937.

[52a] Boe, "Verlustlose Rektifikation von fluessiger Luft," *Z. Ver. deut. Ing.* **78**, 578 (1934)

[53] German Patent 542,400.

[54] Klar, "Fabrikation von absolutem Alkohol zwecks Verwendung als Zusatzmittel zu Motor-Treibstoffen," Halle/S., Knapp. 1937. This book contains considerable data on additional literature on the subject.

[55] Fritzweiler, "Verfahren zur Herstellung von wasserfreiem Spiritus," *Z. Ver. deut. Ing.*, **82** (1938)

[56] Fenske, *Ind. Eng. Chem.* **24**, 1932.

[57] Lewis and Smoley, *Proc. Am. Pet. Inst.* (Tenth annual meeting) **9**, 70 (1930)

[58] Lewis and Matheson, "Design of rectifying columns for natural and refinery gasoline," *Ind. Eng. Chem.*, **24**, 494 (1932)

[59] Cope and Lewis, "Graphical method of computation for rectifying complex hydrocarbon mixtures," *Ind. Eng. Chem.*, **24**, 498 (1932)

[60] Kirschbaum, "Belastungsbereich von Austauschboeden," *Verfahrenstechnik*, no. 5, p. 131, 1938.

[61] Kirschbaum, "Wirkung von Rektifizierboeden bei Unterdruecken und Berechnung des Saeulenquerschnittes von Rektifizierapparaten," *Verfahrenstechnik*, no. 4, 1940.

[62] Carey, Griswold, Lewis, and McAdams, *Trans. Am. Inst. Chem. Eng.*, **30** (1933–4).

[63] Siegel, *Chem. Met. Eng.*, p. 493, Sept. 1937.

[64] Souders, Mott and Brown, "Design of fractionating columns, entrainment and capacity," Refiner Natural Gasoline Mfr. **13**, 32 (1934)

[65] Landolt-Boernstein, *Tabellen*, Vol. 1, p. 209.

[66] Souders and Brown, "Design of fractionating columns," *Ind. Eng. Chem.*, **26**, 98 (1934).

[67] Holbrook and Baker, "Entrainment in a bubble-cap distillation column," *Ind. Eng. Chem.*, **26**, 1063 (1934)

[68] Ashraf, Cubbage, and Huntington, "Entrainment in oil absorbers," *Ind. Eng. Chem.*, **26**, 1068 (1934)
[69] Strang, *Trans. Inst. Chem. Eng.*, **12**, 169 (1934)
[70] Sherwood and Jenny, "Entrainment in plate columns," *Ind. Eng. Chem.*, **27**, 265 (1935)
[71] Claude, Peavy and Baker, "Efficiency and capacitv of a bubble-plate fractionating column," *Ind. Eng. Chem.*, **29**, 1056 (1937)
[72] Kirschbaum, "Untersuchungen ueber die Wirkung von Austauschboeden," *Verfahrenstechnik*, no. 3, 1936.
[73] Volland, "Untersuchungen ueber den Wirkungsgrad von Rektifizierboeden," Thesis, Karlsruhe 1934.
[74] Kirschbaum, "Wirkung von Rektifizierboeden und zweckmaessige Fluessigkeitsfuehrung," *Forsch. Gebiete Ingenieurw.*, p. 245, Sept.-Oct., 1934.
[75] Hausen, "Wirkungsgrad von Rektifikationsboeden und theoretische Bodenzahl." *Forsch. Gebiete Ingenieurw.*, **7**, 177 (1936)
[76] Kirschbaum & Quade, "Ein neues, genaues und einfaches Verfahren zur Bestimmung der Trennwirkung von Austauschboeden," *Forsch. Gebiete Ingenieurw.*, **8**, 63 (1937)
[77] Huette, 26th ed., vol. I, p. 185.
[78] Chillas & Weir, "Design of fractionating columns," *Ind. Eng. Chem.* **22**, no. 3 (1930)
[79] Kirschbaum, "Die Theorie der Rektifikation in Fuellkoerpersaeulen," *Chem. Fabrik* **4**, 38, 51, 63 (1931)
[80] Kirschbaum, "Neue Fragen und Erkenntnisse aus der Rektifikationstechnik," *Z. Ver. deut. Ing.*, **75**, no. 39 (1931)
[81] Weimann, "Untersuchungen ueber die Rektifikation in Fuellkoerpersaeulen," Thesis, Karlsruhe 1932.
[82] Baker, Chilton, and Vernon, "The course of liquor flow in packed towers." *Trans. Inst. Chem. Eng.*, 1935.
[83] German Patent 592,906.
[84] Mach, "*Druckverluste und Belastungsgrenzen von Fuellkoerpersaeulen*," *Z. Ver. deut. Ing.*, Forschungsheft 375 (Nov.-Dec. 1935)
[85] German Patent 620,997.
[86] Fawcett, "Molekular-Destillation," *Kolloid-Z.* **86**, no. 1 (1939)
[87] Langmuir, *Phys. Rev.*, **8**, 149 (1916)
[88] Tyrer, "Latent heats of vaporisation of mixed liquids," *J. Chem. Soc.*, **101**, 81 and 1104 (1912)
[89] Fritzweiler and Dietrich, "Der Azeotropismus und seine Anwendung fuer ein neues Verfahren zur Entwaesserung des Aethylalkohols," Beih. Z. Ver. dtsch. Chem., Angew. Chem. u. Chem. Fabrik 1933, No. 4. Berlin W. 35, Verlag Chemie.
[90] Cummings, Stones and Volante, "High pressure rectification, II, *n*-pentane-*n*-heptane-system," *Ind. Eng. Chem.*, **25**, 728 (1933)
[91] Rhodes, Wells and Murray, "Vapor-composition relationships in the systems phenol-water and phenol-cresol," *Ind. Eng. Chem.*, **17**, 1199 (1925)
[92] Scheeline and Gilliland, "Vapor-liquid equilibrium in the system propane-isobutylene," *Ind. Eng. Chem.* **31**, 1050 (1939)
[93] Kirschbaum and Gerstner, "Gleichgewichtskurven, Siede-und Taulinien von Aethylalkohol-Wasser-Gemischen bei Unterdrucken," *Verfahrenstechnik*, **1**, 10 (1939)
[94] Stockhardt and Hull, "Vapor-liquid equilibria and boiling-point composi-

tion relations for systems *n*-butanol-water and isobutanol-water," *Ind. Eng. Chem.* **23**, 1438 (1931)

[95] Rhodes and Boscom, "Vapor-composition relationships in the system bromine-water," *Ind. Eng. Chem.* **19**, 480 (1927)

[96] Othmer, "Composition of vapors from boiling binary solutions," *Ind. Eng. Chem.* **20**, 743 (1928)

[97] Furnas and Leighton, "Vapor-liquid equilibrium data," *Ind. Eng. Chem.*, **29**, 709 (1937)

[98] Brunjes and Furnas, "Vapor-liquid equilibrium data for commercially important systems of organic solvents. Binary systems *n*-butanol-*n*-butyl acetate and *n*-butanol-acetone," *Ind. Eng. Chem.*, **27**, 396 (1935)

[99] Wucherer, "Messung von Druck, Temperatur und Zusammensetzung der fluessigen und dampffoermigen Phase von Ammoniak-Wassergemischen im Saettigungszustand," *Z. ges. Kaelteind.*, **39**, 97, 136 (1932)

[100] Dodge and Dunbar, *J. Am. Chem. Soc.* **44**, 608 (1927)

[101] Perry, *Chem. Eng. Handbook*, p. 1127, New York and London, McGraw-Hill Book Company, Inc., 1934

[102] Tongberg and Johnston, "Vapor-liquid equilibria for *n*-hexane-benzene mixtures," *Ind. Eng. Chem.* **25**, 733 (1933)

[103] Bromiley and Quiggle, "Vapor-liquid equilibria of hydrocarbon mixtures," *Ind. Eng. Chem.*, **25**, 1136 (1933)

[104] Perry, *loc. cit.*, pp. 1127, 1128, 1130, 1131.

[105] Bromiley and Quiggle, *loc. cit.*, p. 1136.

[106] Tyrer, *loc. cit.*, pp. 81, 1104.

[107] Rhodes, Wells and Murray, *loc. cit.*, p. 1199.

[108] Cornell and Montonna, "Liquid-vapor equilibria in the systems ethanol-water, methanol-water and acetic acid-water," *Ind. Eng. Chem.* **25**, 1331 (1933)

[109] Othmer, *Ind. Eng. Chem., Anal. Ed.*, **4**, 232 (1932)

[110] *ibid.*, **20**, 743 (1928)

[111] Zeisberg, "Thermal considerations in sulphuric acid concentration," *Chem. Met. Eng.* **27**, 22 (1922)

[112] Bromiley and Quiggle, *loc. cit.*, p. 1136.

[113] Hausbrand, *Die Wirkungsweise der Rektifizier-und Destillierapparate*, pp. 209 and 226, Berlin, Julius Springer, 1921.

N. REVIEW BY THE TRANSLATOR OF PROGRESS MADE SINCE THE ORIGINAL PUBLICATION

As in the case of other industrial equipment, the commercial success of distilling apparatus is largely governed by the ability of the designer to predict its performance under conditions anticipated in the project stage.

Since Dr. Kirschbaum's original publication appeared in print, pressure and temperature ranges used in distillation have expanded, as has the physical size of the units. Also, new techniques have been developed. Theoretical methods available for dealing with the separation of mixtures of a limited number of pure components have been improved. New correlations have been proposed, and efforts are being made to achieve thermodynamic consistency between various properties and a satisfactory agreement between generalized correlations on the one hand, and data derived from spectroscopic measurements and calorimetric data on the other. Likewise, the distillation of multicomponent mixtures has been further investigated.

A basic review of the subject of distillation is presented by Othmer in *Chem. & Met. Eng.* of May 1942, pages 5 to 85. The same author and his associates (*Ind. Eng. Chem.*, **32**, 841, 1940, and **36**, 858, 1944) have developed a method of correlating vapor pressures, vapor compositions, relative volatilities, activities, and other functions involved with mixtures of volatile liquids and their separation by distillation. Although this may be developed algebraically from the Clausius-Clapeyron equation, it is most commonly used as a simple geometric chart which has come to be known as an "Othmer plot." This is made as follows for the vapor pressure of any material, or for many different functions of vapor pressure, such as adsorption pressures or partial pressures out of liquid solutions of other liquids, gases, or solids: 1. Select a sheet of standard logarithmic paper and calibrate on the X axis even values of temperatures corresponding to the vapor pressures of a reference liquid (as indicated by the logarithmic graduations). Water is used most frequently. 2. Erect temperature ordinates at these points. 3. Plot on these superimposed temperature ordinates the corresponding values of vapor pressure data using the standard graduations indicated on the Y axis. Lines which are substantially straight over wide ranges are always obtained.

The equation of this line is as follows:

$$Log\,P = L/L' \quad Log\,P' + C$$

P represents the vapor pressure of the liquid or other material being plotted, P' the vapor pressure of the reference material (such as water), L the molar latent heat of the liquid in question, L' the molar latent heat of the reference liquid (all of these at the same temperature); and C is a constant. Since this line is straight, the ratio L/L' is a constant and this is the slope of the line. Thus the value of the latent heat may be obtained for the material in question at any temperature if the value of the latent heat of the reference substance is known; or if the value of the latent heat of the compound and only one point of vapor pressure are known, the line itself may be drawn to indicate the vapor pressure throughout the entire temperature range. Such a plot is particularly useful to interpolate, extrapolate, and evaluate experimental data. An example of this method of plotting is given in Figure 228.

Latent heats at various temperatures may be determined for a given substance from a logarithmic plot against a reference substance, such as water, whose latent heats at different temperatures are known. The angle of the line against the abscissa is measured, and the tangent is taken from trigonometric tables, to give directly the ratio of latent heats. (Alternately, the tangent may be determined directly by measuring on a millimeter scale the length of the side opposite this angle of the right triangle which is given by a horizontal base of 100 millimeters; the number of millimeters is divided by 100.) From the latent heat of the reference substance at various temperatures, those of the given substance are found within a few percent by direct multiplication with this ratio.

A plot of the total vapor pressure of an immiscible substance and water versus the vapor pressure of pure water gives a straight line. Thus, if one variable of a steam-distilling operation is fixed, such as the total pressure, the temperature and partial pressures can be determined. As in the case of single liquids, there is a relation between the slope of the line expressing the vapor pressures and the latent heats. The molar latent heat of the vapor, based on the average molecular weight of the vapor equals the slope of the curve times the molar latent heat of water at the same tempera-

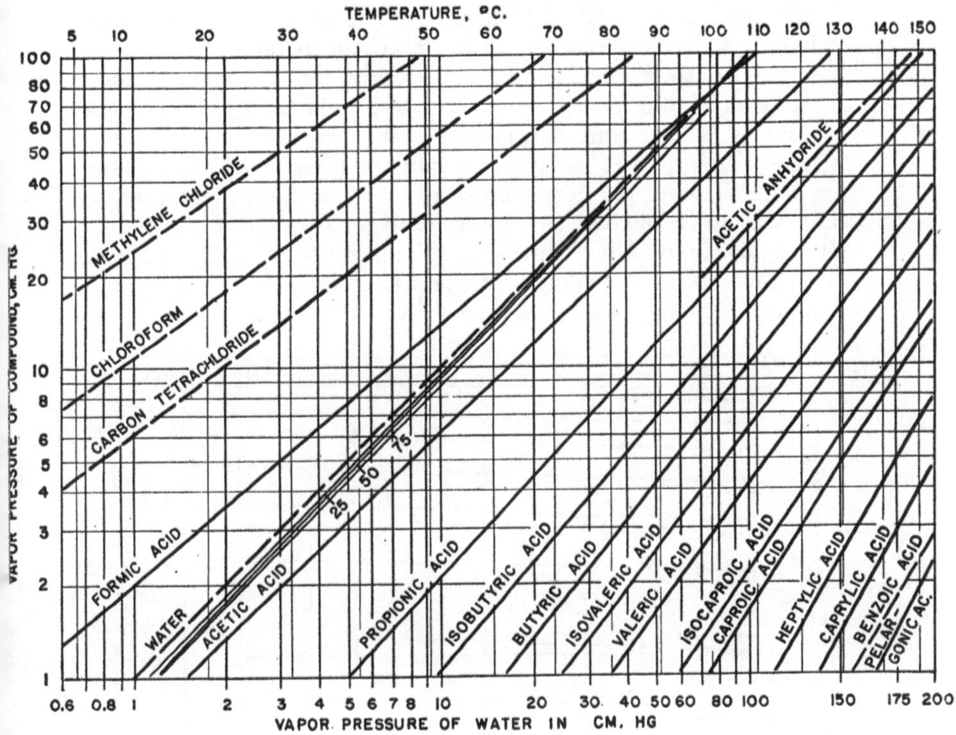

FIGURE 228

Plot of vapor pressures of three chlorine compounds of methane, of lower aliphatic acids, and of several mixtures of water and acetic acid, against vapor pressures of water at same temperature.

ture. Also, the mol fractions of the immiscible substances in the vapor are proportional to the partial pressures used in plotting the individual lines.

In the case of a solution of two volatile liquids, the plot may be made in exactly the same way for the total pressure; for the partial pressure of either component; for the vapor composition (y); for the equilibrium constant ($K = y/x$); for the relative volatility ($\alpha = K_1/K_2 = y_1x_2/x_1y_2$), or for the activity coefficient ($\gamma = p_1/p_1^0x_1$). These functions are of major use in all distillation calculations, and the slopes of the straight lines obtained are functions of the latent heats of the individual components and of the heats of solution.

Still another method of plotting these functions of mixtures of volatile liquids is by plotting any one of them on the *Y* axis of logarithmic paper versus the total pressure on the system taken directly along the *X* axis. Here again straight lines are obtained, the slopes of which are functions of heat quantities. This second method of plotting is particularly valuable because all distillations are done at a constant total pressure and sometimes, temperature data are not available. A representative plot by this method is shown in Figure 229 for the vapor composition of aqueous acetic acid solutions versus total pressures of system at the same temperatures.

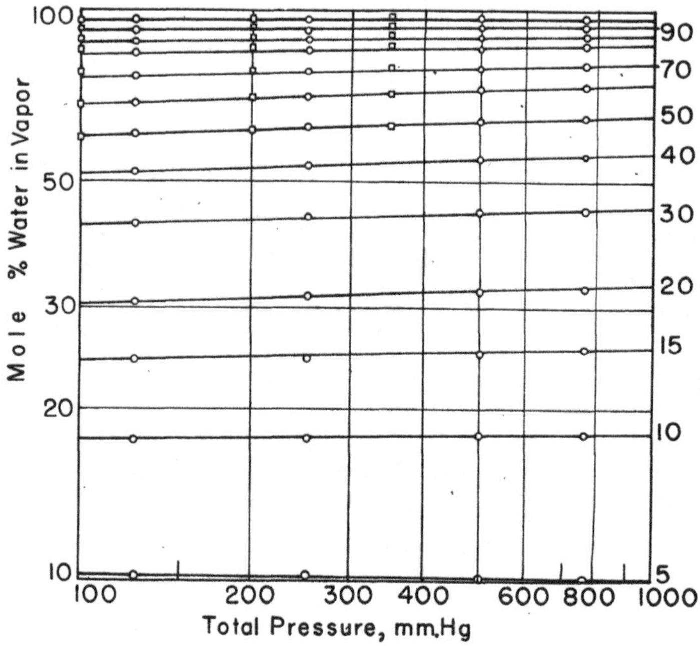

FIGURE 229

Plot of compositions of vapors over aqueous acetic acid solutions vs. total pressures of system at same temperature.

Correlations for heat capacities of ideal vapors and liquids, and heats of vaporization of petroleum hydrocarbons and their mixtures have been presented by Watson and Fallon at the American Chemical Society meeting of Sept. 6, 1943. A summary review of petroleum-processing thermodynamics is given by Edmister in *California Oil World* of Nov.–Dec. 1943.

Small-scale equipment for testing, analysis, pilot-plant development, and manufacturing in limited quantities has been improved by Fischer, Othmer, and Podbielniak. Laboratory data gained from precise analytical fractionators are essential to industrial-scale multicomponent distillation. Automatic distillation equipment operating at low expenditure has been described by Smith (*Ind. Eng. Chem.,* Anal. Ed., **17,** 47 (1945)) and Steffens and Heath (American Chemical Society meeting of Sept. 6, 1943). Such equipment operates at high reflux ratios. A combination of a Podbielniak column with an automatic distillate collector measures successive fractions in a float chamber and empties them into bottles. The boiling temperatures of the distillates are recorded continuously, so that the setup may be left to itself for as long as 16 hours.

In *Trans. Am. Inst. Chem. Eng.,* **39,** 1, Garber and Lerman discuss stripping and steam distillation. General expressions are presented for the stripping medium required to perform a prescribed distillation under given operating conditions for bath, continuous countercurrent, and continuous counterflow. Batch and continuous stripping are compared, and the better economy of the continuous countercurrent process is demonstrated. Its superiority is particularly marked where the concentration of the volatile component in the residue is to be very small. A graph indicates possible savings in steam, depending on feed composition and final composition. Another plot permits determination of the number of actual plates required for a continuous flow stripper, if phase properties, pressure drop per plate, pressure at stillhead, and vaporization efficiency are known or assumed.

Walsh (*Ind. Eng. Chem.,* Jan. 1946, p. 8) discusses high-temperature distillation. Metals are refined by means of simple batch stills, using Cr-Ni steel retorts and water-cooled condensers.

Efficiencies in ethanol distillation have been studied further by Byman and Keyes (*Chem. & Met. Eng.,* July 1945, p. 7–85). Chilton and Colburn's transfer unit (Colburn, *Ind. Eng. Chem.,* **33,** 459, 1941) has been discussed by several authors.

The number of transfer units required in a column in order to obtain a given product from a given feed is the integral of the reciprocal of the difference between the equilibrium and actual vapor compositions. The HTU (Height of Transfer Unit) is the total height divided by this integral value. For two-component systems, Scheibel and Othmer (*Trans. Am. Inst. Chem. Eng.*, **38,** 339, 1942 and *Ind. Eng. Chem.*, **34,** 1200, 1942) have simplified the calculation of the HTU, but multicomponent systems yield either impossible or tedious solutions.

Gerster, Koffolt and Withrow (*Trans. Am. Inst. Chem. Eng.*, **41,** no. 4, 393, Aug. 25, 1946) demonstrate the relationship between HTU and theoretical plate methods in column calculations. Their method is helpful in correlating the operating characteristics of plate columns. A new concept (number of actual plates equivalent to one transfer unit) permits evaluation of column performance in terms of vapor-liquid ratios and slopes of vapor-liquid equilibrium curves.

Flow characteristics of bubble-cap plates of two commercial fractionating columns were studied by Gardner, Pike, Winsche and Urbon. Various reflux ratios and superficial vapor velocities from 0.7 to 3 feet per second were used. The pressure drops observed were compared with values computed according to various authors; the agreement seems satisfactory below the entrainment range. (American Chemical Society meeting, Pittsburgh, Dec. 1946.)

Dodge, in his book on *Chemical Engineering Thermodynamics* (1944) discusses the minimum-work concept and its use in determining separation efficiency. Smoker elaborated a method for determining the number of plates, and ultimately heat requirements.

For a separation of multicomponent mixtures, such as petroleum hydrocarbons, Jenny and Cicalese present new data (*Ind. Eng. Chem.*, **37,** 965). They use the "group" concept in their calculations, i.e., a limited number of hydrocarbons having boiling temperatures close together.

In precision fractionation, the quality of the product can be controlled by suitable adjustment of the overhead stream, with constant heat input, as shown by Verity (*Natl. Petroleum News*, **37,** no. 10, R 169 (1945)).

Good and collaborators (*Ind. Eng. Chem.*, **34,** 1445, 1942) investigated the liquid gradient across bubble-cap plates. Davies defines as a stable plate one that stays balanced under unsteady flow of vapor and liquid, a flexible plate as one that stays balanced under a wide range of vapor and liquid loads. A balanced plate is characterized by satisfactory efficiency under fluctuating conditions when operated at design loads. The balance is largely controlled by the liquid gradient, which is the difference in liquid depth between inlet and outlet side. The author gives an equation for computing the liquid gradient for different vapor throughputs. These relations were established from values obtained from commercial plates using water and air, and checked on a commercial depropanizer plate by means of a gamma-ray detector. The accuracy was of the order of 4%. (American Chemical Society Pittsburgh meeting, Dec. 30, 1946.)

The effect of liquid flow upon bubble-plate economy is particularly noticeable for large-diameter columns. Designs which have the downcomer in the center of a plate, with overflow weirs around the periphery have been proposed to eliminate stagnant areas. Some aspects of the mechanical design of a column of 33 feet diameter are presented by Brummerstedt (*Natl. Petroleum News,* Feb. 3, 1943).

The theoretical capacity of commercial stills can be determined by pressure-drop calculations. Kirkbride (*Pet. Refiner,* **23**, 9, 321, 1944) analyzes the mechanics of fluid flow in bubble-cap columns, and establishes the limit tower concept. Flexibility is balanced against throughput. Also, Walsh (*Ind. Eng. Chem.,* **38,** 1, 1946) indicates the significance of proper determination of the operating pressure in a column for a given design and degree of separation.

A general treatment of multistage separation processes is presented by Benedict in *Trans. Am. Inst. Chem. Eng.,* **43,** no. 2, p. 41 (February 1947).

The number of columns customarily required to effect close fractionation is reduced by the Brugma layout (Figure 230), in which the number of columns is reduced to the power to which 2 must be raised in order to equal the desired number of fractions. This layout supplies multiple sidestreams, and offers improved heat economy. The number of reboilers and condensers is also reduced, although more actual plates are necessary than in the conventional arrangement.

Where low pressure drop and high velocity are given, shower plates (Figure 231) have been used, which consist of baffle-like partitions which are partially perforated. The ascending vapors pass through a series of liquid screens. Further elaboration of this principle has led to the construction of a spray column (Figure 232) which has spray chambers instead of interchanger plates. Refluxing is effected by a number of pumps. In this case, diffusion to and from liquid droplets controls, and determines the limits within such columns are practicable.

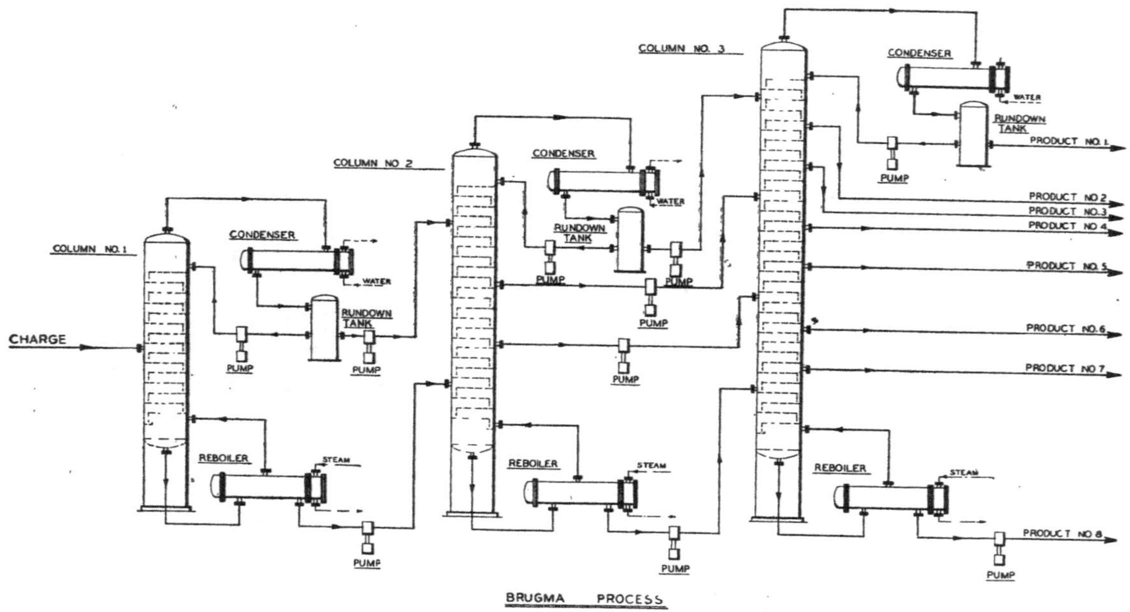

FIGURE 230
Brugma layout.

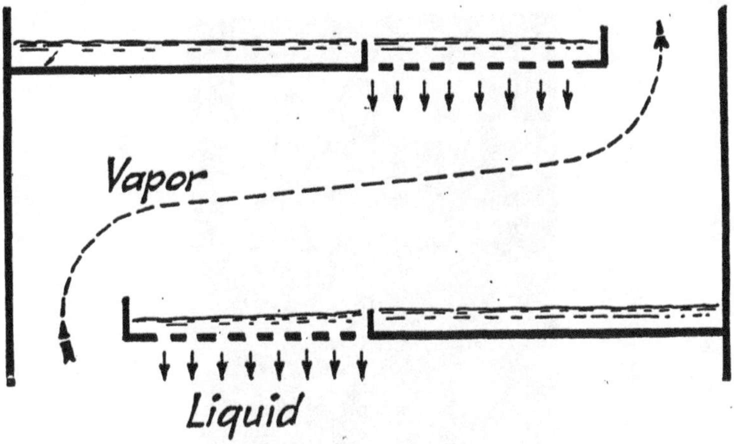

FIGURE 231
Shower plates.

For packed columns, several new types of packing have been developed. General performance data are published by Duncan, Koffolt and Withrow (*Trans. Am. Inst. Chem. Eng.*, **38**, 2). In *Trans. Am. Inst. Chem. Eng.*, **39**, 6, Minard, Koffolt, and Withrow investigate the efficiency of alcohol columns packed with Fiberglass. Bowman and Briant base a new theory of packed-column performance on the hypothesis that the interphase transfer rate of a component is proportional to the amount of that component which would have to be transferred to bring the phases into equilibrium. (American Chemical Society Pittsburgh meeting, Dec. 30, 1946.)

A novel wire-net packing, the Stedman type (Figure 233), is discussed by Bragg in *Trans. Am. Inst. Chem. Eng.*, **39**, 6. In the closely spaced wire-mesh screens liquid is held in film form due to capillary action. This type of packing may be regarded as a modified plate arrangement with close spacing between screens; corresponding HETP values can be as low as 2 inches. In similar designs, the pressure drop is greatly dependent on the load, and the flexibility is correspondingly reduced.

The McMahon packing is made of 100-mesh wire cloth shaped into saddles; its maximum capacities and holdup values are stated to be comparable to those of the Stedman type. However, it is claimed to be easier to install and to maintain. Forsythe and collaborators present performance data obtained in a column of

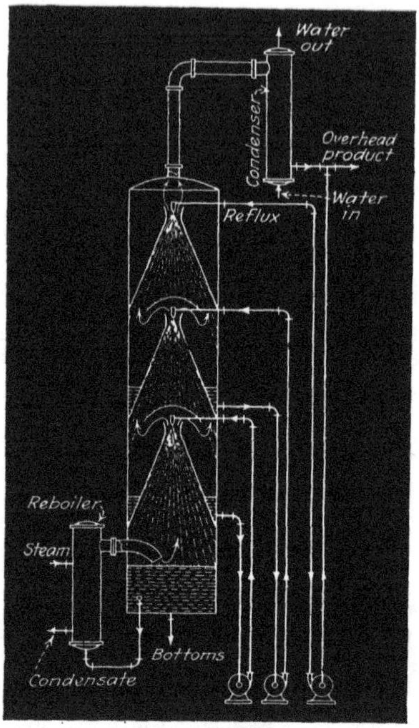

FIGURE 232

Spray column. The conical spray distilling tower developed by the Vulcan Copper & Supply Company for the fractionation of mixtures at low pressures because of very low pressure drop.

6-inch diameter. The weight of the packing is of 25 pounds per cubic foot. (American Chemical Society Pittsburgh meeting, Dec. 30, 1946.)

Much study has been devoted to columns used in the rectification of air. Rushton and Stevenson review general developments in oxygen production in *Trans. Am. Inst. Chem. Eng.*, **43**, no. 2, p. 61 (February 1947). Lobo and collaborators give data on oxygen recovery and fractionating ability of both bubble-plate and packed towers of various sizes, in terms of HETP and plate efficiency. Packings used include $\frac{1}{4}$-inch Raschig rings, $\frac{1}{4}$-inch saddles and Stedman packing, for columns of 6, 8 and 14-inch diameter, with varied spacings.

Weedman and Dodge employed a stripping column of 2-inch diameter and 12 different types of packing (Stedman, Haydite,

a. Perspective view.

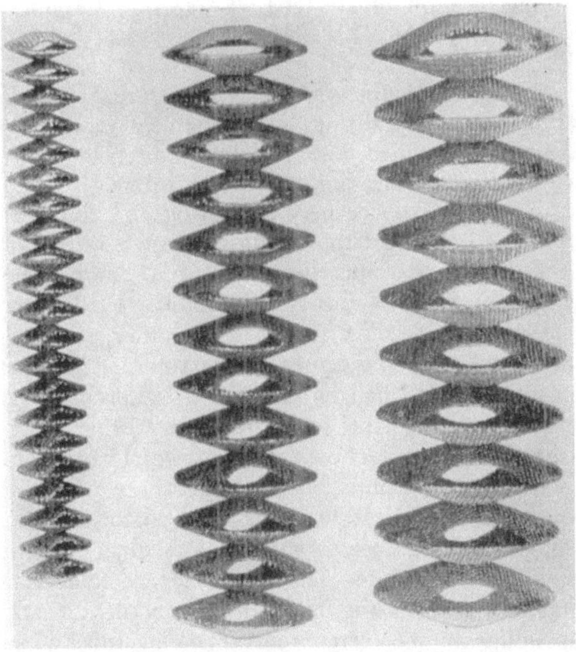

b. Enlarged detail of a vertical section.

FIGURE 233

Stedman-type wire-net packing.

rings and saddles of different sizes). Liquid air was fed at the top and nitrogen gas was drawn off at the top, while oxygen was recovered at the bottom. Beyond 30-inch packing height, extension gave little improvement in separation. Any inclination of the column resulted in diminished separation efficiency. Increased size of glass rings increased the column capacity, but reduced its efficiency. 60×40-mesh Stedman packing gave best results.

D. A. Williams points out that end effects, such as entrainment at the top, and irregular drainage at the bottom, have to be eliminated before consistent column performance can be obtained. Entrainment is reduced by centrifugal separation, distribution of liquid over the cross-section of the column below packing surface, and by means of an ample boundary area of well-drained top packing. The bottom region of the column was drained by liberally dimensioned drain tubes. The variables investigated comprise pressures, pressure drops, vapor velocities for various column lengths and diameters, and various packings, with and without auxiliary refrigeration of air feed. (American Chemical Society Pittsburgh meeting, Dec. 30, 1946.)

Several types of centrifugal fractionators are available or under development. In the Podbielniak design (Figure 234), a horizontal fractionating rotor is built up of a stainless steel strip several hundred feet long, coiled into a spiral, and held between solid head plates. Vapor from the still kettle enters through a vapor pipe into the annular space between casing and rotor; the reflux passes from the bottom of the casing through a return line to the kettle. A thin film of reflux liquid is forced against the path of the moving vapors when the spiral rotates. Feed enters at the one end of the shaft; reflux leaves at the other, passes to a condenser, and returns to the center of the rotor. The shaft is supported on water-cooled ball bearings, and leakproofed against the atmosphere by rubbing ring seals. It is stated that a similar unit accomplishes the same effect as a bubble tower 15 times its volume. Distilling capacities (reflux plus product) are of 30 liters per hour or more. One application of this fractionator is in the separation of isotopes, such as heavy water from ordinary water.

Piazza (Figure 235) uses a flat disk attached to a vertical shaft which rotates in a flat casing, and carries concentric heating tubes. A corresponding number of concentric cooling tubes is suspended from the flat top of the casing; the individual rounds have coolant

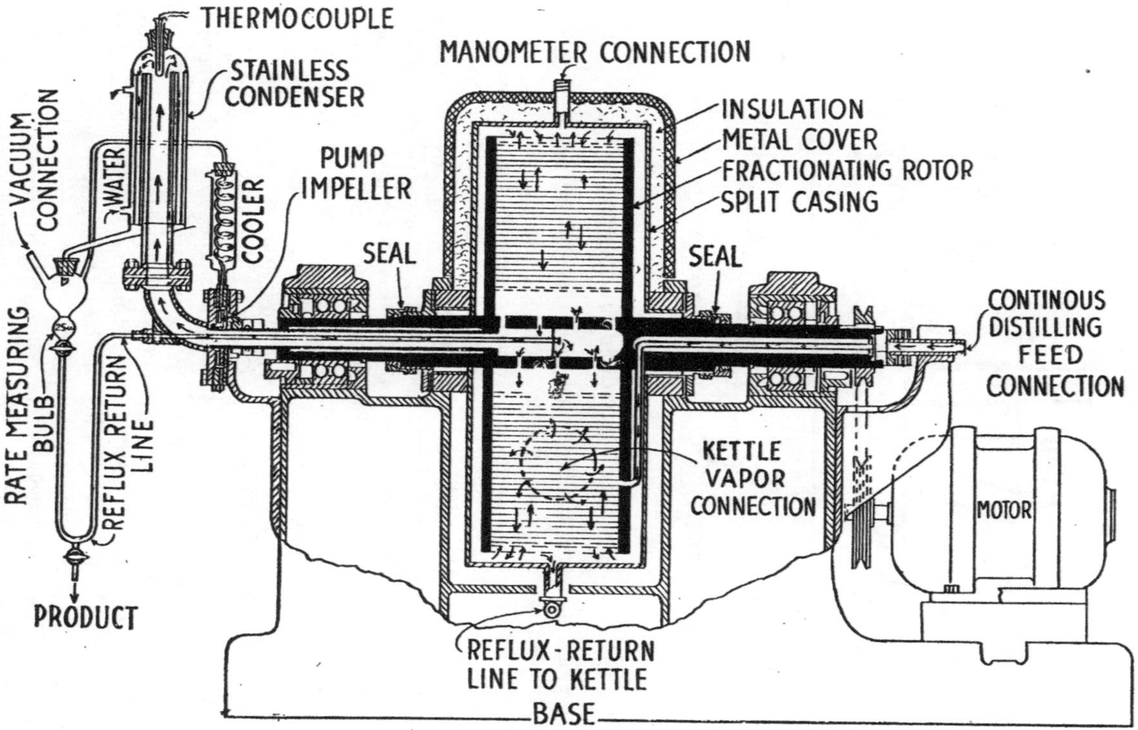

FIGURE 234
Podbieiniak-type centrifugal fractionator.

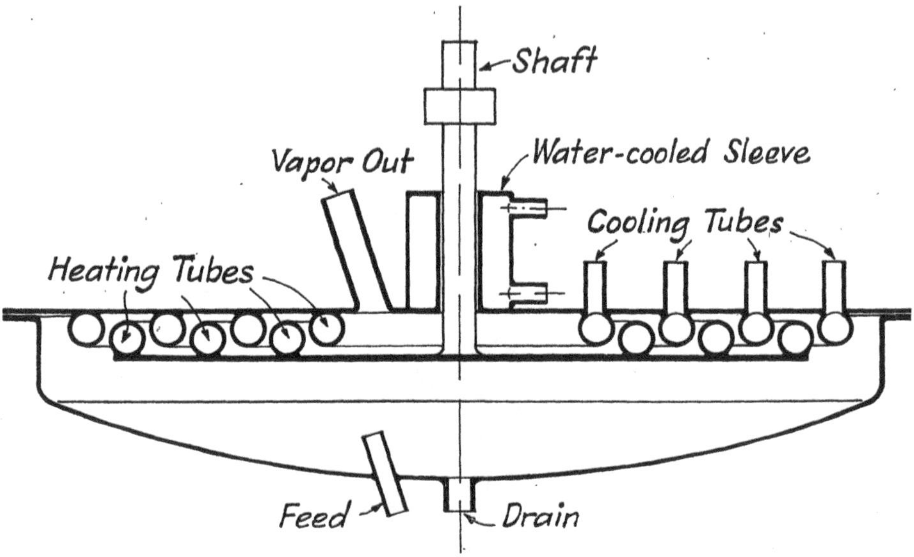

FIGURE 235
Piazza-type centrifugal fractionator.

connections to control reflux quantities. Distilland vapor ascends from the bottom of the casing, enters at the periphery of the rotor, passes through the clearance between casing and rotor, moving to the center. When the disk rotates, a countercurrent motion between the liquid and the gaseous phase occurs together with a series of condensation and reevaporation processes. The distillate is withdrawn at the center on top. In this design, use of a condensation seal is claimed to obviate the use of a stuffing box. Recently, disks with attached tubing have been replaced with spun disks featuring circular corrugations. (Piazza, *Sesiones Quimicas Argentinas de la Ciudad de Rosario,* Nov. 1941).

Hickman and Mees (*Ind. Eng. Chem.*, Jan. 1946, p. 28) report on their design which is particularly suitable for high-vacuum applications. Their molecular still (Figure 236), with outputs of 300 to 1000 kilograms per hour, operates at pressures between 50 and 1 μ. It has a cast aluminum rotor on which a thin film of distilland evaporates. The sides of the pot-shaped rotor, whose diameter is up to 5 feet at the top, slope upwards under 10–25° to the vertical. Feed is introduced through a jet at the bottom of the rotor operating at approximately 400 revolutions per minute,

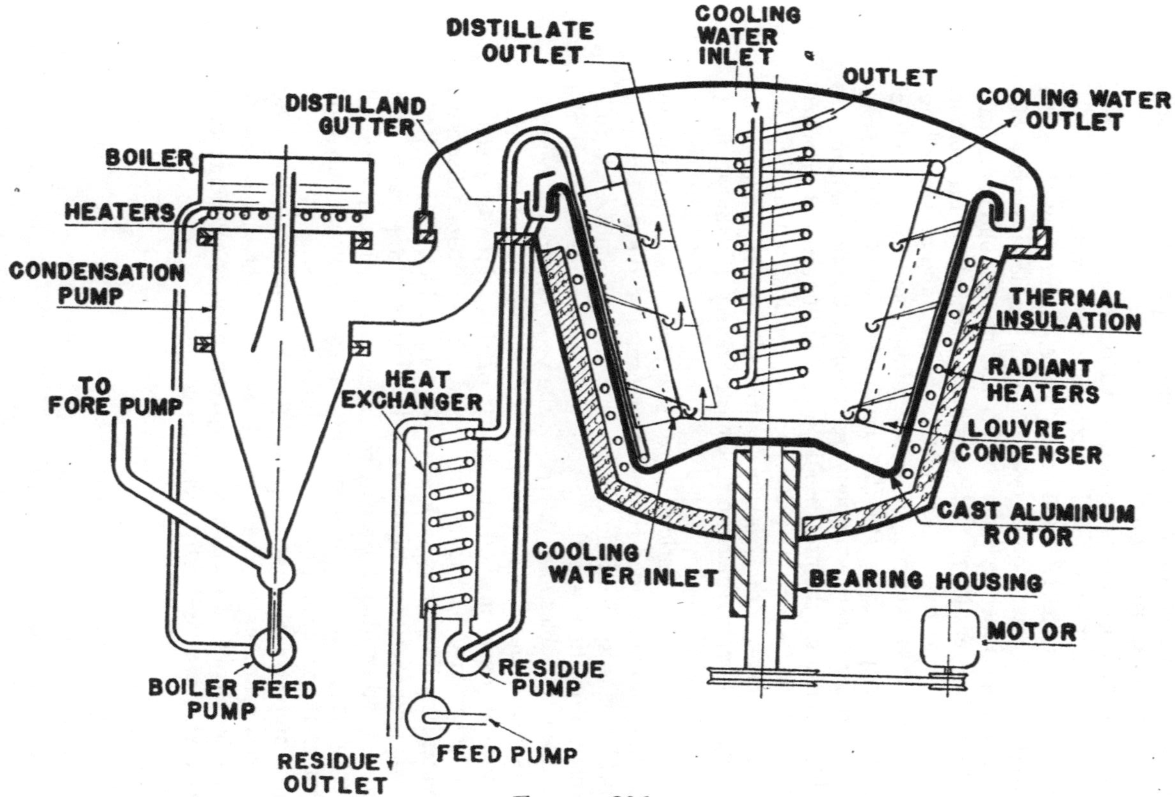

FIGURE 236
Hickman and Mees molecular still.

flows up the hot sides, and passes over the top into a gutter from which it is removed by a pump. Heat is provided by electric resistance units surrounding the rotor.

Willingham, Westhaver, Sedlak, and Rossini describe details of a concentric tube rotary distilling column whose rectifying section is an annular space 1.1 millimeters wide, formed by the inner surface of a stationary outer cylinder, and the outer surface of a rotating closed inner cylinder of 75 millimeters outer diameter and 584 millimeters length. Results are presented for the number of equivalent theoretical plates (at total reflux) and pressure drops at speeds up to 4000 revolutions per minute, and throughputs of 0.6 to 4.7 liters per hour. Per 1000 millimeters length of rectifying section, and for 4000 revolutions per minute, the throughput was of 3 liters per hour, the pressure drop of 2 millimeters of mercury, the number of theoretical plates 62. The calculated holdup for these conditions was of 17.5 cubic centimeters per meter length of rectifying section. (American Chemical Society Pittsburgh meeting, Dec. 30, 1946.)

In all fields of distillation, there is a trend toward continuity of operation and extended runs. This means that there is a widening field for instrumentation and automatic regulators. A fully automatic operation would, however, require control of all the independent variables encountered in the process under consideration. Gilliland and Reed (*Ind. Eng. Chem.* **34**, 551, 1942) have developed an exact method of analysis of the degree of freedom of contact systems. The actual application of this system is discussed at greater length by Perry (*Chem. & Met. Eng.*, Oct. 1945, p. 108). However, distillation equipment finds increased application for operations other than straight rectification, particularly in oil refineries as a means for combining materials chemically and separating the reaction products physically.

An example of this is the production of aviation gasoline by continuous synthesis of C_8 hydrocarbons from lighter components such as butylenes. Reaction and separation of components take place in what are essentially rectifying columns.

SUBJECT INDEX

AUTHOR INDEX

www.ingramcontent.com/pod-product-compliance
Lightning Source LLC
LaVergne TN
LVHW101454200926
848483LV00001B/129
* 9 7 8 0 8 2 0 6 0 0 1 8 5 *